T0231540

AGGREGATES

Aggregates

Geology • Prospection • Environment
Testing • Specifications • Extraction
Processing Plants • Equipments
Quality Control

Editors
Louis Primel • Claude Tourenq
Laboratoire Central des Ponts et Chaussées
58 Boulevard Lefebvre
75732 Paris Cedex 15
France

CRC Press
Taylor & Francis Group
Boca Raton London New York

CRC Press is an imprint of the
Taylor & Francis Group, an **informa** business
A TAYLOR & FRANCIS BOOK

Published by arrangement with "Presses de l'Ecole Nationale des Ponts et Chaussées, Paris

Aidé par le ministère français chargé de la culture.
Published with Financial Aid from the French Ministry of Culture.

Translation of: *Granulats* 1990, © Presses de l'Ecole Nationale des Ponts et Chaussées, Paris. Updates provided by authors for the English edition in 1998.

Translation team: S.A. Velou
G. Ramaseshan
Y.R. Phull
R. Sudha
Margaret Magithia

A.A. Balkema, P.O. Box 1675, 3000 BR Rotterdam, Netherlands
Fax: +31.10.4135947; E-mail: balkema@balkema.nl
Internet site: http://www.balkema.nl

Distributed in USA and Canada by
A.A. Balkema Publishers, Old Post Road, Brookfield, VT 05036-9704, USA
Fax: 802.276.3837; E-mail: Info@ashgate.com

ISBN 90 5410 795 2

Introduction

During the last twenty years, in France as in several other European countries, there has been an increasing trend in the field of extraction of aggregates towards:

— a consolidation of companies, a trend which is still continuing.

— an appreciable increase in environmental problems, which has brought about new legislative measures and change in mining practices.

The increasing attention to quality control and the gradual automation of plants also merits a mention.

The objective of this book is to present the "aggregates" field in a manner as comprehensive as possible, from natural deposits to finished products.

Chapter I deals with the aggregates market in France which is shared by building construction and roadworks and which has remained to a large extent a local one, with average distances of transport by road of 30 kilometers.

Chapter II shows why the relatively recent legislative measures seeking to govern opening of new quarries led French professionals to seriously take into account the problems of environment, from impact study to reclamation of the sites after exploitation.

Chapter III and IV describe how the use of geological information and modern methods of prospecting of deposits as well as the laying down of a set of standard geotechnical tests, make for exploitation in a more rational manner and thus enable the users to obtain a constant quality for the aggregates they use.

Chapter V explains why the exploitation of new resources (rocks not yet used, various wastes) is more difficult as it requires specific studies on the one hand and a change in the habits of users on the other.

Chapter VI, VII and VIII deal with the exploitation of sands and gravels deposits. These require installation designs, methods of extraction and machinery which are quite different massive rocks. These aspects are extensively discussed in chapters VII and VIII with emphasis on the essential role of water in alluvial deposits and that of quarrying of massive rock deposits.

In chapter IX, X, XI, XII there are descriptions of manufacturing equipment even though they have not undergone any revolutionary change in respect of the basic steps of crushing, screening, washing and feeding-batching, which form the unavoidable essentials in a plant.

Chapter XIII explains that though dedusting is linked essentially with the safety of personnel it also has a bearing on environment safety, particularly in the vicinity of urbanized areas. It can also contribute to making the aggregates cleaner.

In chapter XIV, XV, XVI all aspects associated with the movement of materials are dealt with. These operations always entail a huge investment, generally more that 50% of the total. As they cannot be dissociated from the feeding-batching functions described in chapter XII or from the choices mentioned in chapter XV (what are we trying to obtain and with what facilities?) the flow of materials in a plant finds its apotheosis in chapter XVI where the optimization of the functioning of the machinery and the control of the flows leads to improvement in productivity and quality.

Chapter XVII: The limited intervention on new sites over a period of time for taking up heavy site jobs enables synthesization of the various aspects dealt with in this book. Geological reconnaissance is more important in this type of quarry to guard against geologically unforeseen events which may turn out to be catastrophic. This is also the case for many sites in developing countries (Chapter XX).

Chapter XVIII explains that quality control of aggregates serves as an indicator of the proper functioning of a plant. It should become integrated with automation processes; this is a field where considerable progress is yet to be made, particularly for reduction in response time of all controls.

Chapter XIX covers safety and describes a large number of risks of diverse nature. The kind of risks involved, their identification and impact, and the preventive measures are all outlined to provide an overall view of this important aspect. To illustrate the above problems, conveyor belts are dealt with in detail.

Chapter XX deals with export of known technologies to developing countries and their adaptation to local conditions. This chapter provides a link to chapter VI which deals with new materials (laterites etc.) and chapter XVII which discusses planning and setting up from scratch.

1st November 1999

Louis Primer
Claude Tourenq
Paris

Contents

	Introduction	v
I.	Aggregates Market in France	1
II.	Quarries and Environment	24
III.	Properties and Prospecting of Rocks and Aggregates	52
IV.	Properties of Aggregates—Tests and Specifications	109
V.	Non-conventional Aggregates	143
VI.	Presentation of a Quarrying Operation	154
VII.	Stripping, Extraction, Loading and Transport in Quarrying Massive Rocks	164
VIII.	Stripping, Extraction, Loading and Transport in the Exploitation of Sands and Gravels Deposits	200
IX.	Crushing and Grinding	228
X.	Screenings	269
XI.	Feeders and Batchers for Crushers	293
XII.	Washing of Aggregates and Treatment of Fine Aggregates	301
XIII.	Dedusting	344
XIV.	Handling, Stockpiling, Mixing, Weighing, Dispatch and Transport of Aggregates	375
XV.	Engineering Problems in Aggregate Processing Plants	443
XVI.	Automation of an Aggregate Processing Plant. Definition and Methods	492
XVII.	Mobile Plants for Production of Aggregates	504
XVIII.	Quality Control of Aggregates	530
XIX.	Safety: General Rules, Example of Conveyor Belts	549
XX.	Aggregates in Newly Developing Countries	564

I

Aggregates Market in France

Georges Arquié, Alain Camus, Michel Charreau, Jean Louis Dubus, Pierre Dupont and Christian Piketty

Aggregates represent an important stake for:
— the national economy and the Building and Civil Engineering sector;
— the quality of the works and the cost of the projects.

The present chapter, concerned with the consumption, production and flow of aggregates in France, enables determination of this stake.

1. NEEDS

Aggregates constitute the essential raw material for building construction and public works, without which it would presently be impossible to carry out such work.

1.1 Qualitative Analysis of Demand

Aggregates form the skeleton of hydraulic concretes, the skeleton of pavements etc. and the quality of housing, roads and so forth depends in part on their quality. Not all of the rocks available are usable. Just as earlier man selected only those stones most suitable for the construction of buildings and pavements ('when struck with a hammer, the pavement should produce a clear sound'), today's engineers expect from aggregates characteristics suitable for meeting the needs of the material used, the type of work, the site conditions and conditions of use (see Chapter IV).

1.2 Quantitative Analysis of Demand

The French consumption of aggregates may be estimated at 337 million tonnes for 1996 i.e., 6 tonnes per capita. Thus, after water, aggregates are the most consumed mineral resource.

	Total (Mt*)	Per capita (t)
Germany	574	7.1
United Kingdom	231	4.0
Belgium	42	4.2
Spain	197	5.1
Italy	185	3.2

* Mt = million tonnes

1.2.1 DISTRIBUTION ACCORDING TO CATEGORIES OF WORKS

Public works utilise nearly three-quarters of the total tonnage consumed; the rest goes to building construction (see Figure I.1).

It is also interesting to note a few figures of unit consumption:

1 house	120 t
1 hospital	4800 t
1 secondary school	2800 t
1 factory	1700 t
1 hotel	1100 t
1 shop	870 t
1 km autoroute (highway)	30,000 t
1 km road	12,000 t
1 km railway (double track)	16,000 t
1 nuclear plant	6 to 12 Mt

Compared to 1990, consignments of aggregates meant for building construction have decreased by 15% (due to the slump in the construction of houses and office buildings) and those for public works by about 2%.

1.2.2 DISTRIBUTION ACCORDING TO CATEGORIES OF USE

Roadworks absorb nearly two-thirds of the total tonnage consumed; the remaining quantity is taken up in uses associated with hydraulic concretes (see Figures I.2 and I.3).

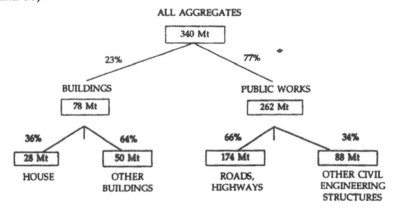

Fig. I.1. Consumption of aggregates (1996).
Distribution according to categories of works. Estimate in millions of tons.

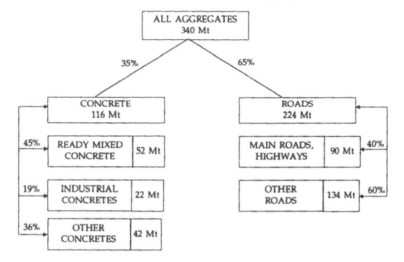

Fig. I.2. Consumption of aggregates (1996).
Distribution according to categories of use. Estimate in millions of tonnes.

However, there is some doubt about the estimated share of hydraulic concretes due to uncertainty about the cement content of site concretes. According to available data, this share may vary between 30 and 40%

1.3 Perspective Trends

1.3.1 PRELIMINARY REFLECTIONS

Analysis of the balance to be maintained between the needed qualities of aggregates, the corresponding quantities consumed and the price levels is very complex and certainly requires in-depth studies, especially if one is interested in the medium term.

In the absence of such a thorough perspective study, here are a few ideas for thought, relevant in particular to the field of pavements:

— Performance of aggregates. For instance, for wearing courses strength and microrugosity have strong chances of remaining strong necessities in the long run due to: a) development of technology; b) probable development of vehicles: private cars will have in the future a total weight lighter than that of today's cars, which will no doubt necessitate ensuring a very good road surface roughness and a quick evacuation of water between the tyre and the pavement.

— Homogeneity of delivery of aggregates. Development of pavement techniques, whether of very low thickness or high performance, but finer than the traditional ones, may widen the difference that already exists between the various quality requirements.

— Required rates of delivery of aggregates at construction sites. If some present-day cases no longer require the very high rates which were justified in 1975, the other situations, such as works under traffic conditions or works for

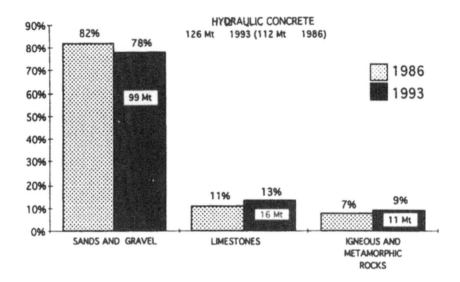

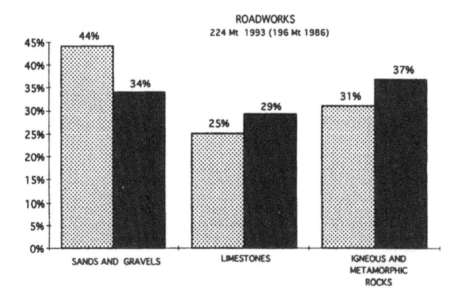

Fig. 1.3. Trends in consumption of aggregates from 1986 to 1993 (according to categories of use and types of aggregates).

strengthening highways with very high traffic, will continue to justify highly concentrated time-oriented supplies in order to minimise discomfort to users.

— Required quantity of aggregates. Some thinking is certainly necessary for defining the medium term development of new techniques of recycling as well as the possible application of strategies of strengthening pavements to combat the effects of freeze-flow on local networks.

— Lastly, competition. In a context which has become more difficult, regional officials are zealous about preserving local activities. They thereby contribute to strengthening regional protectionism and increase the difficulties of the aggregates sector or its means of transport when such are external to the particular region

1.3.2 MEDIUM TERM REQUIREMENTS*

• **For the housing sector,** the major uncertainty results from changes in interest rates. In fact the entire housing sector depends on these rates: households through their solvency and borrowing capacity; investors for their calculation of profitability and their decision of selection between building constructions and financial investments; the State on account of the budgetary cost of the grants that it gives. While the long term interest rates should slightly drop, they will possibly continue to remain high in real terms. Other constraints will get added to this: a slowing down of the economic growth, a limited increase in the purchasing power of households due to transfers of resources to the advantage of the State and a rate of unemployment greater than 10%.

Nevertheless, some factors are conducive to more optimism, particularly the good savings position for housing and the possibility of government measures favourable to a very social rented accommodation and a financial system more favourable to investments by tenants. All these factors should sustain new constructions till 1998. Beyond that the gradual withdrawal of support to house building plan and the fall in population may slow down the rate of constructions.

• **As regards non-residential buildings,** the carefulness of the "financiers" and the level of the occupancy rates will slow down in the short run the revival of office building constructions. Construction of industrial and storage buildings should increase by 5% every year on an average during the next five years with a slowing down at the end of this period whereas agricultural and commercial buildings will increase by 1 to 2% on an average.

As for improvement maintenance, it should increase at an average rate of 3% from 1994 to 2000 but at a lower rate at the end of the period.

• The recession in *public works* which began in 1992 with the completion of big construction works, and the slowing down of government orders, continued till 1995.

Between 1996 and 2000, the average annual growth of civil engineering works should be 2.6% in volume. It will take place through civil engineering investments made by the Administrations (State and local authorities) and private companies. On the other hand, investments from Big National Companies will stagnate but at a high level (two TGVs—high speed trains—will be under construction, the Eastern TGV and the Mediterranean TGV).

---

*Source: Bureau of economic information and projections.

Between 1994 and 2000, the acceleration of the implementation of the highway master plan will redynamize the investments made by civil engineering concessionary companies.

• Greatly affected by the recession of recent years, the *construction materials sector* is in a position to face a revival of activities without major capacity investments. On the whole the growth of this sector will be moderate over the period 1994–2000. Although on a growing trend, production levels would still remain in the year 2000 lower than those of the 1989 peak.

Between 1994 and 2000, the annual production of aggregates should be, on an average, around the present volume with however some differences according to the type:

— In the case of alluvial aggregates, the recession will be due to the unfavourable situation of building construction and to the pursuit of the trend towards their substitution (with aggregates from massive rocks) which is already well under way in some regions.

— In the case of aggregates from massive rocks, the production will be well oriented, particularly for igneous and metamorphic rocks with the TGV and the highways; limestones should in addition benefit from the effects of substitution.

— Moreover, the offer of recycled products should develop. Thus, from the present 3 million tonnes, the production of aggregates obtained from the recycling of products of demolition might exceed 5 million tonnes.

2. RESOURCES

The geology of France determines the location of production areas according to the nature of the terrain (see Figure I.4).

2.1 Qualitative Analysis of Supply

The geological map of France to a scale of one million shows a large wealth and a large diversity of formations (see Chapter III), which results in a wide variety of aggregates.

Sands and gravels are actively exploited along rivers and their tributaries; towns were built and developed in the neighbourhood of rivers and in valleys. Thus alluvial deposits generally constitute the nearest traditional construction materials; these are also the easiest to extract.

On the other hand, the diversity of the sediments (petrographic composition, grain size etc.), when they exist and subsist, enables meeting almost the entire demand.

The proximity of the places of consumption, the large range of qualities, the moderate production costs given the prewashing, precrushing and prescreening work carried out during the geological phases of deposition, account for the significant growth in exploitation of these materials and the dominant place which they occupy in the aggregates market (see Section 2.2).

Public opinion becoming more and more sensitive to problems concerning the environment, the present competition for taking up soils, land speculation,

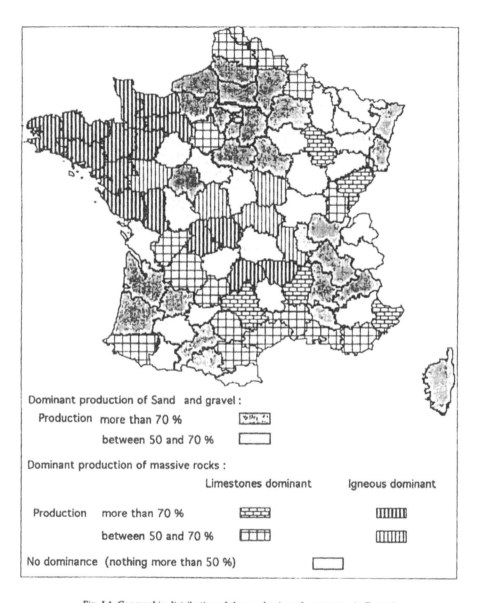

Dominant production of Sand and gravel :

   Production more than 70 %

             between 50 and 70 %

Dominant production of massive rocks :

                                Limestones dominant   Igneous dominant

   Production more than 70 %

             between 50 and 70 %

No dominance (nothing more than 50 %)

Fig. I.4. Geographic distribution of the production of aggregates in France*

strengthening laws relating to quarries—all these constraints work towards limiting extractions in alluvial sites, and even sterilising certain deposits.

    Limestones show essentially on the surface of sedimentary basins and recent folded belts. Limestone aggregates meet the technical requirements of the current

---

*France is administratively divided into 95 so-called "departments", grouped into 21 "regions" of 2 to 5 "departments". The map in this figure shows the geographic distribution of aggregates on a "department"-wise basis.—Ed.

demand: hydraulic concretes and roadworks. Most of these are good as road materials but their use in road surfacing remains limited due to the fact that they are sensitive to smoothening and become slippery.

Igneous and metamorphic rocks are mostly exploited in ancient mountain massifs. They are suitable for manufacturing hydraulic concretes and for carrying out routine roadworks. Many are the preferred constituents of road techniques and more particularly of those specific to wearing courses. Some may be used for ballasting railway tracks.

Most of the aggregates are therefore likely to meet the specifications presently in force.

But it must be noted that contractors preferably use aggregates which are closest to the work sites because the constraints of distance and cost of transport are generally greater than those of the techniques employed. Moreover, the flux of aggregates over long distances is extremely low: this pertains almost exclusively to particular needs which require specific aggregates.

The aforesaid confirms the regional and most often local character of the market of aggregates and explains the importance of short-distance road transports as well as the low concentration of places of production (see Section 3).

2.2 Quantitative Analysis of Supply

French production of aggregates was 367 Mt in 1994, including
  50% of alluvial origin,
  24% from limestones
  26% from igneous and metamorphic rocks.

11 million tonnes of recycled aggregates are also produced including 2 million tonnes of carboniferous schists, 6 million tonnes of blast furnace slags and 3 million tonnes of demolition materials.

Aggregates represent more than three-fourths of the annual production of mineral substances in France.

2.2.1 NATIONAL PRODUCTION

• *Overall production trend*

Figure I.5 traces the trend in production of aggregates from 1980 to 1994 and enables a bird's eye view of the share represented by each category of aggregates. The structural modifications are analyzed below.

Development over the last fifteen years is characterized by three important periods:

From 1980 to 1985, an initial period of decline ending with a production of less than 300 million tonnes (Mt) i.e. 80 Mt less than in 1980. The entire sector of aggregates reacting directly to orders from the Building and Civil Engineering sector, this development is naturally linked with the crisis in the construction sector since 1980.

From 1986 to 1991, a period of growth is registered with a peak production of nearly 400 Mt in 1991 i.e. 100 Mt more than in 1985. In 1985, and especially in 1986–87, there was a revival of activities in the public works sector due to an

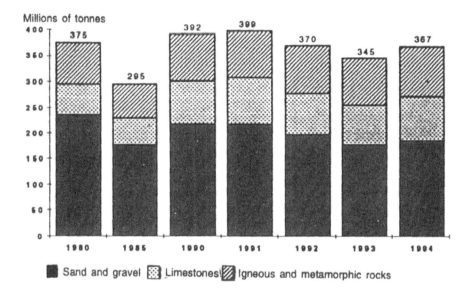

Fig. I.5. Production of natural aggregates, 1980–1994

increase in investments from the local bodies and the execution of large operations at the national level in the field of building construction and civil engineering.

Finally, from 1992 to 1994, the development is irregular with a sharp decline during the first two years (–7% per year) due to an ill-oriented activity in the Building and Civil Engineering sector. The year 1994 records a growth of 6% resulting from a revival plan, an increase in the number of houses constructed and a stabilization of public works activities.

• *Production trends*

From 1980 to 1994, the tonnages of natural aggregates according to their geological nature, were distributed as follows (in Mt):

In million tonnes (Mt)	1980	1985	1991	1992	1993	1994
Alluvial aggregates	238	178	219	197	178	186
Aggregates from limestones	57	53	88	81	77	87
Aggregates from igneous and metamorphic rocks	80	64	92	92	90	94
Total	375	295	399	370	345	367

Compared to the global trends observed since 1980, the various categories of aggregates have not developed in a homogeneous manner.

Developments in %	85/80	91/85	92/91	93/92	94/93
Alluvial aggregates	– 25%	+ 23%	– 10%	– 10%	+ 4%
Aggregates from limestones	– 7%	+ 66%	– 8%	– 5%	+ 13%
Aggregates from igneous and metamorphic rocks	– 20%	+ 44%	0%	– 2%	+ 4%
Total	– 21%	+ 35%	– 7%	– 7%	+ 6%

In terms of limestones and igneous and metamorphic rocks, the production volumes reached in 1991 constitute the highest levels. During the 1980–1985 crisis period, the sharp fall in production is due to the alluvial aggregates losing 60 million tonnes and massive rocks losing 20 million tonnes. These trends result in major changes in their relative share structure:

Relative share structure (%)	1980	1985	1991	1992	1993	1994
Alluvial aggregates	64%	60%	55%	53%	52%	50%
Aggregates of massive rocks	36%	40%	45%	47%	48%	50%
— Limestones	(15%)	(18%)	(22%)	(22%)	(22%)	(24%)
— Igneous and metamorphic rocks	(21%)	(22%)	(23%)	(25%)	(26%)	(26%)

Already begun between 1970 and 1980, the movement of substitution of alluvial aggregates continued during the 80s at the rate of 1 to 2 points per annum. It took place to the advantage of limestones (+9 points from 1980 to 1994) and to a lesser extent to the advantage of igneous and metamorphic rocks (+5 points). Alluvial aggregates represented only 50% of the production in 1994.

2.2.2 REGIONAL SITUATIONS

• *Trend in activities*

If the increase in activity from 1985 to 1991 was not uniformly felt for all categories of aggregates, it also masks some wide regional disparities (see Figure I.6a). Among the regions which have registered a growth of more than 40% between 1985 and 1991, we must mention Auvergne and Lorraine.

The increase in demand in Ile de France also resulted in a big increase in production in Picardie and Haute Normandie. Over the same period, regions like Champagne and Nord registered respectively +65% and +57% but this high growth is brought about by a very good year 1991.

Contrarily, the trend in certain regions has been less than or equal to 25%: Centre and Limousin. The slow growth of Aquitaine and Midi-Pyrénées is due to a particularly bad situation in 1991 in these regions.

From 1991 to 1994, a large number of regions were affected by the recession. Out of nineteen regions, fourteen registered a recession, sometimes a slight one (Alsace, Pays de Loire and Bourgogne, Franche-Comté), sometimes more makedly: Provence, Lorraine, Nord and the two leading consumer regions, Ile de France and Rhône-Alpes Among the few regions which registered a growth, we must mention Limousin and, to a lesser extent, Poitou-Charentes, Bretagne, Aquitaine and Midi-Pyrénées.

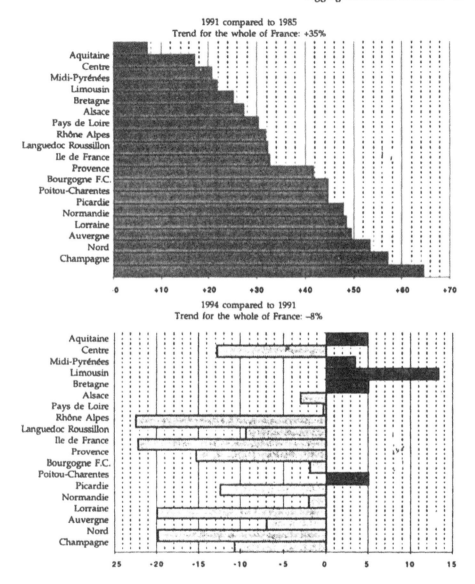

Fig. I.6. Trends in regional productions of aggregates

For all these regions, the year 1994 is different as compared to the preceding two years. From 1991 to 1993, except for Limousin and Bretagne, all the regions were affected by recession (from −2% to −30%).

• Trends in substitution (trend of transfer from production of alluvial aggregates to that of aggregates from massive rocks). At the national level, from 1985 to 1994, the share of alluvial aggregates in the total production of natural aggregates lost 10 points: 5 points in the 1985–91 phase whereas the actual production was increasing for all types of aggregates; 5 points from 1991 to 1994, a phase where the production of only alluvial aggregates decreased.

Over this period, the phenomenon has not been uniform over the entire territory (see Figure I.7).

Regions	% Alluvial aggregates/Natural aggregates		
	1985	1994	Deviation in 9 years
Poitou-Charentes	24%	26%	+2
Alsace	97%	97%	0
Bretagne	15%	13%	-2
Limousin	16%	12%	-4
Picardie	98%	94%	-4
Ile de France	93%	89%	-4
Nord	25%	21%	-4
Rhône-Alpes	80%	75%	-5
Normandie	62%	55%	-7
Centre	68%	61%	-7
Midi-Pyrénées	69%	60%	-9
Pays de Loire	32%	22%	-10
Champagne	69%	58%	-11
Aquitaine	72%	61%	-11
Auvergne	52%	40%	-12
Provence	51%	39%	-12
Languedoc Roussillon	45%	32%	-13
Bourgogne F.C.	48%	34%	-14
Lorraine	82%	62%	-20
Entire France	60%	50%	-10

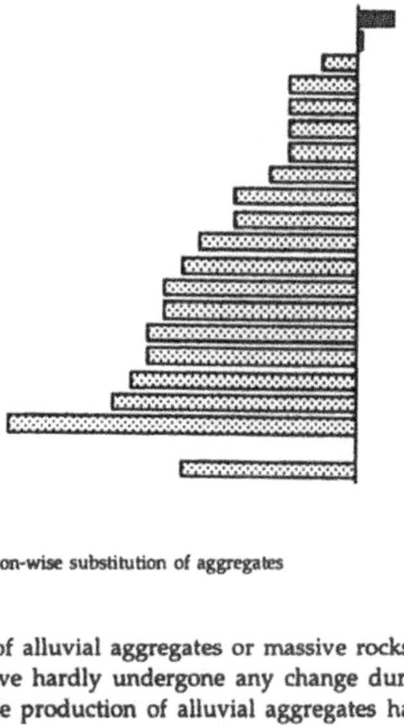

Fig. I.7: Trends in region-wise substitution of aggregates

Regions with high dominance of alluvial aggregates or massive rocks (more than 80% of natural aggregates) have hardly undergone any change during the last ten years except Lorraine whose production of alluvial aggregates has gone down from 82% to 62% of natural aggregates.

The shift was most felt in the regions where the dominance is less marked (between 60 and 80%). It is the case particularly of the Champagne and Aquitaine regions whose deviation from 1985 to 1994 is greater than the national average (–10 points). Only the share of alluvial aggregates in the Poitou-Charentes region increased whereas it is itself massive rock dominant.

It is in the "mixed" regions (dominance not greater than 60%) that the substitution has been most spectacular particularly in Languedoc-Roussillon (–13 points) and Bourgogne-Franche-Comté (–14 points).

2.3 Means of Production

2.3.1 TRENDS IN TURNOVER

With an overall turnover of about 15 billion francs exclusive of taxes in 1994 and a total manpower of 15500 employed in 2100 companies, the industry of aggregates occupies a significant place in the building and public works sector in France.

Being on the upstream of this branch of activities, the entire aggregate sector has always felt the economic movements and the structural changes of these markets right from their very inception.

Figure I.8 shows with a base index of 100 in 1985 the progress in francs of the turnover of the three branches of the profession.

2.3.2 TRENDS IN PROFITABILITY AND INVESTMENTS

• **The industry of aggregates** is a heavy industry which has a high rate of investment.

A plant often costs one-and-a-half to two times its annual turnover. Two thirds of the investments concern fixed installations and one third mobile equipment.

The magnitude of these investments entails a long duration of depreciation. For the product to be considered economically profitable, the plant should generally be written off over a period of more than ten years.

• **In terms of financial results** the profitability of the industry of aggregates has dropped markedly during the 80s.

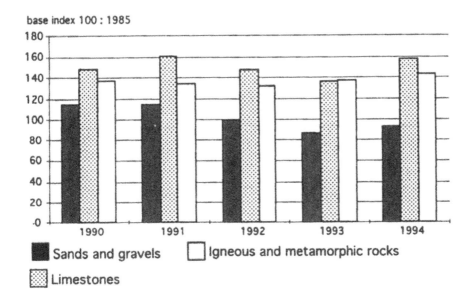

Fig. I.8.

From 1985 to 1990 on the other hand we notice a very marked increase in the result (multiplied by nearly 4 if we compare 1990 to 1985) caused by the resumption of activities.

In 1991 and 1992 the profitability of the companies dropped (with a net result highly negative in 1992). In spite of the reduction in investments, financial charges got multiplied by 3 in two years.

A stabilization of investments registered thereafter and financial charges returned to a more modest level. Since then a fresh spurt of activities in 1994 enabled to close that year with a net positive result.

2.3.3 MANPOWER AND PRODUCTIVITY TRENDS

It is certainly in this field that changes have been most spectacular. Compared to 1985, for a production increased by 23%, the profession employed 17% less manpower in 1994. The productivity gains are as much striking in terms of hours worked; in fact, these latter have fallen by more than 18%. Yet 17.1 tonnes are produced per hour today as against 11 tonnes in 1985 i.e. an increase of 55%.

The following table illustrates these data:

		1985	1990	1991	1992	1993	1994
Number of companies	Alluvial deposits...	1711	1466	1393	1314	1230	1047
	Massive rocks...	920	947	947	934	930	834
Manpower	Alluvial deposits...	10 750	10 178	9 708	9 021	8 473	8 016
	Massive rocks...	7 826	8 207	8 190	7 797	7 605	7 522
Hours worked (in thousands of hours)	Alluvial deposits...	15 645	14 607	13 684	12 449	11 635	11 029
	Massive rocks...	11 491	12 366	12 198	11 465	11 109	11 105

The mechanisation and rationalisation of production equipment has not taken place in a linear manner during the last ten years, however. Productivity gains, in terms of manpower, have rapidly increased between 1985 and 1991 (from 16000 tonnes to 23000, per person). Then this productivity dropped in 1992 and 1993 and finally rose again in 1994 (24 400 tonnes); it may be noted that this increase coincides with that of investments.

2.3.4 INDUSTRIAL STRUCTURES

The industry of aggregates in France is characterised by a multitude of small and medium enterprises (with a national average of 8 jobs per firm) distributed over the entire territory (see following table).

There is no doubt, however, that since the end of the 80s we have been witnessing a trend towards industrial concentration, particularly strong initially around great consumption centres (Paris, Marseille, Bordeaux, Toulouse, Lyon) and which extended, afterwards, through capital sharing or buying back enterprises towards centres of lesser importance.

With the recession which began in 1991 this trend in concentration has slightly decreased. Today, the first five groups control nearly one third of the market.

1994	Enterprises		Productions			
Class of annual production in thousands of tonnes	Number of enterprises per class	Cumulative	Millions of tonnes per class	Cumulative	Percentage per class	Cumulative
More than 1000	61	61	116.9	116.9	32	32
500 to 1000	111	172	74.3	191.2	20.2	52
250 to 500	214	386	75.3	266.5	20.5	73
100 to 250	414	800	66.8	333.3	18.2	91
1 to 100	1081	1881	33.8	367.1	9.2	100

3. ADEQUATION NEEDS RESOURCES

3.1 Operation of Aggregates Market

3.1.1 GEOGRAPHIC PROTECTION

The most striking aspect of this market is the fact that the price of an aggregate supplied at the place where it is to be used includes a very high transport price (at least in percentage) because the aggregate is a heavy product, relatively very cheap when it comes out of the plant.

Besides, geological resources are rather well distributed in the national territory. Of course, certain qualities of materials are missing in certain regions and this poses some specific problems. But for most uses, a technique capable of accepting materials found regionally is available.

The two facts, relative importance of the transport cost and wide distribution of the resource, ensure in favour of each producer a relative geographic protection (but concomitantly an important obstacle to expansion of his quarry).

Let us take the following very simple imaginary model: one producer has succeeded in lowering his cost to 20 francs per ton; a second producer has a cost of 40 francs per ton. Let us assume a centre of consumption located at x_1 from the first producer and at x_2 from the second producer. The latter will deliver his aggregates at a lower price as long as $x_1 - x_2 > \dfrac{20}{p}$ (p being the transport cost per tonne per kilometre).

The second producer enjoys geographic protection, therefore, compared to the first; the location of the deposit with respect to the centre of consumption is thus of considerable importance, so much so that it has prompted producers to set up new plants or encourage clients to do so, and even to induce the opening of quarries in the case of important works, and especially to search for new deposits themselves.

This geographic protection has even come into play at the national level in the past, with producers of aggregates looking for deposits close to Paris and its suburbs. Even today the big quarry development companies are looking for deposits of hard materials nearer to Paris, which are likely to provide aggregates for surface courses of roads.

Geographic protection has two limitations, however. First, quite often the possible zones of deposits are sufficiently vast for several producers to install themselves in the vicinity of one another. In such a case, while all the producers enjoy a protection zone, they have either to come to an agreement or else indulge in fierce competition.

The second limitation results from the fact that the industry of aggregates is one with an increasing yield; it entails very high investments while the marginal cost of production is low. Each producer is therefore tempted to widen his zone of action by selling an additional quantity of aggregates below his average cost but above his marginal cost. The producer is thereby able to compete with another producer in an area which the latter thought to be protected by transport distance.

As it does not appear that during the last thirty years the producers suffered ruin, it must be assumed that clients as well as producers were able to restrict the competition.

As an example of this phenomenon, let us mention the following: producers of limestone materials in northern France have for a long time noticed that Belgian producers could efficiently compete with them and thereby increase their production, which resulted in a fall in price and a further worsening of the conditions of competition. It must be noted, however, that in the present context some producers are compelled to make some productivity gains and therefore to lower their sale price in order to survive.

3.1.2 CONSUMPTION CENTRES

Analysis of demand has revealed that requirements are concentrated in urban areas. Certainly the big construction works of communication routes (highways, roads, new railway lines, ports of any type) or of production of energy (nuclear plants, large dams) involve large consumption of aggregates and are located outside urban areas. But construction and maintenance of houses, urban structures and industrial construction concentrate the requirements in the cities or on their periphery.

Each urban area is therefore a consumption centre and big metropolises control the market of aggregates. This is obviously the case of the Paris region for which the requirements can be met only by using materials from far-off fields and consequently with a high transport cost. It should be noted, however, that a major part of this transport is through waterways at a very low price per kilometre.

3.1.3 ROLE OF TRADERS

This role is not to be neglected. There are, in fact, a large number of small consumers who do not go directly to a producer of aggregates but to a wholesaler, who naturally exploits the competition between quarry owners and intensifies it. Of course, this game sometimes breaks geographic protection.

To some extent, manufactures of ready-mixed concrete play a similar role. However, experience has shown that certain big companies in this profession are branches of cement and sand and gravel companies and this obviously complicates the market game. It is also true for road contractors who own more and more quarries.

3.1.4 MARKET FLOW

Another important element that delays diffusion of technical innovation, results from the habits of the customers.

We shall only mention one example: traditionally a good concrete sand should be siliceous and with round grains. But more than thirty years ago in the South of France people learned to manufacture concrete and even quality concrete with calcareous sands. There are regions in France, poor in alluvial materials and rich in limestone, which are only now discovering this solution. Other regions show reluctance to use concrete sands obtained from crushed igneous or metamorphic materials, so these sands pile up on the floor of quarries, which studies have demonstrated to be a solution.

3.1.5 SEASONAL NATURE OF DEMAND

Roadworks are subject to climatic hazards and are therefore seasonal. Building construction works are better distributed but they too experience significant seasonal variations, the ups and downs of which coincide with those of the road industry. The demand is therefore highly seasonal and the producer has perforce to stock a major part of his production.

Everyone knows that this is very costly. We were surprised that while passing through the period of recession, producers did not propose significant concessions to those of their customers who could have bought materials from them during the lean seasons.

3.1.6 UNSOLD AGGREGATE SIZES

The demand for various aggregate sizes is obviously not based on the production of the quarry; rather, it is the producer who should try to produce the gradings required. He cannot always succeed, in which case he has an excess of gradings, which are actually a by-product, which he attempts to dispose of by lowering his sale price.

During the high-demand years of macadam, certain producers of aggregates nevertheless produced large quantities of manufactured sand, a part of which they sold at a low price but a major part of which they dumped. They found themselves at the head of a huge capital the day new methods valorised this sand.

Full use of the various materials produced in his plant is therefore the main worry of the producer of aggregates.

3.1.7 STERILISATION OF CERTAIN RESOURCES

Not every aggregate resource is exploitable.

From the economic point of view, there are already serious limitations, such as the large thickness of overburden, transport distance up to the centre of consumption, major difficulties of processing etc.

It may also happen that some real bans sterilise a field: this is the case if town planning regulations have imposed a living area or an industrial area at its site. It is also the case if the deposit lies in a protected site.

Even without a real interdiction, constraints on a deposit may be such that exploitation is not advisable. Such is the case, for instance, of a rich agricultural area, which it would be advisable to reserve in the general interest for cultivation.

Lastly, extractions in the bed of streams cause wide imbalances in aquatic life by destroying spawning grounds.

It may thus happen more and more often that for various reasons the Administration or local authorities raise objections to the use of a particular field.

Since it is necessary to continue to supply the market with aggregates, some fields should be reserved and safeguarded from sterilisation. The Commissions on Quarries (several per region) would play an important role in this regard (see Chapter II).

The authorities should also intervene in the case of obvious obstruction and in the public interest.

The mining code has, moreover, provided for zones of exploitation and co-ordinated development which should give rise to the establishment of a plan opposing third parties; this solution is complicated and difficult to implement, however (see Chapter II).

3.1.8 CONCLUSION

Resources are generally adequate, quantitatively speaking, to meet requirements: adaptation takes place essentially at the local and even the regional level.

It may happen, however, that certain qualities of aggregates are not available in an area: the demand then adapts itself technically, or resorts to external supplies for particularly demanding uses (see Section 3.3).

3.2 Transport of Aggregates

3.2.1 SHARE OF VARIOUS MODES IN THE TRANSPORT OF AGGREGATES

• Roads take care of the bulk of the transport:
> — 92% in tonnage;
> — 71% of kilometre tonnes.

• Water transport carries 4% of the total volume, but 11% of the kilometre tonnes.

• As for railways, they also carry 4% of the volume but their share in kilometre tonnes is more significant at 18% of the total.

Over these last ten years, we notice a very clear increase in the market shares of the various modes of transport. If rail transport has remained more or less stable, water transport on the contrary has fallen to the advantage of road transport.

It is interesting to note that the average distance of transport by road has increased and is today of the order of 30 km. That of rail transport is 200 km and that of water transport remains stable at 95 km.

In fact, road transport has not only gained in tonnages, but more significantly, in competitiveness.

Let us incidentally note the privileged relations between modes of transport and sources of aggregates. Water carries only alluvial deposits, essentially meant for hydraulic concretes. This of course is due to the obvious relationships which exist between certain sand and gravel pits and the waterways.

As for the railways, they carry in particular aggregates of igneous and metamorphic rocks and cater mainly to road works.

3.2.2 CHARACTERISTICS OF VARIOUS MODES OF TRANSPORT

• **Roads** are unquestionably the most flexible mode of transport. The rates of supply can be modelled on those of the site. Roads carry aggregates directly, without transhipment of consignment, from the place of production to the place of consumption.

Transport can be equally ensured by trucks run by the producer, the contractor or transporters. However, trucks are also mobilised by concurrent activities, e.g. for the sugar beet season.

Road transport can be very damaging to pavements due to the weight of the loads transported and the number of trucks.

There is therefore the risk that transport may prematurely damage the roads, which are most often the regional or local roads of light and, hence, fragile structure. Furthermore, in view of the low unit pay-load tonnage, meeting the high traffic flows required by big works poses a problem.

Lastly, road transport consumes more energy than its competitors. However, appreciable progress has been achieved in the field of motorisation.

The truck is still the privileged means for short distances. It could, however, prove competitive on distances exceeding 150 km when a return freight is arranged.

• **Rail transport** enables carrying full train loads of aggregates with a very high regularity in special wagons managed by the SGW. This is no doubt a big advantage for carrying out big works. This regularity is due to a strict programming of the trains, which requires setting up adequate means.

In any case, an installation should be available near the site for receiving the trains and unloading the aggregates.

The SNCF has made laudable progress in this regard by making platforms available in its stations. Hence a terminal transport is generally required with the disadvantages inherent in transhipment of consignment (additional cost ...).

The average distance of transport by rail is around 200 km. However, the rail solution could under good conditions prove competitive for distances of less than 100 km.

• Broadly speaking, **waterways** are quite suitable for the heavy goods transport in bulk. Their low cost of operation corresponds well to the low added value of the aggregates.

The places of production and consumption should be connected by waterways. This has been the case of nuclear plant construction sites. This is also the case of extraction zones of the Seine, for instance, whose production is mostly sent to the concrete plants set up 'on water'. Except for these cases, transhipment of consignment is inevitable, which considerably reduces the efficacy of this mode of transport.

Regularity confers on waterways an interesting advantage, albeit programming sometime proves difficult.

Waterways handle very few supplies for roadworks. The few attempts made under the programmes of rehabilitation were not successful. Relaxation of the freight regulations could enable successful implementation of a few suitably located operations in future.

3.2.3 IMPACTS OF COST OF TRANSPORT ON SELECTION OF AGGREGATES AND TECHNIQUES

The cost of transport will unquestionably assume an increasing importance in the coming years and there is risk that the use of a particular type of aggregate will more and more depend on the transport distance between quarries and work sites.

The search for savings thus compels owners to use at best locally available materials and to resort, if need be, to materials which are not very much used because they do not always meet all the specifications in force (see chapter V).

This recourse to local materials should never lead to lowering of the quality of the works, however. This can be precluded only by developing specific methods suitable to the geotechnical characteristics of the materials. If we take the example of pavements, it may be a question simply of:
- either new techniques,
  — processing of aggregates,
  — manufacture and use of treated materials.
- or of techniques avoiding the use of 'fresh' aggregates:
  — in-situ treatment of the soils,
  — in-situ recycling of pavements and coated materials,
  — recycling of demolition concretes and bituminous concretes.
- or new pavement structures,
- or the development of chipping-embedment or mixes of bituminous or hydrulic concretes or even high-performance bituminous mixes.

Finally, we measure how much the transport parameter weighs in the development of such techniques.

An increase in this parameter would motivate owners to evaluate local materials on the one hand, and conserve the use of aggregates from large quarries for high-performance techniques on the other. Such an increase would ultimately reduce the transport market.

Along these same lines, if the short-distance transport cost increases faster than the long-distance, on could consider simultaneous transportation of aggregates for surfacing and road foundation*. Obviously, railways would benefit from this.

3.2.4 CHOICE OF MODE OF TRANSPORT

In practice the mode of transport is not chosen *a priori*. What is important to the client is the overall cost of the work. This cost includes the price of the aggregates, the price of their transport, the price of laying and, of course, the quantities involved, which may in certain cases vary with the quality of the materials.

At best, when the quality of the aggregates is straightaway fixed, it is the delivered price which is decisive.

---

*Road foundation comprises the internal structure of the road, over the formation and under the surfacing, and consists of subbase and base courses.—Tech. Ed.

It is therefore very clear that transport is merely one among many parameters. The choice of the mode of transport will therefore be influenced by competition, which it is advisable to encourage, according to the size and type of works:

— among aggregates;

— among techniques;

— among modes of transport;

— among various possible combinations of the above, by trying to get the most out of them.

3.3 Flow of Aggregates

The demand for aggregates is at the root of transport flow, the magnitude of which is linked with the intrinsic characteristics of the aggregates market, the geological or technical specificities and certain geographic peculiarities.

The abundance of different resources spread over the entire territory, the technical progress which tends to make local products usable, the bulky nature of the aggregates the price of which doubles after 40 to 50 km covered by road: all these play a limiting role regarding the movement of materials over long distances.

Results from a study carried out by the Economics Department of UNICEM on parafiscal tax on aggregates revealed that if local adequation between supply and demand for aggregates is generally ensured, real traffic currents exist which are summarised below.

3.3.1 NATIONAL FLOWS

The figure given below indicates the percentage of production of each region with respect to its requirements; most of the regions ensure at least 90% of their supplies from their own resources. The main exception is Ile-de-France where the deficit exceeds 40% of the consumption but even in this case the deficit is filled by adjoining regions.

This is also the case with the Nord region, although of lower magnitude.

In addition to these flows associated with quantitative deficits, due to the exhaustion of resources or more often limitation of the possibilities of opening new quarries (especially in the field of alluvial deposits), large flows associated with qualitative phenomena are observed at the national level. It is a question either of mutual exchanges between neighbouring regions enabling a rebalance of resources with respect to requirements (case of Picardie which delivers alluvial materials to Nord and receives in turn limestone aggregates) or of flows over long distances corresponding more often to specific requirements for road applications (surface course) and concerning generally aggregates from massive rocks.

• Sands and gravels from alluvial deposits

These flows concern neighbouring regions, the most important being as follows:

— Ile-de-France, which receives 5 Mt from Upper Normandie, 1.5 Mt from Picardie, about 1 Mt from Bourgogne and 1 Mt from Centre. Most of these outside supplies take place through rivers but the other modes of transport are also used.

— Nord, which imports about 1 Mt of alluvial aggregates from Picardie (in exchange for limestone aggregates for an equivalent amount) and also about 1 Mt of marine aggregates.

— Lorraine, which imports more than 0.5 Mt from Alsace.

• Limestones

Nord dispatches 1.5 to 2 Mt to Picardie as well as Ile-de-France and a lesser amount to Champagne-Ardenne. These aggregates are used in roadworks.

The other movements are not significant and correspond rather to border (e.g., inter-regional) phenomena. At most, we can mention an export of a little more than 500,000 tonnes of limestone aggregates from Charente to Aquitaine and some exchanges from the Centre region to Ile-de-France (limestone aggregates from Beauce).

• Igneous and metamorphic rocks

The main regions which import aggregates from igneous and metamorphic rocks are:

— Ile-de-France with about 3 Mt essentially from Poitou-Charentes, Lower Normandie and Loire.

— To a less extent, we can mention Upper Normandie, Picardie, Champagne-Ardenne and Centre.

These deliveries correspond to hard aggregates used in the surface layers of pavements in regions lacking these type of aggregate. Deliveries are generally made by rail, on a regular basis with full trains.

3.3.2 FOREIGN TRADE

It is also worth mentioning the movements of aggregates with the foreign countries. On the whole, the foreign trade with European countries is surplus (13 Mt exports against 6 Mt imports). Compared with the total quantity of tonnage transported, these flows are low (4% for exports and 2% for imports). However, these exchanges, essentially frontier (inter-country), are quite significant for the "departments"* and regions concerned.

• *Imports*

Imports of aggregates represent slightly more than 6 Mt per annum. Aggregates are mainly imported from EEC (more than 90%) and particularly from Belgium (limestone aggregates from Tournaisis, igneous and metamorphic aggregates from the environs of Lessines, and sands from the Valley of Escaut). 'Alluvial aggregates' from Great Britain and the Netherlands also enter France.

These imports are mainly transacted for the Nord-Pas-de-Calais region (5 Mt) and Lorraine but can also enter the more far-off regions.

These flows are significant, especially from Tournaisis towards Lille and its environs, because the French quarries (in the Avesnes and Boulogne regions) are geographically less well placed to withstand foreign competition.

*See footnote to Fig. 1.4.

• *Exports*

The plain of Alsace is a vast reservoir of alluvial deposits which are easy to access and often exceed thirty metres in thickness; hence the production of sand and gravel there has considerably increased over the years. In fact, the navigable Rhine river constitutes a remarkably route for dispatching materials extracted in France to Germany, the Netherlands and Belgium.

Exports also take place by road to Germany and Switzerland.

Alsace exports nearly 9 Mt of aggregates including more than 5 Mt to Germany and 2 Mt to Switzerland. These 9 Mt represent two-thirds of the total French exports of aggregates.

Lorraine exports nearly 1.4 Mt (as much to Germany as to Belgium), the Rhône-Alpes region nearly a million (essentially to Switzerland) and Franche-Comté 700,000 tonnes (also to Switzerland).

3.3.3. CONCLUSION

Unlike other basic materials (gypsum, silica, clay, stones etc.), the geographic adaptation of consumption to production of aggregates is rather satisfactory.

The market of aggregates continues to be a local market with some notable exceptions related to geology and the requirements of certain jobs as well as the peculiarities of some large consumption centres.

We observe a relative reduction in flows over long distances from region to region (less than 10% of the needs). On the other hand, we notice a fairly large number of inter-"department"* movements. The administrative division of the "departments" does not necessarily correspond with the outlines of the fields defined by geology. In terms of volume these exchanges from 'department' to 'department' represent a flow of slightly more than 60 Mt, i.e., 17% of the national consumption.

The development of decentralisation and the evolution of rules and regulations concerning quarries which have given increased powers to local authorities, are well suited to the specific nature of aggregates. It is necessary to avoid too narrow territorial protectionism, however, because if the local level is the dominant factor, it appears quite obvious that aggregates constitute an open market which necessarily requires exchange flow between 'departments' as well as regions.

---

*See footnote to Fig. 1.4.

II

Quarries and Environment

Rémi Galin, Louis Primel and Jean-Michel Sionneau

A theoretical and practical treatise on aggregates should devote a chapter to problems of environment caused by their extraction. These are many and varied and so, since 1970, French authorities have gradually laid down a set of constraining rules.

These rules and their evolution, the problems of environment arising before, during and after the life of a quarry and the means now known to restrict their impact on our environment will be taken up one by one.

The environment of quarries is a vast subject. We shall therefore voluntarily restrict ourselves to a few aspects, summarising our knowledge of them.

The reader should not be surprised if sometimes our remarks go beyond the subject of aggregate quarries because problems arise in an identical manner for all types of quarries.

INTRODUCTION

Many are the interactions between quarries and the environment. Quarries are most often felt by the public through the damages they cause to the environment as a result of the various nuisances and soil disruptions which they produce.

The other aspects associated with quarries, in particular their economic function or the fact that solutions do exist to reduce or eliminate the disadvantages of their exploitation, are less known.

Quarry materials are used by all sectors of our economy (building construction and civil engineering works mostly, but also iron and steel, pharmaceutical, electrical, food industries etc., agriculture) and are the most consumed products after air and water. Nearly 30 km^2 of our land are taken up every year for their production. One can readily understand the problems this may pose.

The legal constraints laid down by the authorities and the state-of-the-art in matters of environment of quarries are described in the following two sections.

1. LEGAL ASPECTS

1.1 Developments up to 1970

Before 1970, it was necessary only to make a simple declaration to the Mayor about quarries with a copy to the Chief Engineer of Mines; the sole purpose of the declaration was to enable administrative surveillance vis-à-vis observance of rules and regulations.

Such a situation could not continue in the general context of a major development in the exploitation of quarries consequent to increasing needs, more and more severe competition among the various possible uses of soils and awareness of the problems of environment.

In 1970 the requirement for declaration was replaced by that for obtaining prior permission with the following objectives:

— to permit a more rational exploitation of the deposits and to avoid wastage or sterilisation of the resources required to provide the country with materials;

— to fix the obligations of the operators with respect to restoration of the soil at the end of exploitation;

— to give the Administration the power to prohibit the opening of quarries whose exploitation might prove an obstacle to the application of a measure of general interest or on the grounds that the applicant lacked sufficient technical and financial capacity;

— to synthesise all the administrative decisions required under the various rules and regulations.

But it very soon appeared, due to the more and more marked sensitivity of public opinion regarding the problems associated with safeguarding the environment (and one of the results of which was the creation in 1971 of a Ministry of Environment), that it was difficult to leave matters at that and not to give to the authorities the legal means to intervene more efficaciously for the protection of the environment.

This concern of the community to protect its surroundings naturally led the Parliament to considerably strengthen the legislative machinery.

1.2 Legislative Texts at the End of the 1970s

These are four in number:

— modified Code of Town Planning (1976),
— law pertaining to the protection of nature (1976),
— law on classified installations for protection of the environment (1976),
— modified Mining Code (1977).

These legislative texts considerably strengthened the legal means to protect the environment. Nevertheless, it may be noted that they do not systematically result in more obstacles to the opening and exploitation of quarries just because the environment is taken into account; on the contrary, possibilities of exploiting deposits are created; similarly, a better organisation of mining operations has been sought through these texts.

For the convenience of the users (Administration, public, quarries) and in order to ensure the principle of uniqueness of procedure and decision, the authorities opted for a single decree (1979) which lays down for the use of quarries the conditions of application of the last three laws mentioned.

1.3 Present Legal System

1.3.1 MINES OR QUARRIES?

Before describing the broad outlines of the legal system applicable to quarries, it seems useful to define what is a quarry in the sense of the Mining Code: 'a quarry is a deposit of a mineral or fossilised substance existing on the surface of the earth' (1972).

Substances whose deposits constitute quarries are negatively defined in the Mining Code; they are those which do not figure in Article 2 of this Code, of which aggregates constitute the main ones (which are not defined in the general introduction to the work), stones, slates and other building materials, limestone used in the manufacture of lime, cement and hydraulic binders, clays and other materials used in the ceramic industry, enriching agents, barite and peat.

Deposits of those substances restrictively mentioned in Article 2 of the Mining Code are considered mines; these are essentially energy substances, metal ores, certain salts and metalloid ores and certain gases.

The demarcation between mines and quarries does not therefore result from the mode of exploitation (there are underground quarries and open-cast mines), but from the substance itself.

The demarcation between mines and quarries originates from the lawmaker's desire to give, as it were, the character of a public property to the mine substances because of their scarcity and consequently their high value. The more common quarry substances are subjected to less constraining conditions.

In fact, two widely different legal systems are applicable to mines and quarries.

In the case of quarries, Article 105 of the Mining Code specifies that their disposal be left to the owner of the land. This is not so for mine substances over which the owner of the land has no right. Permission to exploit them is given by the government. The permit holder (permit to exploit or concession) enjoys an exclusive right, as opposed to the owner, who is compensated however for the loss he suffers due to the exploitation.

The exploitation of a quarry by the owner or his assignees is subject to permission granted by the Prefect.

1.3.2 OPENING A NEW QUARRY

It is necessary in any case, whatever the size of the quarry, to obtain authorisation from the Prefect.

• *Résumé of procedure*

1) Submission at the Prefecture of a file comprising:
    — an application,
    — maps and plans,
    — financial guarantees,
    — an impact study (see Section 1.3.3),

— a study of the risks.

2) Conducting a public inquiry for a minimum duration of one month.

3) Consultation:

— with various administrative departments,

— with the concerned communes (districts),

— with the departmental* commission on quarries (see Section 1.3.4).

4) Administrative decision and publicity of the decree.

5) Declaration as to when quarrying operations will commence.

The total duration of this procedure is about twelve months.

1.3.3 IMPACT STUDY

From the 'environmental' point of view, the obligation to carry out an impact study prior to the opening of a quarry represents an essential step in taking into account environmental problems. It is a question, right from the beginning, of characterising the nature and extent of the expected impacts of exploitation and defining the means to minimise or eliminate them, and specifying the subsequent method of rehabilitating the grounds.

The impact study is therefore, from the very beginning, a key factor in any quarry opening operation. It is a privileged means of compelling everyone concerned (quarry owners, the communities, the administrations, the public...) to look into the consequences of setting up a materials exploitation plant on the environment (in a broad sense).

The preliminary study and analysis of the various components of the environment enable measuring the extent of the impacts, the realistic possibilities of rehabilitation and to assess afterwards, with full knowledge of the facts, the advisability of the project.

• *Initial state of the site and its milieu*

This section of the impact study enables examination, one by one, of the basic data relating to the geographic milieu, geology, hydrogeology, hydrology, pedology, ecology, climatology, socioeconomy and land occupancy, landscape, rules concerning the site and its environs etc.

• *Analysis of the effects on the milieu*

After an outline of the planned mining operations, this section of the impact study analyses one by one the direct and indirect effects of the project on the various components of the natural and human surroundings described earlier. This environmental evaluation describes the nature of the expected nuisances (noise, dust, ecology, socioeconomy...), whether they are negative or positive. It also tries to quantify them.

• *Reasons for the choice of project*

This section describes the motivations and environmental reasons (technical, economic, ecological etc....) leading to the definition of the project.

---

*France is divided into 95 administrative divisions called "departments" (mean surface: 6000 square kilometres).

• Steps proposed for eliminating, reducing and, if possible, compensating for the harmful effects of the project taking into account the environment impacts described in this section of the impact study, the possibilities and means to be used for finding a remedy for the possible impacts on the landscape, natural surroundings, waters, soils, neighbourbood... are studied. An estimate of the cost of these protective steps is made.

• *Rehabilitation*

Depending on the various possibilities associated with the site and the motivations of the concerned partners (owner, producer, public bodies...), this part of the study describes the principles to be adopted for the gradual and final rehabilitation of the lands to be exploited.

1.3.4 DEPARTMENTAL COMMISSION ON QUARRIES

The Commission examines the applications for permission to open quarries. In addition, the Commissioner of the Republic may refer to the Commission any question of general interest regarding the setting up and exploitation of quarries in the department and the latter may formulate in this regard guidelines and recommendations which it may deem useful for reconciling the implementation and development of this activity with protection of the environment.

The composition of this Commission is as follows:

a) a department councillor (elected), appointed for three years by his colleagues;

b) a mayor appointed for three years by the association of mayors of the department or, failing which, by the mayoral college of the department;

c) a representative of the Chamber of Agriculture appointed by the latter;

d) two members representing associations for the protection of nature and the environment;

e) a person qualified in the field of natural sciences;

f) a representative of professions using quarry materials;

g) two representatives of the profession of producers;

h) seven officials representing the Ministries in charge of mines, equipment, agriculture, sanitary and social affairs, environment, architecture and telecommunication stations.

It may be noted that the Prefect may or may not agree with the views of this Commission.

2. EFFECTS OF QUARRIES ON THE ENVIRONMENT AND MEANS TO REMEDY THEM

2.1 Effects of Quarries on the Environment

For the sake of convenience, the impacts on the environment are listed below according to the medium on which they are exerted: impacts on the landscape, the atmosphere (noise, vibration, dust...), the waters (underground and surface), the ecological milieu (flora and fauna), the soils etc. The technical solutions generally adopted for controlling them are indicated. Let us point out, however, that this description is not intended to be exhaustive, given the vastness and complexity of these various subjects. The level of information is therefore of a general nature and the reader interested in some particular aspect is requested to consult the References given at the end of this chapter. It may be noted here that joint actions taken in recent years have significantly improved knowledge in the field of interactions between quarries and their environment by virtue of the numerous works and studies carried out.

2.2.1 EFFECTS ON THE ATMOSPHERE (NOISE, DUST, VIBRATION...)

• *Noise*

One can differentiate in quarries:
   — sounds emitted by site equipment (processing of materials, crushing, screening) causing a relatively continuous and repetitive noise level;
   — pulsating short sound emissions, generally of much higher decibel intensity (in the case of exploitation by mine charges and blasting);
   — sounds arising from movement of vehicles carrying materials.
   These various sources of noise have effects of a distinctive nature:
   — The charging of mines in a quarry is generally not very frequent (maximum of a few blasts per week) and an administrative permission is required for stockage and usage of explosives. Charging methods and their execution (prior warning to the public, use at regular hours, firing in deep holes...) combine an obvious concern for the optimum pursuit of safety of the workers and reduction of nuisances, in particular sound nuisances.
   — Noises caused by site equipments (drilling equipment, motorised compressor sets, motorised site equipments) and processing plants (crushing, screening, washing...) remain relatively confined to the mining site and its peripheral area (a few hundred metres at most). Noises caused by the processing and site equipment can be reduced by proper selection of machinery, covering or hooding their sensible parts, proper maintenance, but also and especially by a proper organisation and planning of exploitation (see Table II.1).
   Generally the sound levels emitted at the source by site equipment or processing plants approximate or exceed 100 dB (A). This high intensity requires moving away two or three hundred metres if one wishes to have an undisturbed ambient noise level (of the order of 45 dB (A)).
   Appreciable reductions in the sound level are possible by simply setting up the plants in a judicious manner, taking advantage of the natural topography (hollows, mounds) or such 'designed' specially to act as a screen (plant-covered earthworks, stocks of materials...) (see Fig. II.1).
   Transport of materials outside the quarry may cause an intermittent rise in sound level during conveyance depending on the increase in vehicular traffic induced by the quarry.
   Rules and regulations specify a noise level for various areas (urban, semi-urban, rural, industrial areas), which should not be exceeded.

• *Vibrations*

These concern essentially massive rock quarries whose mode of exploitation (mine charges) is likely to produce vibratory movements in the soil.

Table II.1: Means of reducing noise is quarries [1]

Reduction principle	Level of application of means of reduction	Description of means
Equipment technology	Source	• Use of rubber: — screen cloths — wearing components in rod mills — inside covering of chutes
	Source	• For movable equipment: designing less noisy engines — treatment of air inflows — special exhaust silencers
Buildings or entirely closed hoods	Source	• For movable equipment: hooding of engines
	Nearby field	• Hooding of fixed equipment: either unit-wise (case of cladding screens) or overall (confining the plant in a building)
Screen	Nearby field	• Earth mounds or stocks of materials placed at periphery of the plant
	Nearby field	• Provision of annexes (offices, garage etc.) between source and receiver
	Nearby field	• Placing equipment in pits
Screens	Nearby field/distant field	• Placing equipment at bottom of quarry
	Distant field	• Plant located at lowest and deepest possible point in the site
	Distant field	• Various screens (earth mounds and others) near receiver
Distance	Distant field	• Keeping noisy equipment as far away from receiver as possible

In addition to aerial waves, charges induce vibrations in the surrounding ground which are defined by their amplitude, speed and acceleration against time.

Two types of movements essentially characterise the vibrations generated by a quarry:
— stationary movements associated with the activity of crushing units;
— transitional movements associated with mine charges.

As for the first type of movements (stationary), their propagation depends largely on the geological nature of the terrain traversed and their frequency falls within a range of 5 to 10 Hz. The possible movements associated with this type of vibration are almost insignificant.

The second type of movements (transitional) involves accelerations of several orders of magnitude which are lower than the levels which would be critical.

A certain number of simple measures make it possible to tone down the effects due to vibrations. It is generally recommended that charging be carried out on fixed days and at certain hours.

The use of microdelay action electric detonators (staggering by a few tens of milliseconds) ensures that vibrations will be felt separately without increase in maximum amplitude. Efforts should also be made to achieve a preferential orien-

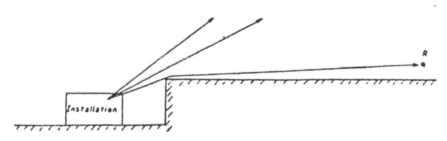

Plant located at bottom of quarry (diffraction due to face)

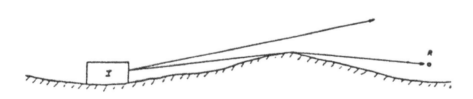

Plant located in a depression (diffraction due to topography)

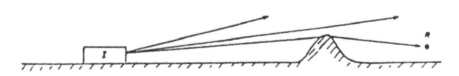

Receiver protected by an earth mound

Fig. II.1. Principal means of reducing noise in a distant filed [1].

tation of the extraction plan, permitting an energy dissipation towards a no-risk zone.

• Dusts

The main sources of dust in a quarry are the crushing-screening plants, conveyor belts and movement of vehicles (incidentally drilling). The flight and movement of dusts into the atmosphere and the environment close to the quarry depend on their grain size, meteorological conditions and topography. These emissions are perceived as sources of nuisance and may, in some exceptional cases, be a health hazard for the operating personnel, cause disturbances to nearby crops (vineyards, orchards...) or, more simply, aesthetic changes to the landscape.

A certain number of measures safeguard against these various effects:

— hooding and placing, if need be, at a lower level the processing units which produce the maximum dust (crushing-screening and conveyor belts);

— use of drilling equipment fitted with an independent dedusting device which enables concentration of dust and its removal;

— regular watering of the tracks during dry whether to tamp the dust on the ground, or even surfacing the tracks;

— if need be, incorporating in the quarry mining plan provisions relevant to the meteorological conditions (direction and force of prevailing winds) favouring the dissemination of dusts (topographical layout favouring the trapping of dusts, quarrying faces serving as screens...).

2.2.2 EFFECTS ON WATERS

The impact of opening a quarry on groundwaters differs from that on surface waters. Generally, the more the waters and materials interpenetrate, the greater the probability of interactions. In this respect it should be emphasised that clogging appears to be a key parameter in the exchanges which occur in all fields of interaction.

The following are the various fields to be studied:

— *hydrodynamic:* modification in the water flows, in the piezometry of the groundwaters and in the productivity of the water catchments;

— *hydromechanical:* transport of suspended matter, change in the banks and beds due to erosion and sedimentation or reduction in exchanges due to clogging;

— *physicochemical:* change in the chemical or bacteriological quality of the waters with or without intervention of pollutants;

— *hydrothermic:* temperature variations in groundwaters or rivers caused by gravel pits;

— *hydrobiological:* modification of the biological balances in rivers or gravel pits, in particular their eutrophy.

The resultant consequences will be more or less extensive and lasting depending on the type and mode of exploitation and subsequent evolution of the excavations.

• *Hydrodynamic field*

Extraction of materials from the bed of minor water courses entrains a remodelling of their profile lengthwise and crosswise, which modifies the water flow. This results in an overall lowering of the water line of the river and the associated alluvial groundwater.

Overdigging the bed may also compromise, due to regressive erosion, the stability of certain structures (banks, bridge supports...) upstream. All these phenomena are known and the present law prohibiting extractions in the bed of minor water courses reduces their importance. Nonetheless, certain amenities (weirs) do make these exploitations compatible with the necessary stability of the bed.

Extraction of materials in a major water bed entrains the opening of an excavation in an aquifer system; this locally modifies the parameters which govern the flow of groundwater, namely, the limits of the aquifer (nature, geometry), the associated conditions (nature and distribution of the exchanges), its hydrodynamic characteristics (distribution of permeabilities and storage coefficients) and finally the resultant piezometry.

The extent of the impact will depend on the geometry of the excavations, on the mode of exploitation and redevelopment and finally on their age.

So, as soon as the excavation touches the groundwater, a system of exchanges takes place between it and the water plane, the intensity of which is the key factor for most of the interactions between the quarry and its environment.

The resultant main effects relate to one of the diagrams in Figure II.2 and underscore the importance of the clogging factor.

This local modification in flows may also have some effects on the productivity of the water catchments, whose performance will be degraded or, contrarily, improved (as the case may be).

Except for the problems of accidental or chronic pollutions, only those exploitations in water other than the bed of a minor water course are of concern. In fact, quarries in the bed of a minor water course have only a negligible influence on the chemical (except for suspended matter) and bacteriological quality of these waters.

We generally observe a change in the groundwater in the gravel pit, which leads to a reduction in the total mineral content, hardness, alkalinity and pH. These mechanisms reflect precipitation of the bicarbonates and hydroxides of iron and manganese in particular, and an increase in dissolved oxygen content. These modifications persist downstream but subside rather rapidly either by dilution or, whenever possible, also be reaction with the aquifer, which always rapidly imposes its facies on the waters which impregnate it.

Nitrate contents also decrease regardless of whether they are accompanied or not by an increase in ammonia and nitrogen, and organic carbon, due to biological activity.

The sulphate and phosphate contents in the gravel pit can also decrease under the action of sulphate-reducing bacteria or other micro-organisms.

Chlorides remain practically unchanged, probably because of the high solubility of this anion under all circumstances and its low involvement in biological activity.

From the bacteriological point of view, the various research works carried out (Alsace and Haute-Normandie in particular, see References) have shown no appreciable contamination of the water in gravel pits by conventional pathogenic bacteria. Consequently, the impact on groundwaters and, for greater reason, on water catchments, can only be low or nil due to the filtering role played by alluvial deposits.

Sulphate-reducing bacteria, as in the upstream and downstream groundwaters, are also observed in the gravel pits.

The risks of pollution of waters in the gravel pits however remain real due, most often, to unauthorised dumping into them. An antagonism between two mechanisms thus occurs:

— a momentary improvement in the quality of the groundwater downstream of the gravel pit over a distance of about a hundred metres;

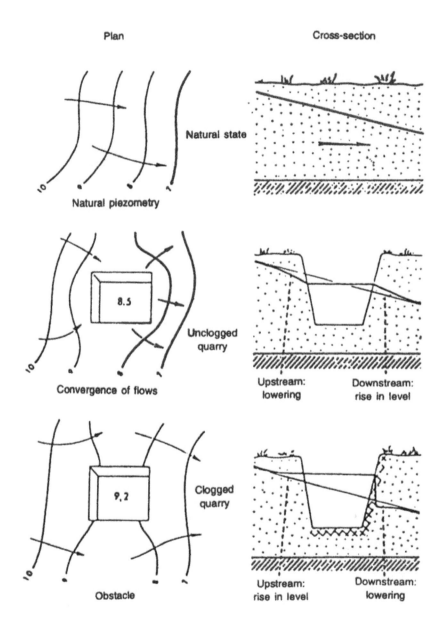

Fig. II.2. Piezometric effects associated with clogging of an exploitation (after P. Peaudecerf, BRGM).

— exogenous contamination (accidental pollution, malevolence).

Subject to precautions against accidental pollution, the uncovering of groundwater which results from mining the materials generally has a neutral and even beneficial (denitrification) effect on the quality of the groundwaters.

• *Hydrothermic field*

Groundwaters generally show a constancy in temperature which results from the absorption of climatic variations in the soil. The opening of a quarry reduces or eliminates this heat protection, which in turn increases the water/atmosphere exchanges: groundwater warms up more in summer and likewise cools down in winter just below the quarry. The temperature variations thus caused have repercussions downstream on both underground or surface waters.

The climate and temperature of the groundwater condition the direction and intensity of the heat exchanges, the degree depending on the area of the water expanse. The depth of the quarry exerts an effect on the temperature distribution within it. According to the seasons, we notice a more or less marked homogenisation or stratification which may, moreover, be reversed from summer to winter or may be perturbed by the exploitation.

The thermal impact of the quarry on the groundwater depends directly on the hydrous exchanges: a 'bubble' of hot water is injected in summer (or the opposite in winter) which will increase according to the rate of flow of the groundwater. If it is desired to restrict this impact, the exchanges should be stopped by voluntary clogging of the quarry, for instance.

The presence of gravel pits undoubtedly modifies the temperature of the groundwaters downstream. These disturbances take place in proportions and at distances which vary greatly according to the local situation.

As for surface waters (water courses), exploitations which may have a thermal impact on these surroundings are, a priori, quarries in the bed of a minor water course and those situated in alluvial groundwater. The thermal effects of these exploitations are generally negligible or nil.

The temperature of the outlet may exceed by 10°C that of the river and at a few metres downstream of the communication, the thermal impact may reach 4 to 5°C. These effects subside very rapidly downstream by dilution over a distance of a few tens of metres. The cumulative effect of the unit influences is not felt in the river.

Should the need arise, outlets with siphons drawing the cooler water from the bottom of the borrow pits may be installed to reduce the thermal impact. The remoteness of the borrow pit from the river and clogging of the exploitation are also factors which will help restrict this impact.

• *Hydrobiological field*

The natural balance of the water courses undergoes the various impacts already reviewed, especially in the hydromechanical and hydrothermal fields. In gravel pits a specific flora and fauna will develop which will tend towards a proper balance, which will modify the conditions of the surroundings (clogging, denitrification, desulphation, day/night consumption of CO_2, production of O_2 etc.). In the case of open gravel pits, the existence of free communication with the river favours the introduction into the water courses of competitive or predatory species (which may prove harmful).

The increase in turbidity (due to lack of light) and suspended matter (due to clogging of the tissues) reduces the photosynthetic activity and restricts the growth of plants, especially seaweeds: primary production then decreases significantly.

The impact on the fauna is discernible downstream of the excavation sites by the decrease in biomass of invertebrates over several kilometres (without reduction in species diversity).

Fish are directly and indirectly affected by excess suspended matter, which may cause lesions and which does exert an abrasive and clogging action on their branchiae, with a more or less marked inhibition of the respiratory function. More indirect effects are felt on the reproduction and development of eggs, by disappearance of the spawning grounds (clogging by suspended matters or erosion), and by asphyxiation of the eggs and fry, if the suspended matters contents are too high.

As for excavations in water (other than the active bed of rivers), mineral inflows (nitrogen and phosphorus in particular) of the groundwater and run-off determine to a great extent the development and biological evolution of the surroundings.

The variable mining depth of quarries in water and their morphological characteristics exert great influence on the biological qualities of these surroundings and their future development. Size thus has an impact on the stability of the ecosystem. The larger the latter, the richer it may be in plant or animal species, and the more diversified and complex it is, the better it can withstand external aggressions. The processes of penetration of water by light are linked with depth and influence the thermie and dynamics of this mass of water and consequently the productivity of the entire water expanse. Finally, the facies of the banks is an element of the first order because their development and their profile determine the diversity of the surroundings and also influence their productivity. Their profile (slope) is an essential factor in the recolonisation by aquatic and semi-aquatic plants whose various floristic categories are distributed according to depth.

An analysis of the various populations present in gravel pits makes it possible, however, to derive a stable common specific association consisting of species characteristic of this type of closed environs, comparable to shallow ponds or lakes: roaches, rudds, bleaks, breams, tenches, associated with perch and pike. Around these basic populations, other species become associated (gudgeon, dace, catfish, eel...), a result of voluntary or accidental introductions (communications).

2.2.3 EFFECTS ON THE LANDSCAPE

The setting up of a quarry leads to a change in the landscape (contrast of shape, colours...) with impacts which vary according to the types of exploitation (appearance of a water expanse, a mineral coal face, deforestation etc.). From another point of view, land constraints also pose the problem of "moth-eaten" landscapes which arises from the proliferation of exploitations in alluvial valleys and the very rectilinear contour of excavations.

Changes in the landscape caused by the opening of quarries should be thoroughly analysed right from the stage of the impact study in order to define a landscape treatment which may be resorted to at several levels:

— at the level of reconditioning the soils directed towards reducing the chaotic look of the terrain and permitting their reuse;

— at the level of their rehabilitation, which involves supplementary work for enhancement.

The landscape objective should enable either reconditioning for economic development or the best integration possible into the terrain. We may thus be led to conceal or 'heal' a site into the surrounding landscape or, why not, exploit its architectural or aesthetic potentialities.

Figures II.3 and II.4 present examples of reintegration of massive rock quarry sites into the environment.

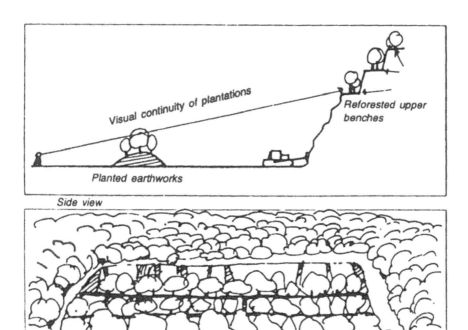

Fig. II.3. Creation of wooded fronts.

In matters of landscaping, the intervention of a landscaper will generally be most useful for defining such basic data as:

— the choice of location of the exploitation, if the deposit permits (land and technical aspects);

— the direction of exploitation, which will leave a rough or smooth face depending on the desired reconditioning as well as the final height and width of the face;

— the subsequent possibilities of modelling and ensuring safety.

At the end of exploitation, one should not forget to anticipate the long-term consequences of the mode of rehabilitation as well as the problems of maintenance and management. Intervention of the landscaper may guide the final works for the sake of landscape integration.

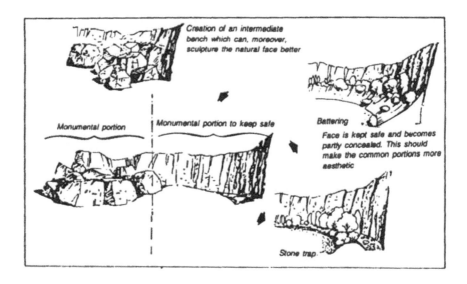

Creation of an intermediate
bench which can, moreover,
sculpture the natural face better

Monumental portion

Monumental portion to keep safe

Battering

Face is kept safe and becomes
partly concealed. This should
make the common portions more
aesthetic

Stone trap

Fig. II.4A. Additional alternatives for treatment of landscape

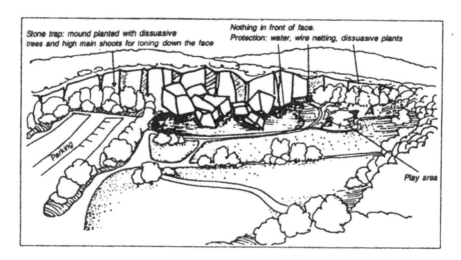

Stone trap: mound planted with dissuasive
trees and high main shoots for toning down the face

Nothing in front of face.
Protection: water, wire netting, dissuasive plants

Parking

Play area

Fig. II.4B. Simplified diagram

2.2.4 EFFECTS ON NATURAL ENVIRONMENT (FLORA, FAUNA)

These effects are exerted mainly because of the profound perturbations caused by
setting up a quarry in a natural environment. Here again, the impact study plays
an important role because it is very much the preliminary analysis of the natural
environment which will enable determination of ecological value and biological

potentiality. A knowledge of these elements should enable selection of the site of extraction according to the environmental constraints while avoiding areas of major ecological interest. Additionally, a study of the potentialities of the environment should guide the principles of its rehabilitation. In certain specific cases, the potential ecological interest of the area after rehabilitation may even prove greater than that before exploitation.

Let us recall here the main questions we should ask ourselves while assessing the ecological consequences of quarrying:

— nature and degree of rarity of the existing milieu (ecosystem);

— effects of opening the quarry (disappearance of the milieu, diminution of ecological diversity...);

— geographic and climatic position (migration corridor);

— evolutionary dynamism of the milieu.

2.3 Rehabilitation of Sites

This section presents the main rules and rehabilitation techniques for the principal modes of reutilisation of space.

We have to distinguish:

— rehabilitation of the soils, which is obligatory for the producer as per the statutory texts in force (articles of the Mining Code and decree of 20 December 1979, see § II. 1.3)

— amenities which are sometimes necessary and which involved works additional to those require for rehabilitation;

— redevelopment, which is the sum total of these two operations.

From the technical point of view, the numerous experimental or demonstrative operations have considerably advanced our knowledge of the various methods of rehabilitation. Among the various possibilities, there exists for each quarry a main formula of reutilisation and even, depending on its size, a combination of different formulas. Before taking them up briefly, let us recall the importance of a prior consultation among the main parties concerned (producer, owner, administration, community) regarding the choice of final method of rehabilitation. This dialogue should reflect the motivation and consensus required to enable the producer to carry out his work with a certain objective for redevelopment and guarantee in the long run the use of these lands for the purposes envisaged. During exploitation the importance of this consultation remains paramount since it will influence adaptation of the project to take into account the various modifications or constraints which are likely to arise during mining.

The main possibilities for developing quarries after (or during) extraction are shown in Table II.2.

2.3.1 RESTORING TO AGRICULTURAL CONDITION

In France, about 1500 hectares of quarries are opened in agricultural areas every year, which have to be reconditioned. The techniques followed, as of today, have been well mastered. They enable, after reconditioning, realisation of a soil whose agronomic potentialities are equivalent to those which existed before extraction, and this right from the first harvest. For this purpose, restoration to an agricultural

Table II.2: Main possibilities of development of quarries after extraction of materials

| Type of exploitation | Particular conditions | | Development possibilities | Remarks |
	Quarries	Environment		
Quarries in water	Depth of water—low	Rural	Ornithological reserve	Small or medium areas
			Aquatic game hunt	
			Artificial lake, basin	Impervious quarry bottom, large areas
			Infiltration basin	Large or medium areas in communication with the water table
			Water reclaimed for agricultural or silvicultural use	Assure that groundwater not polluted by fill
		Semi-urban and urban	Discontinuity in urbanisation	No significant fluctuations in water level
			Partial or total filling for utilisation:	Problems of fill quality (chemical, geotechnical)
			• green zones and leisure zones	
			• constructible zones	
	Depth of water—medium to high	Rural	Fishing for recreation	Small areas
			Fish farming	
			Bathing place	Water temperature
			Boating	Adequate connection with navigable channel
			Pleasure resort	
			Infiltration basin	Permeable quarry bottom
			Water-storage basin	Significant utilisable volume of quarry
		Semi-urban and urban	Housing development on waterfront	Facility for development of river banks
			Industrial port	
			Leisure base	
			Multipurpose	
Dry quarries	Pits	Rural	Reconstitution of agricultural terrain	Medium to large area, good drainage assured
			Reforestation	
			Infiltration basin	Permeable substratum

Contd.

Table II.2: *Contd.*

Type of exploitation	Particular conditions		Development possibilities	Remarks
	Quarries	Environment		
		Semi-urban and urban	Controlled filling	Possible pollution problems
				Keeping in view ultimate utilisation of fill area
		Green belts		Drainage assured
		Park zone		
		Residential		Small depth
		Industrial zone Artificial lake		Drainage assured after making bottom impermeable
On the hill side	Loose walls	All environments	Bank sloping and vegetation	
	Rocky walls	Distant views	Strengthening and treatment of the wall	Artificial patina for 'camouflage'
		Nearby views only	Talus eventually with plant cover	Provision of plant cover
	Quarry bottom	Rural	Revegetation (grassland, agriculture, silviculture)	Ultimate deposition of soil
				Purpose defined
		Urban and semi-urban	Green park	—
			Residential zone	Orientation to be considered
			Car park	
			Industrial zone	
			Leisure zone	Especially mountaineering school, shooting range

condition must imperatively be studied right from the prelude to the opening of the quarry.

Table II.3 groups the parameters that must be studied in order to arrive at a conclusion regarding the feasibility of restoring a quarry to an agricultural condition.

• *Procedure*

The procedure described below is applicable to all methods of reconditioning and should ensure reconstitution of a plant cover, be it for agricultural, silvicultural, landscape or ecological purposes.

The tasks of restoring a quarry to an agricultural condition 'in advance' start from the moment an extraction has been opened up to the minimum initial excavation permitting mining operations (deployment of machinery, possible installations).

Table III.3: Reconnaissance data prior to rehabilitating a quarry to agricultural condition

	Points to be studied	Eventuality	Objectives
1.	Level of water table (high waters, decennial mark)	Only if risk exists of flooding or water-logging due to oscillations of water table, a nearby water course or water expanse	Fixing a reference mark for level of reconstitution of soil (maintaining upper 0.5 m outside water)
2.	Deposit bottom (underbase): — nature — general topography — characteristics (permeability in particular)	Only if it is established that 1. is situated below the deposit bottom	Designing a drainage plan. Prevention of risks of hydromorphy
3.	Soils in place: pedological characteristics (horizons encountered, thickness, analysis)	Compulsory	Definition of manuring plan, overturning if need be, plus technical rules for reconditioning horizons
4.	Risks of run-off and erosion at site: drainage device	Compulsory	Cleansing and drainage plan
5.	Tailings: characteristics and volumes	Only if deposit study or mining technique reveal materials which cannot be 'reused'	Possible widening of range of technical possibilities of reconstitution of the soil
6.	Materials fill: — nature (refer if need to be 5.) — volume — rate of input	Only if established that 1. is situated above deposit bottom (even if just for part of exploitation)	Feasibility of eventual filling

Preparation of the underbase

A distinction has to be made depending on whether we are dealing with a pit bottom outside water (dry deposit) or under water. In the latter case the underbase is actually backfill brought in to fill the excavation under water. This type of underbase or blocking course sometimes requires over one or two years before the backfill stabilises.

The surface of the underbase should be gently sloped (most often between 0.5 and 1%) by levelling the ground. A tearing of the surface (using a ripper) is generally recommended if the underbase is not very permeable, in order to facilitate cleansing. Ripping is preferably carried out just before depositing the layers of soil cover (overburden).

Soil cycle

The care with which the soil is handled before and after exploitation determines the quality and time limit for resumption of agricultural operations. Usually, above a certain thickness of the soil (50 cm) selective scouring becomes necessary, which expedites restoration to an agricultural condition. While scouring the soil, a certain number of precautions should be taken:

— overburden stripping should not be undertaken when the soil is water-logged or during rainy weather (negative impacts on structural stability...);

— the surfaces to be exposed should not be used for movement of tyre-mounted machinery (harmful settlements...);

— soil should not be pushed over distances of more than 20 m (blade-grading).

The machinery generally used for overburden operations are: a bulldozer, loader or hydraulic shovel.

When transport is required, programming co-ordinated mining and reconditioning of the quarry should be directed towards minimising this factor as costs are high. Transport is most often effected by means of tipping trucks. These vehicles exert great pressure on the ground and their movement over the underbase should be adroitly manoeuvred.

Stockpiling is generally not advised (leaching, plant invasion, additional handlings). Moreover, proper management of the exploitation of the quarry (reduction in the transfer price, space gain, maintenance of soil qualities) recommends soil rotation with immediate laying of the soil cover on the sector just exploited.

If stockpiling cannot be avoided, one should take care not to constitute a dump which is too high; it should be rounded off and sown if possible with grass or leguminous seedlings (protection against erosion or overgrown vegetation).

Covering the soil with earths is the reverse of the stripping operation. Therefore, the lower horizons will be laid in sublayers and then covered with the humus-bearing horizon.

The total thickness of the soil to be reconstituted varies most often from 50 cm to 1 m. During these covering operations, particular care should be taken to avoid any compaction at depth.

For this purpose, trucks and tip carts are stationed outside the area being levelled. The most suitable levelling machines are the hydraulic shovel for the sublayer and the caterpillar loader or bulldozer for the humus-bearing layer (these machines can then function from the final surface).

In matters of programming, several formulas are possible for the reconstitution of an agricultural soil, which take into account the modalities and precautions mentioned earlier. Schematically, reconditioning 'during progress' (reconstitution carried out in liaison with progress of extraction) can be represented according to standard methods which, in practice, allow among themselves various combinations. Two of the most frequently used methods are represented below (Figs. II.5 and II.6).

2.3.2 RECONDITIONING FOR REFORESTATION

All the general principles examined earlier in matters of reconstitution of soils continue to be valid. It is just that reforestation presents stricter constraints, especially since errors, if any, cannot be made good after plantation (unless we start from scratch again).

• Clearing

It is advisable to do this in successive phases and sectionally in order to carry out scouring of the soils (and possible stockpiling) to preserve in so far as possible the soil qualities.

Clearing and scouring operations are therefore carried out in so far as possible according to the strict requirements of exploitation.

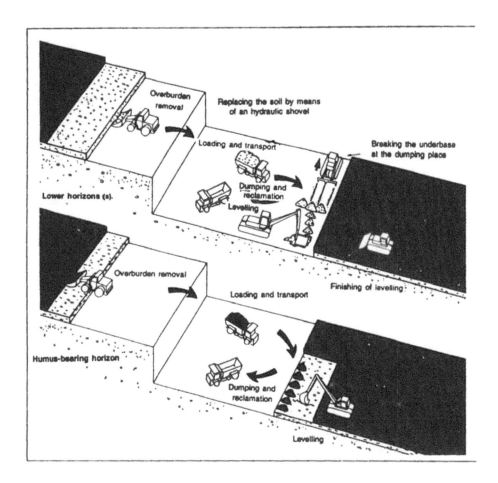

Fig. II.5. 'Bucket and shovel' method (simultaneous scouring and levelling)

Generally, the humus-bearing surficial horizon and the lower horizons are imperatively preserved and scoured selectively if the total thickness of the soil is adequate.

It is difficult to specify the thickness of the loose soil to be reconstituted because the qualities of the materials in each quarry differ.

Figure II.7 summarises the rules presently accepted for the main species of trees, in which the absolute minimum depth is indicated.

On average, the height of the level of the reconditioned soil should be at least more than 1.50 m from the level of the decennial height of the water table.

In fact, soil preparation, choice of trees and plants, techniques of plantation and maintenance works are all interdependent. The choice of trees is the task of specialists; trees are planted either by sowing or through seedlings or cuttings.

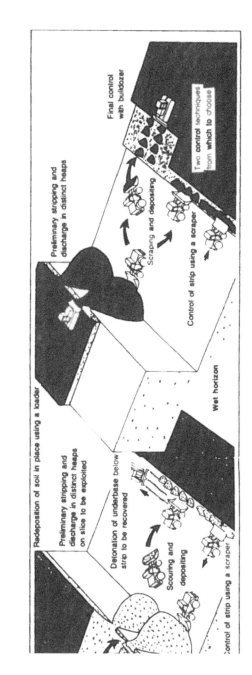

Fig. II.6. 'Loader method'

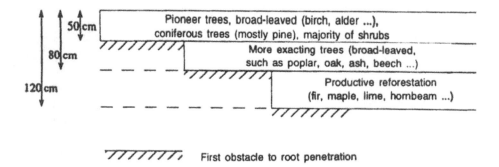

First obstacle to root penetration

Fig. II.7. Minimum depth for a reconstituted loose soil

It may be noted that plantation on plastic film generally gives better results and requires less maintenance (for a slightly longer preparation time).

Protection (fences, wire nets) and maintenance are important guarantees for the ultimate success of the plantations.

2.3.3 ECOLOGICAL RECONDITIONING

The conditions laid down below concern, as the case may be, dry quarries and/or quarries in water. The latter represent the most frequent case of this type of reconditioning.

• *Case of sand pits under water*

In the case of an extraction in groundwater, an essential parameter will be the water level which can be predicted and which will determine almost all the characteristics of reconditioning. The hydrogeological study carried out at the stage of the impact study will enable specifying the average equilibrium level of the future water expanse as well as the lower and upper limits of the water level. These water level variations condition the extent and nature of the surface relief of the banks, shallows, islets etc.

Unless it is a special case, the quality of water poses no problem. Let us note, however, the importance of the limiting factor that turbidity may constitute according to the nature of the deposit (extent of suspended fines which disturb light penetration; ongoing studies suggest that the suspended matters may be more in a silicoclayey substratum environment than in a limestone environment).

The basic principle of the final project should be to establish the maximum diversity in the facies and habitats created (or recreated). The operation will seek to restore or provide in the milieu those types of habitats essential to the life of animal communities, which will once again settle in that milieu.

If possible, a certain number of facies or biotopes favourable to aquatic or terrestrial life will be encouraged:

— shores and muddy grounds (for limicoles, for instance...),
— sand beds (small plover),
— reed beds (shelter and sleeping area for a number of species),

— marshy areas (winter season area),
— emergent grasslands (and submerged, if possible),
— woody hedges,
— islets,
— etc.

The surface relief of the shores and the nature of the banks should be different. The banks should have different types of slopes, from steep to gentler, of the order of 5 to 10°.

The contour of the shores should be so chosen as to attain sufficient length and avoid shapes which are too linear: certain parts will be colonised by vegetation which will favour frequentation of fauna. The creation and surface relief of the shores and beaches should take into account exposure to the sun and the direction of prevailing winds. These various arrangements will bring about, to a certain extent, variations in the temperature of the revering waters, which in turn will facilitate mechanisms of exchanges. Differences in depth of the water expanse will also contribute to this mechanism of alternation between cool and warmer areas.

At the bottom of the excavation, care should be taken to leave some parts with a small thickness of gravel (10 to 20 cm) whereas others (under a shallow water height) will be filled with topsoil.

From the ecological point of view, we must here and now pay great attention to the possible preservation of the existing fringes and plan for the creation of new ones alongside the future water expanse (remodelling and vegetation).

In fact, the surface relief of the banks is going to partly determine the success or failure of the project planned. The extent of marling will be proportional to the difficulty in planning the banks to be created. If the difference between the lower and upper limits of the water level is too great, a very large beach type free-board should be planned in order to ensure a useful fringe from the ecological point of view.

The ideal, while planning a reconditioning, is to be able to reconcile the setting up, design and coherence of ecological projects with the typology of the deposit (generally banks with a gentle slope where the thickness of the deposit is the least, and similarly for an islet or residual shallows, for instance).

• *Case of dry quarries*

Under certain conditions, these workings (massive rock, extraction outside the groundwater) may be of some importance for a worthwhile reconstitution of biotopes or biocenoses: calcicolous lawns, fauna adapted to a particular milieu....

To conclude, let us not forget the advantage of consulting a specialist who will be able to specify for each particular quarry the essential ecological rules to be followed. This can generally be done without fear of major difficulties or excessive cost.

2.3.4 RECONDITIONING FOR RELAXATION AND LEISURE AREAS

An outline of the development project should be prepared either as part of the section 'reconditioning' in the impact study or as a specific report.

On this basis, preliminary consultations with the local bodies should enable reconciliation of implementation of the rules laid down pertinent to reconditioning by the producer, with the subsequent work of development or accoutrements to be carried out by the local bodies.

The use of water expanses for sports necessitates a prior socioeconomic study in order to assess their possible frequentation, since the costs of maintenance and management can sometimes be high.

This observation is valid for any basic leisure project. Local bodies should properly evaluate the annual costs they may entail.

The finalities of reconditioning a relaxation and recreational area are many; they concern:

— Bathing: particularly the condition of being able to create friendly banks and ensure sufficiently clean water. The banks have to be transformed into gentle slopes and covered with sand, the approaches planted with grass; tree stands to create shaded areas are also recommended.

— Sailing: attractive use of a minimal area, for instance 10 ha for an initiation basin, 100 ha for a water expanse close to a town of 200,000 inhabitants, about 700 ha for an olympic water expanse. This new use given to the site implies, even during periods of low water, a water depth equal to at least 1.5 m.

— Other nautical sports: rowing, which necessitates ample development in length, as does water-skiing. In this case also the depth of the water expanse should be at least 1.5 m. The banks should be converted into gentle slopes.

— Sports grounds: Training ground, football field and tennis courts require prior levelling with subsequent careful preparation of the underbase. The surface relief of a hollow quarry may require some original planning.

— Fishing: considering its multiple advantages (recreation, ecology, production...), this topic is discussed separately below.

2.3.5 RECONDITIONING FOR FISH-BREEDING

The exploitation of materials in alluvial valleys leads most often to the creation of relatively shallow (3 to 5 m) water expanses of variable size (between 3 to 10 ha on average). Development of these water expanses for fish has two objectives:

— 'Breeding' fish in the sense of pisciculture per se. This objective is difficult to achieve in a quarry, a place that cannot be cleaned, and hence is dealt with only indirectly here. Studies and experiments defining the modalities that enable attainment of this objective are now drawing to a close.

— Recreational fishing: this is the most common utilisation.

The development of a borrow pit for fish-breeding purposes should enable fulfilling a series of conditions directed towards obtaining a balanced and diversified aquatic ecosystem as well as modifications in the land environment for the purpose of using it as a fishing site. Thus production of fish in sand pits depends to a great extent on the physical and chemical quality of the waters. Certain parameters are decisive for the living (if not reproduction) conditions of fishes. The most important of these parameters concern dissolved oxygen content, pH and temperature.

The last mentioned parameter plays a dominant role in the biology of fishes (food, environment, reproduction...). The main parameters deterministic for the productivity of waters are by and large well known and identifiable. Their prior examination is necessary for determination of the potentialities of a water expanse from the piscicultural point of view (recreational fishing and even extensive pisciculture).

The surface relief of the banks and their configuration will be decisive factors for the success of reconditioning. It is the development plan which should foresee, before commencement of extraction, the conditions which will optimise fish production and fishing. Attempts should be made:

— To develop the length of the banks: productivity of the water expanse will thereby be improved. For instance, a pike is generally introduced for every 10 m of banks.

— The water expanse should have some banks with a gentle slope, others somewhat steeper. The former will permit water to warm up faster during fine weather and will help reproduction of fish. They will be more readily colonised by vegetation, which will make fishing more difficult. This is not a big disadvantage because the reproduction sites should be protected from fishermen and strollers. Moreover, if the dimensions of the water expanse permit, creation of a shallow or an islet in the centre of the pond will provide excellent conditions for a spawning ground well away from passage areas. Proper conditions of reproduction can usefully replace costly and risky stocking of fish. Banks with steeper slopes will be more favourable for fishing.

— Vertical banks, which are particularly dangerous (risk of undermining and cave-in) should be smoothened. Banks with a gentle slope, and only these banks, will be covered with topsoil, already stockpiled before commencement of quarrying. This soil, which is better than the subsoil, will have all the qualities required for the creation of aquatic vegetation. On the other hand, it will reduce the water supply to the pond (settlement, clogging). It is thus advisable to plan before extraction the positioning of banks with a gentle slope. Care should also be taken regarding the location of these banks with respect to prevailing winds.

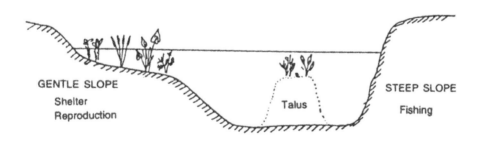

Fig. II.8. Different arrangements of banks for a water expanse

2.3.6 RECONDITIONING FOR A CONTROLLED DUMP AREA

Human activity, which involves consumption of aggregate materials (about 7 t/year/inhabitant), is also a producer of household wastes. This waste has to be disposed of and, in certain cases, quarry activity may be supplemented with redevelopment by exploitation of a controlled dump area.

These two activities are subject to distinct rules which must be followed. In case, after or during quarry exploitation, plans call for its use as a controlled dump, the impact study should have allowed for assessment of the actual feasibility of this plan in the zone authorised for extraction of materials. It should be particularly ensured that:

— that hydrogeological characteristics of the grounds preclude any risk of pollution of the waters;

— all the steps necessary for the prevention of pollutions and nuisances can actually be implemented; besides protection of waters mentioned above, one should check in particular for gas discharges, interfering animals, fires etc.

The decision to fill a quarry as a controlled dump area will be guided by:

— 'the geological situation' (lithological nature and thickness of the bedrock up to the level of the first aquifer at least, structural geology, nature and thickness of the surface formations...);

— the hydrogeological context (vulnerability of the groundwaters, hydrodynamic parameters...);

— and the hydrological data (extent of run-off, relation with the hydrographic network, absorption points of the flow...).

It will then be important to define as early as possible the principle of the subsequent (or even joint) use of the quarry as a controlled dump, as well as the type of final rehabilitation of the dump. This will enable conceptualisation of projects whose function can be useful both for the quarry and dump activity.

2.4 Conclusion

Today, public opinion truly hostile to quarries is a thing of the past. This has resulted from effects of regulations which have become very restrictive and from technological advances which make it possible to restrict damages to the environment. Let us mention once again the studies, research and rehabilitation works which are essentially at the root of the major advances made in the field of quarries.

A certain awareness of the operators should also be noted because, without that, it would not have been possible to make such a satisfactory report.

Nevertheless, one should not forget that quarries do directly undermine our natural heritage and, if they are necessary, they should integrate themselves during and at the end of the exploitation so that their presence is forgotten.

However, rules and regulations, and the techniques of environment are not the total panacea; the dialogue firmly established between the champions of the environment and producers must necessarily continue.

REFERENCES*

1. Sound environment of quarries of massive rocks and alluvial deposits. Main report and annexes. Regional Laboratory of Bridges and Highways, Angers, 1982 (69 pp.).
2. Development of sand pits into fishing sites. Brochure. Special tax on aggregates and Ministry of Environment. Young Economic Chamber of Romily, 1983 (46 pp.).

*All entries in French—General Editor.

3. Qualitative impact of quarries under water on groundwaters. Summary of results obtained in Haute-Normandie and Alsace. Reports BRGM no. 87 SGN 499 HNO and 87 SGN 199 ALS, 1987. Two main reports (160 pp.) and annexes.
4. Continental water and quarries. UNGP Technical Collection no. 5, 1987 (83 pp.).
5. Interactions between quarries and underground and surface waters: Bibliographical summary and technical report concerning them. BRGM Report (60 pp.) + annexes no. 87 SGN 391 PAC, 1987.
6. Ecological potentialities of quarries. Brochure. Special tax on aggregates and Ministry of Environment 1986 (28 pp.) and article by J.M. Sionneau, *Industrie Minérale, Mines et Carrières*, April 1987.
7. Revegetating quarries. General principles applicable to reforestation. Brochure. Special tax on aggregates and Ministry of Agriculture. CEMAGREF, 1984.
8. Quarries of massive rocks: landscaping. Brochure. Special tax on aggregates and Ministry of Environment. CEMAGREF, 1986 (47 pp.).
9. Cost of reconditioning quarries for agricultural purposes. Brochure. Committee of management of the special tax on aggregates, 1985 (24 pp.).
10. Return to earth. Brochure. Special tax on aggregates. Permanent Assembly of Chambers of Agriculture, Ministry of Agriculture, 1987 (17 pp.).
11. Reconditioning quarries and development of sports or recreational centres. Technical information note (20 pp.). Central Laboratory of Bridges and Highways, 1983.
12. Studies of impact on the environment. Technical specifications nos. 1–7. Special tax on aggregates, Ministry of Environment. [n.d.]
13. Piscicultural potentialities of quarries under water. Main report (75 pp.) + annexes. Higher Council for Fishing, 1987.
14. Technical modalities for reconditioning quarries for agricultural purposes. Special supplement, Chambers of Agriculture, no. 671-19. [n.d.]

III

Properties and Prospecting of Rocks and Aggregates

Louis Primel and Claude Tourenq

1. THE EARTH'S CRUST

Aggregates are obtained either from massive rocks, worked in quarries, or from loose rocky material recovered from sand-pits and gravel-pits. They acquire from these rocks a major part of their properties and especially all their intrinsic characteristics, i.e., those which are related to mineralogy and texture.

Moreover, the manner in which these materials occur determines and manner of their extraction and processing, i.e., the profitability of quarrying.

The geology of France is characterised by the presence of a wide variety of rocks over a small area. Before reviewing them, let us recall some general information about the geological history of our Earth and the formation of the rocks which constitute the Earth's crust.

The history of the Earth has been subdivided by geologists into eras and periods. This classification was made in an exclusively relative manner through the study of the succession of species of fossilised living beings. More recently, some modern techniques, such as radiochronology, have enabled a more accurate evaluation of duration of each of these periods (Table III.1).

Evolution of the terrains of the Earth's crust resulted from the combined influence of internal and external dynamic actions. Thus geological phenomena belong to two separate domains.

1.1 Internal Geodynamics

The Earth's crust consists of a certain number of rigid plates, which have a thickness of the order of a few tens of kilometres; they float on an underlying viscous mass (asthenosphere or magma), the temperature of which reaches several hundreds of degrees (Figure III.1). In-between some of these plates there are areas wherein an oceanic crust is created from the upward movement of the magma and areas wherein this oceanic crust is destroyed by penetrating under the plates, thereby causing their displacement, fracture or folds.

Table III.1. Age of geological periods

Era	Period	Age in millions of years
Quaternary		3
Tertiary	Pliocene	9
	Miocene	25
	Oligocene	37
	Eocene	65
Secondary	Cretaceous	135
	Jurassic	190
	Triassic	225
Primary	Permian	280
	Carboniferous	350
	Devonian	400
	Silurian	450
	Ordovician	500
	Cambrian	570
Precambrian	Age of the Earth	4550

All these movements perpetually alter the structure of the Earth's crust and rare are those regions which have not been successively deep pits, high mountains and plateaus flattened by erosion. Crystallisation of the magma during formation of the oceanic crust or during its injection through the deformation of some plates, led to the formation of magmatic rocks (granite massifs, outflows of basalt etc.).

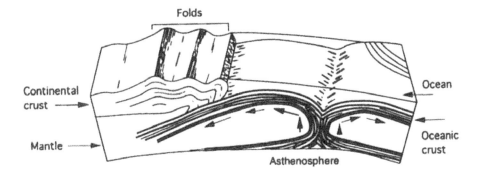

Fig. III.1. Internal geodynamics and theory of plates

1.2 External Geodynamics

External geodynamics is concerned with the phenomena observable on the surface of the Earth's crust and thereby more accessible. It can be schematised by the combined action of three essential factors:

— mechanical actions due to wind, water, ice and the solid particles which they produce;

— chemical actions due to water dissolving mineral salts;

— biological actions due to proliferation of living beings.

In continental regions, mountains and coasts are subjected to gradual wear and tear: this is erosion. The elements torn from the continents undergo transport (run-off, streams, wind, glaciers). Eventually sedimentation takes place on the continents by deposits of river sands and gravels, wind sands, glacial moraines and particularly in the seas by accumulation of the solid fragments from the continents, shells from marine animals and precipitation of dissolved salts.

The aforesaid deposits occur in successive layers. The weight of the upper layers and the internal chemical changes convert (diagenesis) these materials gradually into more or less coherent rocks (sedimentary rocks).

Concomitant with their accumulation at the bottom of the seas, these deposits gradually sink (subsidence) and sometimes reach areas in which the combined action of pressure and temperature converts them into metamorphic rocks, which thus acquire a very particular lamellar structure.

1.3 Materials of the Earth's Crust

The rock which we encounter on the surface of the earth, those which are worked in mines, and quarries belong to several groups.

We shall first distinguish the massive rocks consisting of minerals strongly bonded together. These rocks have three distinct origins:

— If produced from the cystallisation of magma deep in the earth, they are magmatic rocks (granites, basalts etc.);

— If they result from the cementation of loose rocks, they are massive sedimentary rocks (limestones, sandstones etc.);

— If they result from the recrystallisation of any other rock under the effect of pressure or temperature, they are metamorphic rocks.

We shall conclude with a description of loose rocks, which are generally of sedimentary origin, but which have not been cemented (marine or alluvial sands and gravels etc.).

Before taking up the study of rocks, however, it seems pertinent to define their components, namely the minerals.

2. ROCK-FORMING MINERALS

Rocks are formed by mineral assemblages.

The most common minerals (90%) are silicates, the rest consisting of carbonates, metallic oxides, chlorites, sulphates etc.

2.1 Silicates

Silicates are essentially geometrical arrangements of oxygen ions grouped in fours around Si^{4+} ions (tetrahedrons), a certain proportion of Si^{4+} ions capable of being replaced by Al^{3+} ions, the equilibrium of the charges being ensured by K^+, Na^+, Ca^{++} etc. ions.

. The type of arrangement characterises the family to which the silicate belongs. The proportion of Al, Si, various cations and the existence of Mg, Fe etc., introduces into each family a certain number of categories which are more or less clearly differentiated.

2.1.1 SILICA SiO_2

Quartz is the most common representative of this group; it is colourless with a greasy lustre. It is found in certain magmatic rocks (granites, rhyolites), in sedimentary rocks, where it is sometimes the only component (certain sandstones and sands), and in metamorphic rocks (Figure III.2).

Opal and chalcedony, amorphous and microcrystalline respectively, are some varieties of quartz which are found in certain sedimentary rocks (flint).

2.1.2 FELDSPARS

The replacement of a certain number of Si by Al leads of the family of feldspars. They contain potassium (orthoclase), sodium and potassium, sodium and calcium (plagioclase), or are less rich in silica (feldspathoids).

These minerals generally occur in the form of colourless, pink or white tablets. They are found in most magmatic rocks and in many metamorphic rocks (Figure III.2).

2.1.3 PHYLLOSILICATES

All the representatives of this family occur in the form of flakes.

Muscovite or white mica, brown or green biotite, which contains in addition in its formula Mg and Fe, are found in magmatic and metamorphic rocks in very variable proportions (Figure III.3).

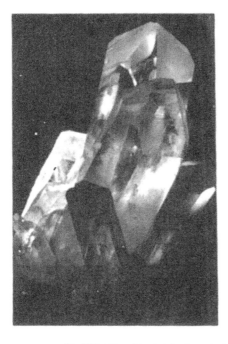 

Fig. III.2. Prismatic crystals of quartz (× 0.4) (left) and orthoclase (× 0.7) (right)

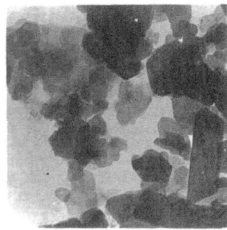

Fig. III.3. Piling up of mica flakes (× 0.7) (left) and clay crystals (kaolinite × 25,000) (right)

Chlorites of green colour are weathered minerals which are found accessorily in all rocks.

Clays consist of very small flakes (Figure III.3); they result from the weathering of other silicates. They are mostly found in sedimentary rocks but may themselves constitute rocks (kaoline).

2.1.4 FERRO-MAGNESIAN SILICATES

Periodots, of which olivine is the most important, are found in magmatic rocks without quartz (basalts, peridotites).

Pyroxenes and amphiboles are alumina silicates: The former, of which augite is the most common, are found in the same rocks as olivine while the latter (hornblende in particular) contain water and are present in magmatic rocks, which are poor in quartz, and in metamorphic rocks. These minerals are often green or black and amphiboles are sometimes fibrous (asbestos) (Figure III.4).

2.1.5 METAMORPHOSED SILICATES—SiO_2–Al_2O_3

These minerals are quite typical of metamorphic rocks: blue kyanite, black andalusite, pearly sillimanite. Some, such as staurolite, cordierite and garnets (Figure III.5), have a more complex chemical composition with Fe and OH. They are very stable minerals.

2.2 Other Minerals

2.2.1 CARBONATES

Calcite (CO_3Ca) is generally a colourless to white mineral. It is found in sedimentary rocks as an essential mineral of limestones and in certain metamorphic rocks (calcitic marbles) (Figure III.6).

Fig. III.4. Amphibole needles (× 0.7)

Fig. III.5. Staurolite (left) and garnet (right) (× 0.7)

Dolomite $(CO_3)_2$, CaMg is found in sedimentary rocks. Actually a variety of intermediate stages occur between pure limestones and dolomites (dolomitic lime-stones).

Fig. III.6. Calcite (x 0.7)

2.2.2 OXIDES

Titanium and alumina oxides are accessories in igneous and metamorphic rocks. Iron oxides, haematite and magnetite are more widespread but always accessory.

2.2.3 SULPHATES

Gypsum (SO_4Ca-2H_2) is an accessory mineral in sedimentary rocks and may constitute an exploitable mass.

2.2.4 SULPHIDES

Pyrite (FeS_2), a golden-yellow mineral, is accessory in all rocks.

2.3 Relative Importance of Minerals

The occurrence of the minerals described above varies markedly. Only a few are really widespread, as can be seen from Table III.2. But in very selective exploitation, occurrence and harvesting are two different things.

As shall be seen later, we have classified silicated rocks on the basis of the most important minerals.

Table III.2: Relative importance of minerals

Minerals	Percentage in	
	Earth's crust	Superficial pits and quarries (France)
Quartz	12	53
Feldspars	60	20
Micas	4	
Silicates Fe-Mg	17	
Metamorphosed silicates	2	
Carbonates	1	25
Other minerals	4	

We have also taken into account other minerals which are rather scarce but typical of a certain tendency. Thus feldspathoids have been retained to characterise certain rocks which are deficient in silica.

The major characteristics of the principal minerals mentioned above, in particular hardness, which conditions to a great extent several properties of rocks such as abrasiveness, polishability and wear resistance, are listed in Table III.3.

Products resulting from the weathering of minerals usually pose problems because of their sensitivity to water and their reaction to hydraulic binders. As far as aggregates are concerned, these products will always be found in sand fines.

3. MASSIVE ROCKS

3.1 Igneous Rocks

Igneous rocks, characterised by a massive structure, come from depths below the Earth's crust. If the magma has cooled slowly, the minerals are greatly developed and we have a rock with a granular texture with minerals of millimetric size quite visible to the naked eye (granites). If the magma has cooled rapidly by outpouring on the surface of the Earth's crust (e.g. outflow of lava), most of the crystals do not have time to develop and are not visible to the naked eye (< 0.1 mm); we then have a microlitic texture (extrusive rocks).

The cooling of extrusive rocks has generally been too rapid for all the minerals to have had time to take shape. A vitreous or very finely crystallised phase thus occurs in the rocks. However, a few large crystals may form, in which case we have a porphyric texture (Figure III.7).

Thus, for an identical chemical composition (same original magma), we may have very different textures.

A very fine texture always leads to a very high mechanical strength but the breaks are smoother and such rocks are more polishable.

If the chemical composition of the original magma varies (Figure III.8), it generally causes crystallisation of different minerals (Table III.4) and therefore the formation of families of different rocks.

Table III.5 gives some important physical and mechanical properties of main rocks.

Fig. III.7. Porphyric texture. Thin section in a rhyolite (× 7)

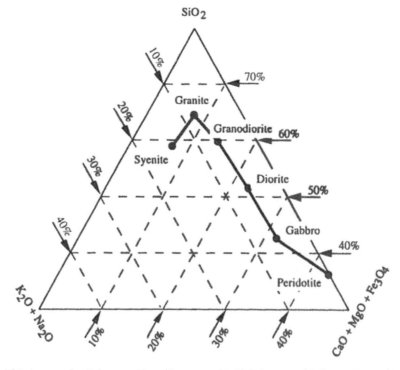

Fig. III.8. Average chemical composition of igneous rocks. Al₂O₃ is assumed to be constant and equal to 20%

Table III.3. Main characteristics of major minerals

Minerals	Apparent density T/m³	Hardness (Vickers) kg/mm²	Weathering	Sensibility to water	Remarks
Olivine	3.5	820	Often weathered		
Metamorphosed silicates	2.5/4.5	800/1300			
Pyroxenes	3.4	680			
Amphiboles	3/3.4	730	Moderate		
Muscovite	2.8	85	Not liable to weathering		Shape unfavourable for adhesiveness
Biotite	2.7/3.3	90	Often deferred	High	Alkali silica reaction with cements
Chlorite	2.6/3			High	
Clays	2/2.6			Very high	Very hydrophilic Unfavourable for adhesiveness
Quartz	2.65	1280	Not liable to weathering		Alkali silica reaction with cements if structure is cryptocrystalline Problems of adhesiveness with bitumen
Chalcedony					Low alkali silica reaction with cements
Opal					Very strong alkali silica reaction with cements
Volcanic glass	2	600			Strong alkali silica reaction with cements
Feldspar	2.57/2.75	690/720	Often weathered		
Calcite	2.71	110			
Dolomite	2.87	350			
Haematite	5.2			More or less hydrated forms of iron may be very sensitive	
Pyrite	4.9	1050		High	

For information, hardness: Steel 550, Carbides Si 1800, Carbides Bo 3300, Corundum (Emery) 2000.

3.1.1 GRANITE FAMILY

Granites and granodiorites are the most widespread among igneous rocks in France. The rock is granular and contains obligatorily a potassic feldspar and some quartz. It may contain sodium plagioclases and micas (0 to 30%) and accessorily some amphiboles (Figure III.9).

Granites may form at great depths from sedimentary and metamorphic rocks or nearer the surface (a few kilometres), associated with tectonic zones. They appear on the surface through folds and erosion. Because of erosion, the more ancient the region, the more abundant they are on the surface: Massif Central, Massif Armoricain for example.

Table III.4: Igneous rocks

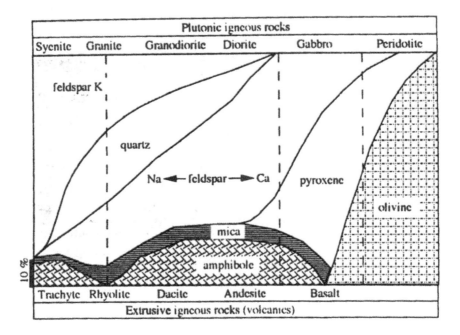

The physical and mechanical properties which we shall be delineating always refer to sound rocks located outside the zone of surficial weathering (Table III.5).

Microgranites and rhyolites have the same mineralogical composition as granites and from the point of view of texture, all the intermediates exist between them. Like granites, they are found in ancient massifs but occupy more limited areas. Their very fine texture confers on them (see Figure III.7) a double advantage: on the one hand, they are weakly permeable and hence withstand weathering very well and, on the other, due to this, possess high strength.

Certain products from volcanic projections, when they have a well-developed cement, can have a porphyric texture and are often of excellent quality: rhyolitic tuffs, cinerites etc.

3.1.2 GRANODIORITE FAMILY

These rocks hardly differ from granites. Plagioclases are more important than orthoclase and many rocks called granites are in reality granodiorites or intermediates.

3.1.3 SYENITE FAMILY

These are granites without quartz but with the same accessory minerals.

The microlitic equivalent or trachyte, though fairly wide-spread in France, is not much used because of its high porosity.

Table III.5. Major physical and mechanical properties of main rocks

Particle density, T/m³	Porosity, %	Speed of propagation of longitudinal waves, m/s	Compressive strength, MPa	Modulus of deformation, 10 MPa	Los Angeles, %	Micro-Deval in presence of water, %	Polished stone value	Abrasiveness	Linear thermal expansion, 10^{-7}/1°C	Thermal conductivity, K 10^{-5} cal/s/cm	Specific heat, 10^3 cal/g
2.65	< 0.5	5800	150	80	20	10	0.55	1400	90	970	190
2.65	< 0.5	5800	250	80	13	8	0.50	1500	80	840	190
3	< 0.5	6200	180	90	16	12	0.48	1000	70	550	190
2.9	< 0.5	6200	200	90	13	8	0.43	1100	70	600	180
3	< 0.5	6500	200	100	14	13	0.50	900	50	720	170
2.9	< 0.5	6500	300	100	10	10	0.45	1400	50	700	200
2.65	2	6000	180	75	20	14	0.35	50	110	800	240
2.40	10	4600	80	40	40	20			90	500	200
2.10	20	3800	35	28	65	30		15	70	400	180
1.80	30	2700	10	12	85	60			50	300	160
1.50	40	2000	5	10	100	100		0	25	200	150
2.80 2.60	< 0.5 7	7000	160	80	25	16	0.45	600 220	100	1200	220
2.40	10	2500	60	14				800	110	1100	220
2.10	20		30		80			200	100	600	160
1.80	30		3							240	170
2.56	< 0.5	5500	260	90	18	6	0.50	2000	115	1920	225
2.6	< 0.5	5800	220		14	10	0.52	800	90	900	210
2.7	< 0.5	5000	100	50	60	90	0.38		70	890	210
2.6	3	3500	100	70	18	14	0.55	20	90	780	185
2.6	1	4000	200	80	20	10	0.55	1700	70/100	800	195

Fig. III.9. Thin section of a granite (× 7)

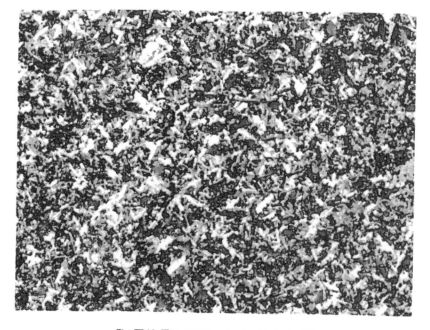

Fig. III.10. Thin section of a microdiorite (× 100)

3.1.4 DIORITE FAMILY

There is no quartz in these rocks and the dominant feldspar has a sodium tendency. Amphiboles are frequent in these rocks and sometimes pyroxenes. They are poorly represented in France.

Microdiorites are much more appreciated. They have sometimes undergone deep changes, probably hydrothermal, developing secondary minerals, and often very abundant: chlorites, epidotes (Figure III.10). The corresponding microlitic equivalent are the andesites, rocks which are often porous and therefore seldom quarried for aggregates.

3.1.5 GABBRO FAMILY

These rocks contain neither quartz nor orthoclase and plagioclase in richer in anorthite than in albite. Pyroxenes are always abundant and sometimes so is olivine. Amphiboles are often present.

Gabbro is often weathered and some French deposits are not free from this rule (green rocks).

Dolerites, diabases and ophites belong to this family. They have a special texture due to intergrowth of plagioclases, which confers on them excellent qualities (hypabyssal rocks).

Basalts with a microlitic texture are very widespread. They are black and massive rocks which often display large crystals of pyroxenes and olivine, and periodotite nodules of a few centimetres, which are granular rocks consisting mainly of pyroxene and olivine (Figure III.11).

3.2 Massive Sedimentary Rocks

We shall classify these according to their genesis. Table III.6 gives the main sedimentary rocks under this classification in which all the intermediates exist and duality of origin occurs quite often. As a matter of fact, only a few of these rocks are quarried and we shall describe just the siliceous (sandstones, quartzite sandstones) and the carbonate (limestones and dolomitic limestones) rocks.

3.2.1 SILICEOUS ROCKS

Some coherent rocks are formed through cementation of sand grains, namely sandstone to quartzite, according to the degree of filling of the voids left between the grains by the silica which binds them (Figure III.12).

The strength of these coherent rocks depends primarily on their porosity (see Table III.5).

3.2.2 CARBONATE ROCKS

Limestones are generally of organochemical origin: organic due to the accumulation of fragments of shells or the colonial activity of corals, and chemical due to the precipitation of carbonates, which generally form the cement. The structures may be very varied depending on the size, shape and proportion of the shells, extent of cement, percentage of other minerals (quartz, clays, organic matter) and pores (Figure III.13).

As for sandstones, the essential factor is porosity. Table III.5 gives the characteristics of limestones containing at least 90% calcite. These are mean values

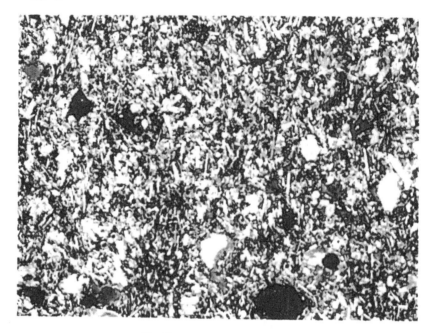

Fig. III.11. Thin section of basalt (x 7)

Fig. III.12. Sandstone-quartzite. Thin section (x 35)

Table III.6. Classification of sedimentary rocks

Origin			Chemical nature
Detrital	Chemical	Organic	
Sandstones	Flint	Radiolarites	
Quartzite sandstones	Siliceous limestone	Lydian stones	
Molasse	Cherts	Diatomites	Siliceous
Conglomerates	etc.	Gaizes	
Breccias		etc.	
Clays			
Shales			Silicoaluminous
Slates			
	Marls	Coral limestones	
	Lacustrine limestones	Algal limestones	
	Oolitic limestones	Foraminiferal limestones	Carbonated
	Dolomitic limestones	Shelly limestones	
	Dolomites		
	Chalks		Carbonated

which may be highly influenced by the presence of particular minerals. For example, a clayey limestone containing only 60% calcite has shown for a porosity of 6% a compressive strength of less than 50 MPa. Conversely, a limestone containing 15% quartz gave for a porosity of 1% and a compressive strength of 240 MPa.

Limestones are rocks which are generally very sensitive to polishing; this is particularly so the finer their paste and the higher their purity. The polished stone value (PSV) may fall to 0.25 in the case of certain limestones. On the other hand, a high quartz content in certain siliceous limestone yields a PSV of 0.60.

The presence of dolomite may also modify somewhat the properties of limestones by slightly improving them. The PSV of dolomitic limestones is 0.48.

Abrasiveness may be increased by the presence of quartz and dolomite (Figure III.14).

3.3 Metamorphic Rocks

The transformation of igneous or sedimentary rocks under the action of temperature and pressure at great depths in the Earth, leads to metamorphic rocks. Other factors are involved in metamorphism also, in particular ion exchanges.

The multiplicity of original materials and actions renders any classification difficult. Table III.7 presents our tentative schematic classification of these rocks according to the nature of the original material and the intensity of metamorphism.

3.3.1 ROCKS OF CONTACT METAMORPHISM

In contact with certain granitic massifs, schists undergo gradual transformations: new minerals (metamorphosed silicates) appear first, forming small darker masses; these are the spotted schists (Figure III.15).

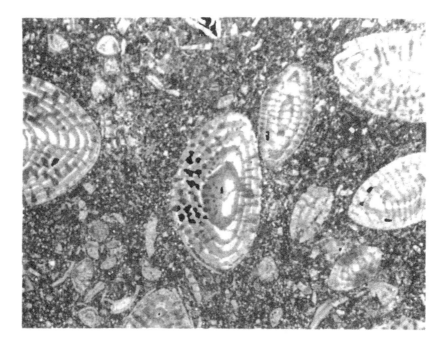

Fig. III.13. Thin section of a fossiliferous limestone

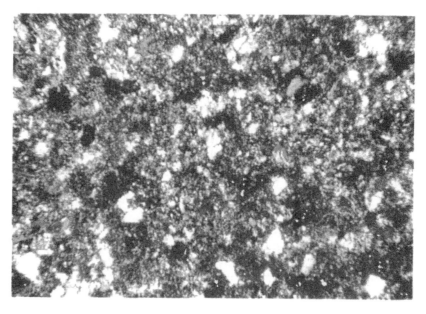

Fig. III.14. Sandy limestone. Thin section (× 50)

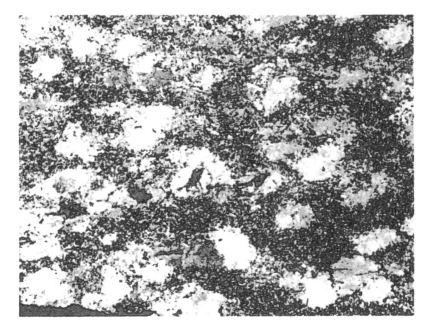

Fig. III.15. Thin section of a spotted schist (× 7)

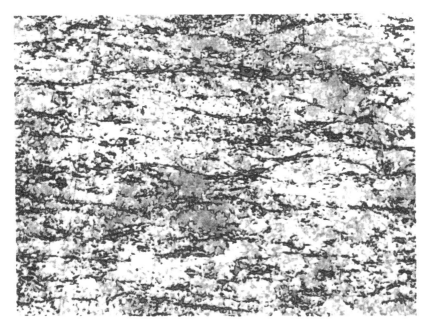

Fig. III.16. Thin section of a gneiss (× 7)

Nearer to the granite the rock gradually loses its schistose structure to become massive with fine grain. These are the hornfels consisting of quartz, feldspars and in very variable proportions, micas, pyroxenes and amphiboles.

Under the same conditions, limestones undergo recrystallisation, often with the appearance of new minerals. These calcitic marbles have the same properties as sedimentary limestones but the presence of larger crystals gives them a lower strength.

Like their sedimentary counterparts, sandstones transform into quartzites but often with a more developed agglomeration of crystals.

3.3.2 ROCKS OF DYNAMOTHERMAL METAMORPHISM

These always have in common a more or less marked lamellar structure (foliated crystalline rocks).

Schists and sericitic and chloritic schists consist of oriented quartz and micas. According to the continuity of the mica beds, the rocks may be strongly anisotropic (Table III.8).

Certain schists with very fine grains (0.005 mm) may have remarkable strengths; 180 to 250 MPa under simple compression. This is particularly true of slates, but the strong anisotropy of the mechanical properties generally renders these rocks hardly usable in the field of public works.

Micaschists, consisting of quartz and micas arranged in layers, in general are more granular in structure. Well-developed black micas give the rocks very mediocre qualities, like those of schists.

Gneisses (Figure III.16) have the same constitutents as granites but are laid in beds. The continuity of the mica beds, their proportion and the size of the rock grain are factors which determine their strength.

Table III.7: Classification of metamorphic rocks

Initial rock	Contact metamorphism (very limited, in contact with granites)		
	Schists	Limestones	Sandstones
Increasing intensity of metamorphism	Spotted schists Nodular schists Hornfels	Limestone with minerals Calcitic marble	Quartzites

Initial rock	General metamorphism (large extension in space)					
	Schists, Clays	Sandstones	Limestones	Marls	Granites	Gabbro
Increasing intensity of metamorphism	Phyllites Sericitic schists			Calcschists		
	Chloritic schists	Quartzites	Calcitic marble	Micaschists		
	Micaschists					
	Gneisses			Amphibolites Pyroxenites	Gneisses	Amphibolites Pyroxenites
	Leptynites Granites					

Leptynites are gneisses with fine grain and without mica. They are found as intercalated layers in gneisses. Their mechanical characteristics are generally better.

Amphibolites consist mainly of green amphiboles and feldspars. The crystals are visible to the naked eye. These rocks are found in beds or in lenses intercalated in gneisses.

Table III.8: Anisotropy of schistose rocks

Orientation / Characterisitcs	In plane of schistosity	Perpendicular to plane of schistosity
Compressive strength (MPa)	49	100
Wave velocity (m/s)	5800	3100

4. LOOSE ROCKS

Loose materials used for construction or roadworks are unconsolidated detrital sedimentary rocks; they occur

— either in the form of deposits belonging to 'conventional' stratigraphic horizons,

— or in alluvial deposits of rivers or streams (lower valleys or terraces),

— or, more rarely, on present-day beaches or dunes.

As a matter of fact, all these deposits, be they new or ancient, are the result of the same process of erosion and sedimentation. These processes may be very briefly schematised as follows:

— An *in situ* decomposition of the various rocks that are more or less liable to weathering according to their nature and the climate. The residues of this weathering are carried away by flowing water, then by rivers and sometimes also by glaciers, or for fine fractions, by wind.

— These agents of transport constitute an initial classification. A major part of the weathered residues carried by rivers or streams gets deposited before reaching the sea; they comprise, among others, alluvial sands, which are almost always associated with coarser elements precisely because of the brevity of their transport. These sands additionally contain elements lifted by run-off from the terrains through which they pass, which explains the more or less apparent heterogeneity of alluvial deposits.

— The elements which best withstand chemical alteration and mechanical wear and tear reach the sea where they mix with fragments lifted by the latter from the shores. This mixture, triturated and agitated for a long time, is eventually deposited at more or less long distances from the shore. This distance depends on the fineness of the sediment. These marine sands are afterwards covered by deposits which may, if bioclimatic conditions vary, be quite different, e.g. limestones or clays.

— A part of the marine sands, littoral sands or beach sands, is frequently picked up by the wind and accumulates in the form of sand-dunes.

— There is no need in a classification to make a distinction between ancient deposits, present-day or recent, since the process of formation has always been the same. It may be said, however, that most of the significant deposits of marine sands in France date back to the Secondary and Tertiary eras whereas most of the alluvial and wind deposits, as well as moraines and screes, are Quaternary. This is explained by the fact that the latter materials were deposited on the continent and have therefore been almost always destroyed by erosion much before protection could be provided by the deposit of another layer, as in the case of marine sands. This protective cover also explains, along with other factors, why most of the sands prior to the Secondary era have been partially or totally converted into sandstone.

— The important point is that all these processes which gave birth to the various types of loose deposits, have also conferred on them special physical properties. That is why careful geological reconstitutions should enable predication, at least at the regional level, of the geometry of the deposits as well as their grading and their petrographic or mineralogical composition.

From the economic point of view, materials of alluvial origin are by far the most important (nearly two-thirds of the French production of aggregates). A description of the major French alluvial deposits will constitute the bulk of this section, complemented by some data on other loose materials used as aggregates (non-alluvial sands and certain other deposits).

4.1 Alluvial Materials in France

French alluvial materials are extremely varied due to the geological diversity of the catchment areas of the water courses which have deposited them. Among the various possible classifications, we shall retain the one based on mineralogical and petrographic composition even though all the parameters are closely linked (the shape of the grading curve depends markedly, for instance, on the petrographic nature of the elements).

A comparative study on the evolution of the main characteristics of gravels (grain size, petrography, mechanical characteristics) was conducted for most of the important rivers in France.

4.1.1 CALCAREOUS ALLUVIAL DEPOSITS

The most abundant are those of the Parisian basin (part of Seine, Marne, Aube, Meuse, Somme) as well as those of some tributaries or parts thereof the Jura (Doubs, Ain, Tilles). In most cases they consist of rather less coarse materials, the percentage of elements higher than 50 mm being generally low. When the catchment areas consist of relatively homogeneous limestones of average mechanical strength (Seine and upstream Aube, tributaries of the Jura), the grading curves are often disjointed. When these areas consist of a mixture of limestones (e.g. limestone of average strength and chalk), the curve is always smoother. It was also observed that the chalk content varies considerably from one granular class to another, the highest percentages occurring naturally, given the friability of chalk, in the finest classes (Figure III.17). These calcareous gravels, which represent in number about 8% of alluvial quarrying in France are most often used for manufacture of ordinary concrete. However, it so happens that they are also used in road technology. Barring some exceptions (e.g. Drôme: 20) the Los Angeles values lie between 23 and 28.

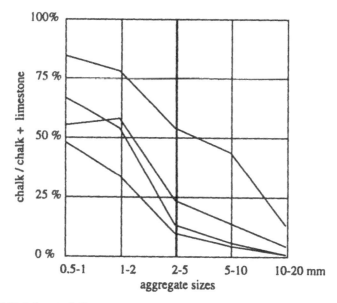

Fig. III.17. Variations in chalk content in the alluvial deposits of the Seine between Montereau and Nogent

4.1.2 SILICO-CALCAREOUS ALLUVIAL DEPOSITS

These represent about 43% of the alluvial quarrying in France. Various materials are included in this category, the 'siliceous' portion consisting sometimes of fragments of igneous or metamorphic rocks (Rhône, Durance, Rhine, for instance) and sometimes of flint (Seine, Yonne). The calcareous portion is also very heterogeneous, the catchment areas of the rivers being very different. A very distinct patterns of evolution of the petrographic composition of these gravels from one grain-size class to another was observed regardless of the river under study (Figures III.18 and III.19). So this aspect should always be taken into account when mechanical tests are carried out on raw material before processing (Los Angeles values on class 4/6 will not give the same results as on 10/14).

Given the heterogeneity mentioned above, test results are very spread out. The Los Angeles value can reach 17 or 18 (e.g. Rhine, Durance) because the tested fine gravels correspond essentially to the crushing of big unweathered siliceous elements. On the other hand, it can go up to 28 or 30 when the large elements are predominantly more or less weathered granites or when they contain limestones of average strength.

Petrographic heterogeneity frequently leads in the case of these alluvial deposits to grading curves with gaps. Taking into account the mechanical performance, many alluvial deposits are widely used not only for construction concretes, but also in roadworks, including the wearing course. Their use at this level poses two essential problems: their Los Angeles value on the one hand (rarely less than 20) and compliance with the crushing ratio on the other, which theoretically compels use only of the coarsest elements, which often represent only a low percentage of the overall material.

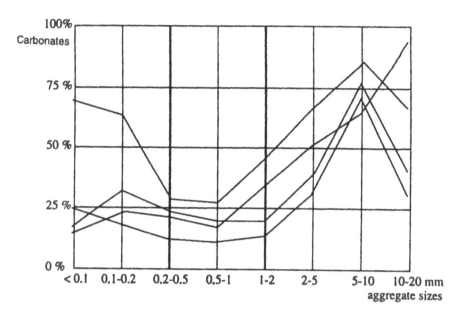

Fig. III.18. Variation in carbonate mineral content of silico-calcareous gravels of the Seine basin

Fig. III.19. 2.1-mm fraction of a silico-calcareous gravel

4.1.3 SILICEOUS ALLUVIAL DEPOSITS

These represent about 48% of alluvial quarrying in France. They are located in the south-west (Garonne, Gaves, Adour), Centre (Loire, Allier), the Parisian basin (Eure, Loing) and in the east (Meurthe, upstream Moselle).

These deposits contain no limestone (or very small quantities) but nevertheless are not homogeneous; some (Eure, Loing) consist of a mixture of flint, which constitutes the coarse part, and quartz which is very widely dominant in sands (Figures III.20 and III.21) whereas most of the others result from the gradual wear and tear of elements of igneous or metamorphic rocks, i.e., they consist of pebbles

Fig. III.20. Siliceous gravel

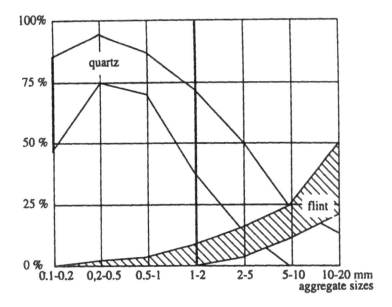

Fig. III.21. Distribution of flint and quartz content in river Loing

and gravels of granites, diorites, gneisses (locally, basalts or other lavas) etc. and sands, mainly of quartz.

The mechanical characteristics of these alluvial deposits are not better than those of the silico-calcareous deposits and in this case also largely depend on the choice of the fraction which was crushed to obtain the fine gravel on which the test is carried out. The Los Angeles value varies from 15 (processed from un-weathered siliceous pebbles and gravels, which corresponds to the values obtained in quarries on sound rocks) to 30 and more, when the constitutents of the gravel are partially or totally weathered. The uses and problems which they pose are the same as for the silico-calcareous gravels. These few indications show therefore that we should never make hasty deductions from the classification of alluvial gravels. Some calcareous gravels have a better Los Angeles value than some siliceous gravels.

4.1.4 SANDS AND GRAVELS EXTRACTED FROM THE SEA

Let us finally point out that about 1% of the sands and gravels extracted in France come from submarine deposits of the continental shelf. These are either ancient river alluvial deposits corresponding to alluvial terraces similar to those quarried on the continent which have been submerged (submarine extension of the Seine, Loire etc.) or hydraulic dunes resulting from the play of sea currents which have accumulated detrital sediments of terrigenous or organogenic origin. These deposits are almost always sands of variable grain size and also very varied petrographic composition, from pure quartz sand to pure shell sand, with all the intermediates.

Reserves of marine sand and gravels are huge but their intensive quarrying poses ecological, economic and technical problems.

4.2 Non-alluvial Sands

These are basically ancient sands, i.e., detrital deposits deposited at different geological periods, at more or less large distances from the shore. Depending on the palaeogeographic conditions, these deposits have been more or less mixed up by currents, covered with other sedimentary formations and are presently prominent over variable stretches and thicknesses, which also differ markedly. Generally, however, these deposits constitute much more extensive and thicker beds than the alluvial deposits. In France the main quarried sandy horizons belong either to the Secondary era (e.g. Triassic sands, Albian and Cenomanian sands of the Paris basin) or to the Tertiary (e.g. sands from Fontainebleau) and even the Quaternary era (e.g. sands from Landes, Orléanais, Picardie etc.).

An attempt at classification of the various sands found in France for the purpose of utilisation revealed that most marine sands are either 'fine' sands or 'medium' sands. The essential characteristic of fine sands is that 75% of the elements higher than 80 microns are less than 0.5 mm. These sands (Figure III.22) have all been very well sifted by the sea or wind and contain by the large only quartz grains (more than 90%).

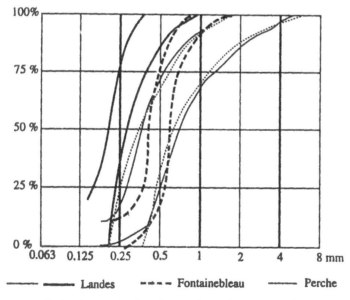

Fig. III.22. Examples of grading envelopes of three fine sands

Medium sands, of which by definition more than 50% of the elements higher than 80 microns lie between 0.2 and 2 mm and less than 20% are higher than 2 mm, include, among others, a part of the sands from Perche, sands from Orléanais, Triassic sands, Pliocene etc. As shown in Figure III.23, these sands have grading curves which make them similar to a certain number of very widespread alluvial sands (e.g. sands from the Loire, Seine). Barring a few exceptions (e.g. shell sands from Blesois and Touraine which are very rich in shell fragments), these medium marine sands are also predominantly quartzose.

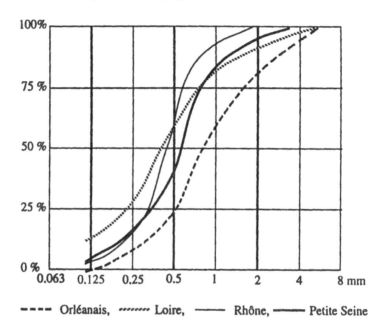

- - - - Orléanais, **·········** Loire, **——** Rhône, **——** Petite Seine

Fig. III.23. Examples of grading curves of medium sands

If the use of these 'medium' sands as aggregates poses no particular problem, simply those posed by many recent alluvial sands, 'fine' sands continue on the other hands to pose problems that are difficult to solve. The extent of the beds and the huge reserves which some of these sands constitute (sands from the Landes, Tertiary sands from the Paris basin) enabled their use for a long time as fills or subbases but hardly as wearing courses. Recent developments have made it possible to consider them for use after treatment (especially with slag) as subbases for medium traffic pavements. Let us remember that many of these 'fine' sands are very widely used in fields other than civil engineering: sands for foundry works, glassworks etc. but the tonnages consumed for these uses continue to be low (a few million tonnes per annum).

4.3. Other Loose Deposits Likely to be Used as Aggregates

Though economically less important than the preceding ones, some other loose deposits should nonetheless be mentioned: these are mainly the granitic sands, moraines and screes. They are generally deposits which are either residual or seldom transported and therefore often poorly classified.

4.3.1 GRANITIC SANDS

These are residues of *in situ* weathering of the most widespread igneous or metamorphic rocks, the most quarried being granite quartz sands. Barring outcrops of hard rocks, in fact all the surfaces indicated as feldspar (granites, diorites, gneisses, migmatites) in geological maps, they are covered with a layer of granitic sand, the thickness of which may reach 15 m in certain regions (Limousin, Bretagne). Thickness variations are very quick and linked with fracturing of the rocks. Those sands, whose grain size is more or less that of coarse sands, pass on gradually to weathered but coherent rock, and then to sound rock. The elements of the granitic sand are those of the mother rock. There is hardly any chemical dissolution in a temperate climate, the main evolution being granular due to fragmentation of the minerals. The mineralogical composition of granitic sands reflects therefore that of the crystalline rocks from which they have been formed. However, most of the igneous rock massifs in France (Massif Central and Armorican Massif in particular) have been subjected since the end of the Primary era to very different climates and chemical alteration is often extensive in them. Illite, chlorite, montmorillonite and kaolinite are frequently found in the fine fraction of granitic sands. The presence of these clayey minerals explains moreover the limited use of these sands; to date they have mostly been used as fills or subbases for roads with low traffic. The extreme heterogeneity of the deposits also explains the fact that granitic sand is generally utilised as a by-product during opening of quarries or excavation of rocky materials.

4.3.2 MORAINES

These are always very heterogeneous deposits and seldom quarried unless picked up by river and redeposited downstream (fluvio-glacial deposits which can be summarily compared to coarse alluvial deposits). The main difficulty in mining moraines comes from their granular heterogeneity. The same bed contains side by side huge blocks whose quarrying often depends on the explosive and silty or clayey fine sands.

4.3.3 SCREES

The constitution of most screes corresponds to the effects of frost and especially of alternation of frost and thaw during various glacial periods, which affected France during the Quaternary. There are therefore screes of very diverse nature on almost all the present catchments. The only screes minable as materials are the calcareous ones. Their potential utilisation depends on two factors:

a) Mechanical characteristics of the rocks from which they come. As such, the most interesting are the screes of the hard limestones of Provence, certain regions of the Alps and the Jura. Since chalk screes can only be used as fills, they cannot be considered as aggregates.

b) Grading of the material, not readily predictable insofar as the screes consist of successive deposits corresponding to periods the intensity of frost could have varied considerably and hence fragmentation of the rocks also.

Finally, let us mention that quarrying screes is limited by the problem of safety, the relative stability of these screes often being compromised by extractions.

5. INVENTORIES OF AGGREGATE RESOURCES

So, as has been seen, a wide range of materials is available to aggregate producers in France.

An extractive industry of such magnitude as uses up 25 to 30 square kilometres of land annually for its operations proper could not possibly be devoid of major problems, the more so since, unlike the majority of industries which enjoy options of location, this one can be located only where deposits exist.

For decades the aggregate manufacturing units were established at places where the resources were 'evident'. These 'evident' sources have become increasingly scarce for various reasons. Thus it has gradually been realised that quarry materials should also be appraised in much the same conscientious manner as an inventory of the resources of metallic ores or those of energy products.

Why inventories of aggregate resources? a) in some regions the traditional local reserves, most often alluvial sands and gravels, will shortly be exhausted (within 10 years or even less) as the present rivers in France are not bringing along fresh coarse deposits. In these regions, which like the Paris region are generally large consumers of aggregates, more economical alternative sources should be expeditiously identified.

b) In all the regions and more particularly in the alluvial basins, the procedure on the one hand for official approval for opening new quarries and the ever-increasing public awareness of environmental problems on the other, have gradually made it necessary to launch regional directive plans with due regard to different land uses.

c) Over the last few years a number of techniques, especially in the field of road construction, have become increasingly exacting with respect to the properties of the aggregates used, which in turn makes a thorough qualitative picture of the regional resources imperative.

To take care of such wide-ranging concerns, inventories have been drawn up for some 20 years now, especially by the regional laboratories of Roads and Bridges and the Geological Survey. Though the series of inventories are not necessarily a very logical chain, they can still be divided into two broad categories:
— inventories on a small scale (1/100,000 on average);
— inventories on a medium scale (of the order of 1/25,000).

5.1 Small-scale Inventories (1/100,000)

These relate primarily to:
— alluvial basins (most of French important rivers);
— departments or regions where all the materials have been taken into account;
— particular geological horizons, hardly used at present, but likely to find use (for example, the soft calcareous limestones of the Paris region).

These surveys have several attractive features.
— They furnish a good overview of the region-wise availabilities, especially from the qualitative point of view, and as such are highly useful to the project managers, for example, highway networks.

— They enable determination of whether some particular geological horizon is suited to serve the role of a substitute aggregate.

— They are not costly since they call for no arduous method (geophysical, boreholding).

— They pose no confidentiality problem.

Their drawback is that their precision is inadequate for formulating a regional quarrying policy, not to mention taking decisions at the level of urban planning.

5.2. Medium-scale Inventories (1/25,000)

The main objective of these more precise inventories is to help arbitrate between the different possibilities of land use after collecting the data which enables doing so with full knowledge of the relevant facts. They must also help to launch realistic regional or local quarrying policies with due regard to all the parameters. Therefore, these inventories of resources should be thought of as the springboard for a more comprehensive data file, which simultaneously takes into account the needs, constraints and diverse economic and socioeconomic data of the region.

The methods employed for drawing up these inventories combine the use of surface geophysical survey (especially resistivity), hand augers, fast boreholes with logging, core drilling naturally completed by geotechnical identifications and often by trials on the various possible utilisations of the materials under scrutiny. Such

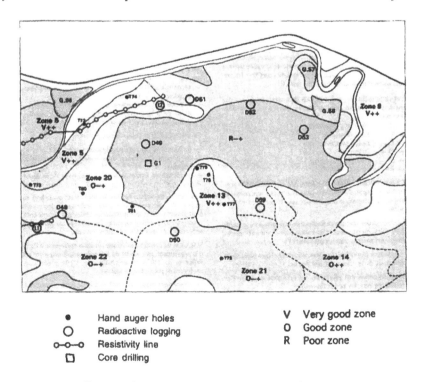

●	Hand auger holes	V	Very good zone
○	Radioactive logging	O	Good zone
o–o–o	Resistivity line	R	Poor zone
☐	Core drilling		

Fig. III.24. Example of orientation map for potential extraction

leads to the demarcation of zones; these are described either 'objectively' (by giving for each its geometric characteristics and the characteristics of the materials it includes) or through 'interpretation'. In the latter case the authors assign to each zone a 'value' coefficient (very good, good ... poor) taking into account the above criteria interpreted in the light of the general economic and regional context (ratio of overburden to material thickness, sand/gravel % etc.). In both cases the simple fact of demarcating the zones on a 1/25,000 scale implies an adequate number of measurements, the authors of the surveys not permitted to reduce the number of measurements below a reasonable limit (Figure III.24).

Thus the main interest behind these surveys is to enable effective comparison of alternative quarrying schemes and problems of land use at the regional policy level. For the surveys to be wholly efficacious, it is necessary to involve all the concerned bodies and services right from the planning stage so that the results can then be effectively utilised. Regional or departmental co-ordination bodies have generally enabled achievement of this objective.

6. DETAILED PROSPECTING OF MASSIVE ROCK DEPOSITS

6.1 Objectives

These comprise arriving at the conditions and modalities of quarrying (stripping, extraction, transport, reclamation of the mined area) and determining the type of machinery and modalities of treatment most suited to the material and the deposit, with due regard to the intended end-uses.

Precise knowledge of a certain number of parameters is indispensable for achieving these objectives.

6.1.1 GEOMETRICAL PARAMETERS AND THEIR EFFECTS ON QUARRYING

In all types of deposits a good knowledge of the total reserves, variations in thickness and nature, is indispensable.

a) Overburden: its geometry

In calcareous deposits: A calcareous geological horizon with homogeneous features generally has no overburden of notable thickness. One may say from experience that when the height of the face is of the order of 15 to 20 m, however, which is the case with a great many deposits in France, the overburden thickness is a very important parameter (Figure III.25). Lack of knowledge of its thickness could have catastrophic economic consequences for the quarry operator. We may cite the example of a calcareous borrow fit having up to 6 m of highly weathered and platy material, the total thickness of the bench worked being 15 m. An inadequate survey forced the enterprise to quarry the plates along with the sound rock. The installation output was 50% below prediction as a major portion of the quarried material had to be discarded at the prescreening stage before primary crushing.

Fig. III.25. Massif bench and "loose" top zone in a calcareous deposit

In igneous (or metamorphic) deposits: Demarcating the overburden in massifs of igneous rocks is more complicated than for calcareous deposits, since the extent of weathering varies progressively from weathered clay *in situ* up to the weathered rock. Weathering may be visible on the rock surface, as it can only exist at the level of feldspar crystals or ferro-magnesian minerals, and thereafter is discernible only under a microscope. It is important to know this progression well since if one can, as a last resort, permit aggregates with a certain degree of weathering to be used for the lower layers of pavements on the basis of specific investigations and taking a certain number of precautions, especially to avoid degradation of stockpiled materials, the same is not the case for surface layers.

b) Overburden: its nature

In calcareous deposits: When the overburden is unpolluted, one may proceed to open a deposit in two stages. Firstly, a rough stripping removes the unusable material. Then the quarry operator completes the stripping of the rest of the overburden. This method can render materials from the prescreening stage usable, for example in shoulders, which would otherwise not be suitable for anything other than fill under other conditions of opening of the deposit.

In igneous deposits: A good knowledge of the distribution of weathering can sometimes allow proceeding in three stages: earthworking the zone of very intense weathering, working of the less weathered zone used as aggregates for subbases and lastly, the truly sound rock for road-bases or surface aggregates. But generally quarrying is done in two stages: after removing the most weathered portion, the rest is added to the quarried material and eliminated at the time of processing (before or after primary crushing).

c) Mineable material

The question now is one of defining the total mineable volume, variations in thickness of the mineable material, and the number, volume and nature of the heterogeneities. This calls for specifying:

For calcareous deposits: The dip and the structure of the deposit since calcareous horizons are often thin with rapidly changing facies.

We may cite the example of a deposit wherein 600,000 t of material were mined from a 12 m thick horizon from the medium Bathonian intercalated between a more or less crumbling dolomite cover and a friable and frost-cracked white oolitic-calcareous substratum. Considering the structure of the massif, the problem posed was ascertaining the homogeneity and continuity of this facies, as faults even of small magnitude could trigger displacements between the benches and changes in dip of the layers. Since the totality of such data directly impinges upon demarcating the mineable zone, the options of the point of opening-up, extraction and stationing of the processing plant, a detailed knowledge of them is indispensable. It might even be necessary to make a three-dimensional model or a block diagram of the deposits in order to follow the quarrying with possibilities for quick corrective action should such be needed.

For igneous deposits: Study of the structure is often more involved, the heterogeneities generally being distributed in a manner which is more difficult to foresee: veins of rocks of a type different from the main rock, faults, diaclases, crushed zones etc., often presenting a larger clayey contamination (Figure III.26).

6.1.2 PARAMETERS DETERMINING QUALITATIVE CHARACTERISTICS OF THE MATERIAL

Four kinds of information are essential in the case of calcareous and igneous deposits:
— petrographic type,
— mechanical properties of the materials,
— state of fragmentation of the materials in place,
— types of contaminations.

a) Petrographic type

This is determined first by visual inspection and then under polarising microscope (see below).

In the case of calcareous deposits it may be useful to complete the mineralogical analysis by a chemical analysis to give the silica and dolomite contents which have a direct bearing on the utilisation of certain processing equipment. The samples for analysis are taken either from outcrops or from cores (borehole samples), or in the form of fragments (percussion drilling with hammer drill).

Fig. III.26. Argillaceous fractured zone (dark).

b) *Mechanical properties of the materials (see Chap. IV)*

These are defined by a certain number of standardised tests (micro-Deval in presence of water, Los Angeles, and possibly completed by freezing tests).

The samples in this case come from the crushing of rocks sampled from the outcrop or in drilled cores. It is even possible to directly use the fragments coming from percussion drilling (impact test).

c) *State of fragmentation of materials in situ*

Knowledge of this determines the blasting techniques and the magnitude of the first stage, that is, the type of drilling, mode of quarrying, type and size of the primary crusher and the aperture size of the scalping screen before primary crushing.

d) *Various contaminations*

(i) Contamination related to the genesis of the rock mass, essentially in the form of:

— Intercalation of clayey or marly beds or lenses of a nature different from the whole facies in the case of calcareous deposits. It happens that both types of heterogeneities related to the conditions of sedimentation coexist in the same deposit.

— Presence of veins which intersect the main facies and which can be considered all the more frequently as contaminations since they are often more weathered than the enclosing rock.

(ii) 'Accidental' contamination. The presence of highly fractured or even crushed zones can considerably hinder quarrying. If this has not been revealed in the preliminary survey, it might happen that the working face runs into such a zone parallel to its maximum stretch. Since these zones are largely materials which are highly weathered and unfit for utilisation, one might be obliged to run through or turn around them and thus lose several days of production.

(iii) Vertical fissures and washouts containing clay. These are relatively frequent phenomena in some calcareous deposits and directly influence the success of blasting plans and the state of the material after quarrying, which in turn determines the mode of recovery and the dimensions of the scalping screen to eliminate the clayey fraction, depending on atmospheric conditions.

A major fissuration accompanied by clayey contamination will constrain the quarrying to be carried out in limited volume, so that the material does not remain exposed to moisture for too long. Prescreening has to be provided at the plants with 'bypass' for the periods of very bad weather or high pollution, a very long conveyor belt, placing of materials upon the pre-screen in a single layer, a disconnectible belt to enable removal of clay lumps etc.

6.2 Methods

What methods enable encompassing all these parameters and how exactly are they employed?

Any study of a deposit commences with a geological investigation. This is greatly facilitated in the case of a quarry under work, the working face making it possible to define:

— Petrographic type and mechanical properties of the different facies;
— State of fragmentation of the rock mass;
— Various causes of contamination and directions of fracturing of the rock mass, the heterogeneities etc.

These various aspects enable preparation of geotechnical map of the quarry for demarcating the unweathered, weathered, crushed etc. zones upon the different working faces (see Chapter XVIII Figure XVIII.1).

In the case of a new deposit, examination of the working face will be replaced by geological investigation of the sector, drawing of samples from the outcrop (with attendant risk of bias in sampling) and taking of drill cores from the site chosen.

Petrographic examination is essential, which is carried out on thin sections that might be cut from fragments of very small dimensions (chips flying off while drilling with a hammer drill) and completed by a 'methylene blue value' (see Chapter IV, Section 2.4.1) for quantifying the degree of weathering.

Thus this phase broadly defines the petrography of the deposit. It now remains to specify the structure of the deposit, for which surface geophysical methods are applied first: electrical and seismic.

6.2.1 ELECTRICAL METHODS

The most often employed electrical method is the resistivity line (at two different lengths of line), for which the results are presented in the form of a resistivity map. Their quality depends mainly on contrasts in resistivity between the overburden, which is generally conducting, and the unweathered (sound) material, whose resistivity is generally much higher. Thus one obtains a picture of the overburden thickness and location of the zones of faults wherever they are of considerable size: such data serves as a guide to subsequent prospecting.

However, this method does not generally enable detection of the fissured rock-unweathered rock boundary nor the local heterogeneities, owning to the characters of the measurement and its discontinuous mode. Precision in respect of overburden thickness remains inadequate also. The main attractions of this method are: ease of application and quickness of execution.

Recourse can also be had to another method of resistivity measurement, the artificial magneto-telluric method. It consists of measuring on the ground surface the horizontal components of the magnetic field H and electrical field E created by radio transmitters.

Application of this method is exceedingly simple. Two capacitive electrodes that measure the electrical field are implanted on a "rubber mat" that can be dragged along and thus make a continuous measurement of the electrical field. Since continuous measurement of the magnetic field poses no problem, an electrical model enables direct calculation of the E/H ratio (possibly the resistivity directly). This method has three advantages vis-à-vis the resistivity one:

— Continuous measurement, enabling establishment of minor features of unevenness;

— Quickness of measurement, the "mat" being moved at a man's walking speed (it might possibly be pulled by a vehicle);

— Finer separating power.

6.2.2 SEISMIC REFRACTION

In the exploration of quarries of massive rocks, the recordings generally point to three layers: a surface zone at 300 m/s which forms part of the overburden, a high velocity zone (between 2000 and 4000 m/s) which represents the sound zone, and between these two a zone of medium velocity (of the order of 1000 to 1500 m/s) which is difficult to characterise since it could be a weathered material that is practically not fractured (compact granitic sand, for example) or a highly fractured unweathered material, as the case may be. This uncertainty can be removed only through observations, or in some cases in conjunction with the electrical method (Figure III.27). Yet another difficulty connected with this intermediate zone is its poor definition on the recordings, which could entail rather large errors in determination of thicknesses.

It should also be mentioned that the seismic velocity reflects the state of consolidation of the zone and not its nature. Thus this method does not help distinguish schists from quartzites, for example.

Whatever the geophysical method employed, two points limit the precision of interpretation:

Seismic cross section, seismic velocity in m/s

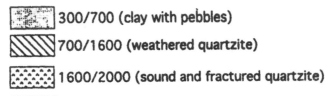

300/700 (clay with pebbles)

700/1600 (weathered quartzite)

1600/2000 (sound and fractured quartzite)

2000/4000 (sound quartzite)

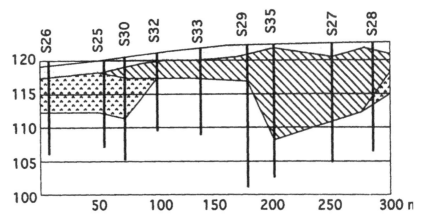

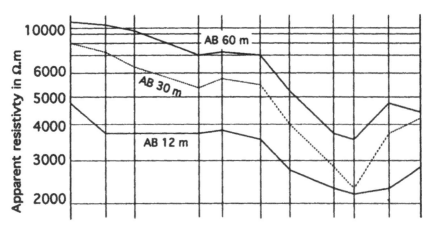

Resistivity cross section

Fig. III.27. Example of seismic cross section and corresponding resistivity profiles

In the case of rock masses it is difficult to locate the overburden-unweathered rock boundary. The geophysical characteristics (resistivity, seismic velocity) vary in a smooth manner and depiction of discrete layers in a diagrammatic scheme could lead to errors.

The fissuration of rock masses is not isotropic. One can generally discern several directions of fractures; the parameters do not have the same value when measured parallel to these directions, or at a right angle (except perhaps by the artificial magneto-telluric method). It is generally difficult to reconcile the interpretations of the two measurements for calculated thicknesses, which underscores the great significance of preliminary structural geology study for fixing the directions of measurement.

To conclude, geophysical study furnishes an 'image' of the deposit which enables more judicious borehole location.

6.2.3 BOREHOLES

Making exploration boreholes is generally understood to imply core drilling. Cores generally provide excellent information. Continuous sampling enables obtaining very precise geological sections (accompanied by petrographic identification), studying fractures and conducting mechanical tests. But such core drilling is very slow in practice and also very expensive. In order to contain the cost within reasonable limits, it is therefore necessary to limit their number, generally resorting to them only when it is absolutely imperative to visualise

Fig. III.28. Percussion drilling machine

the rock (case of new deposits). But then the study of a deposit presupposes a large number of exploration boreholes. To remove this contradiction, resource should be had to another means of drilling. The only one available which satisfies these cost criteria is drilling with a hammer drill (Figure III.28) (which is seen in quarries during exploitation anyway). Therefore the problem posed is one of obtaining sure and precise geological data from rapid destructive drillings.

The following constitute the different stages of study.

a) Follow-up of drilling

This is the fundamental factor in the overall interpretation of results and should be carried out with the maximum precision possible. It is essential that the section be taken by an experienced geologist since there is good reason not only for making visual observations, but also interpreting variations, or new or unexpected appearances of facies or even predicting variations or different anomalies. So it is imperative to observe all that takes place at the time of drilling (colour and appearance of the dust, size of the cuttings, relative volume of fine elements and chips, variation in penetration rate of the drill, noise made by the drill upon rock contact, uniformity in emergence of cuttings) and to take more samplings in doubtful zones. The recovery of cuttings is relatively simple, that is, sample cuttings and dusts at least once every 50 cm. After sieving, a certain amount of fragments over 2 mm in size is washed for petrographic examination. The person carrying this out should observe all the fragments, not concentrate solely on the large ones, and note the variations in petrography, weathering, colours and the presence of contamination.

In this manner the geologist prepares what is termed a 'drill log', which cannot be perfectly precise in every case. Indeed, between the surface observation and the cuttings there may be a deviation due to the time taken for the fragments to be ejected. Observations support the notion that in a dry hole this deviation will increase with depth and depend on the force with which cuttings are blown out.

There is good reason to preserve the cuttings for subsequent examination in the laboratory or subjection to mechanical tests, such as impact test. Identification at the very instant of drilling could well be too hasty and entail errors in assessment. While opening up the 'drill log', it is highly useful to re-examine these chips or their thin sections with a binocular microscope.

b) Rate of penetration of the bit

While making the borehole section, systematic measurements of the rate of penetration of the bit are carried out.

This measurement consists in accurately recording the time taken to drill a given length (Figure III.29). Analyses of graphs showed that the time taken for boring 50 cm had to be adopted for many reasons (quickness of scrutiny, adequate precision, facility of chronometering).

The measurements can be made automatic, enabling sizable reduction in the follow-up work after drilling and a far greater precision in measurements.

The rate of penetration is a function of the zones traversed and the drill bit employed. Hence it is advisable to always use the same drill bit and if possible

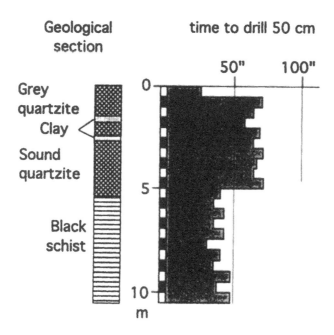

Fig. III.29. Extract of a section from drilling and graph of velocity of penetration of the bit

the same drill man within the same quarry. Caution should also be exercised in interpretation.

6.2.4 LOGGINGS

Loggings consist in continuously recording a parameter of the soil during drilling operations. This could be its natural radioactivity, resistivity, density, water content etc., as a function of the depth drilled.

a) Equipment (Figure III.30)

The apparatus commonly employed comprises an electric winch of regulated speed 0.5 to 5 m/min, a cable with 4 conductors and a bi-trace recorded (movement of the chart of the latter is controlled by the movement of the cable), to provide 1/50 to 1/100 scales of recording.

b) Various logs

Gamma-ray logging: This is due to the presence of radioactive elements, the nature and proportion of which vary with the types of rock. Measurement is done using a probe which consists of a crystal and a photomultiplier. Thereby one hopes to distinguish different petrographic facies, provided the thickness of the layers exceeds 30 cm.

Fig. III.30. Logging equipment assembly

Logging is generally done at the rate of 1 m/min. This probe can be put to work inside a borehole after enclosure or not in a tube empty or full of water, by virtue of the high penetration power of gamma rays. Measurement can be refined through the use of a selective probe capable of distinguishing the various radioactive elements.

Resistivity: Measurement is done using a 'laterolog' to obtain a fine definition of the beds (it is possible to demonstrate the presence of centimetre-thin conducting layers embedded in a electrically resistant body, which is a problem to contend with in quarries). It should be noted that these measurements can be used to reinterpret surface electrical prospections. Low resistivities correspond either to weathered zones or to fissured ones.

The use of a laterolog dictates that the probe be filled with water or mud, encased or not inside a plastic tube to which a strainer is tightly fastened. This poses no major problem in the case igneous rock masses, unlike calcareous ones. Measurements can be taken very rapidly, the rate of advance being of the order of 3 m/min.

Velocity of wave propagation: This can be measured either continuously (base of velocity measurement (20 cm) in a borehole filled with water), or from point to point in a dry borehole (base of measurement 30 or 60 cm). In this case a portion of the definition is naturally lost.

The measurements so made enable assessing fissuration of the medium. This log accords very well with that of the rate of penetration, but has the advantage of being quantitative and not dependent on the drilling machine.

Density and water content: The $\gamma-\gamma$ probe can also be employed, which measures density, but its precision for values around 2.7 is inadequate for studying fissuration. The neutron-neutron probe, which theoretically measures the water content, in our case is mainly sensitive to the chemical composition (presence of iron for example) and hence gives results which are complementary to natural radioactivity.

c) Results

It is difficult to give a table which would summarise the cases wherein such or such method is the most effective since the problem is not one of making quantitative measurements, but rather differentiating the various facies.

Some examples help appraise the results obtained.

— The record of natural radioactivity given in Figure III.31 enables perfect distinction between microgranite and rhyolite and thus corroborates the record of penetration rate. Resistivity pinpoints the intermediate zone best.

— Contrarily, if the natural radioactivity log enables correct distinction of the part of the overburden (limestone and clay), it does not differentiate the schists from the diorites (Figure III.32). Resistivity brings out these two facies. In the case of calcareous deposits one generally comes across one or more distinguishing radioactivity levels, enabling, among other things, determination of the geometry of the rock mass by correlations between the different probes. Furthermore, surface weathering and contamination in layers, pockets and fissures are generally well revealed. Thus it is possible to establish the geometry of the rock mass of the basis of the γ-ray logs (Figure III.33).

The resistivity recordings are generally much more distinctive than those of natural radioactivity, which makes their interpretation easy. High resistivity values are almost invariably synonymous with sound rocks whereas nothing certain can be said of low values: examination of the chips enables clearing any confusion between weathering versus fissuration.

These few examples highlight the wide variety of cases one can encounter as also the hopeless situation in which one often finds oneself when it comes to identifying *a priori* which probes will best differentiate the various facies in a quarry. Since there is no question of trying all the probes one after the other, their discriminatory power is tested on two or three standard boreholes (that is, thoroughly studied from the viewpoint of cuttings). It should be noted that at present the natural radioactivity and resistivity logs, complemented by recording of the rate of penetration, have invariably helped resolve the problem posed, namely, location of the borehole section.

A log is an assurance in this sense that all the facies have been brought to the form of chips (which is not evident, for instance, in fine clay layers in a limestone deposit) and the geologist has identified all of them in the proper order (case of a mixture of chips).

A synthesis of the results (borehole section, logs, laboratory examination of the chips) enables preparation of the definitive geological section, with a precision that almost matches that of a core drilling.

It is essential that the studies keep exploration in focus above everything else. One should never forget that what is needed is to make available to the quarry operator a working scheme that will enable exploitation of a deposit to the best if his ability with regard to the quality and homogeneity of the products.

To this end, it is desirable to clearly demarcate the facies that differ geotechnically, appraise the variations in thickness of the overburden and make available

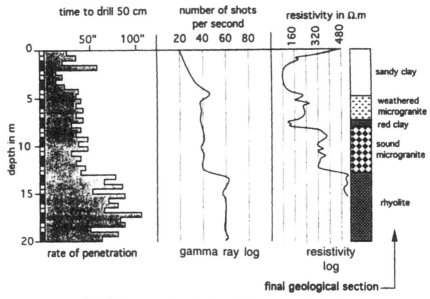

Fig. III.31. Recording of rate of penetration, γ-ray and resistivity

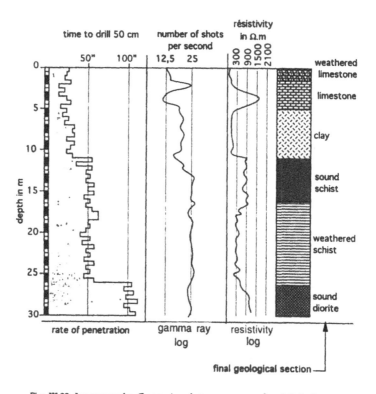

Fig. III.32. Igneous rocks: Comparison between γ-ray and resistivity logs

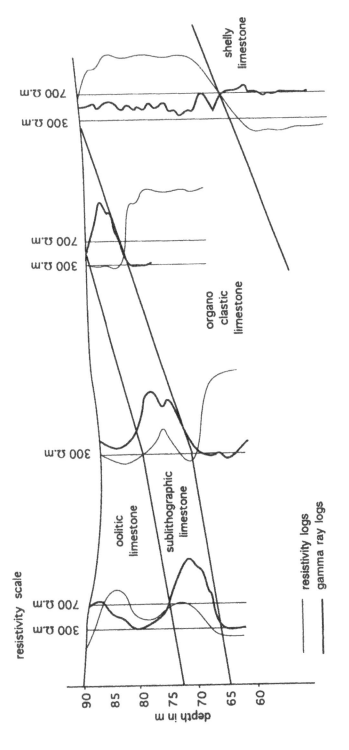

Fig. III.33. Structure of a limestone deposit

the maximum data which will facilitate exploitation through a proper mapping of the working faces.

Exploration of a deposit goes hand in hand with development of rational exploitation, which is reflected in the achievement of a consistent quality of aggregates, judicious acquisition of lands, layout suited to dumping mine wastes and future location of the processing plant.

The methods described above now enable carrying out explorations in an effective manner and at reasonable costs.

7. DETAILED PROSPECTING OF LOOSE ROCK DEPOSITS

As in case of deposits of massive rocks, the stake for quarry operators in having a thorough knowledge of their deposit cannot be overemphasised. The reasons for this are the same (output rates, quality required for the materials, cost of investments etc.) to which may be added the problems of reclamation of the mined area, which are even more acute than for quarries of massive rocks. But then no programme which is clear and viable, both technically and economically, can be laid down and then implemented without a detailed preliminary study of the deposit, which implies a close co-ordination of the phasing of exploitation, movements of soil (overburden) and reclamation jobs proper (see Chapters II and VIII).

In the area of alluvial deposits, which we shall elaborate here, formulation of any scheme as such is difficult given the local peculiarities of each valley. However, as in the case of deposits of massive rocks, it is possible to isolate the main parameters, a knowledge of which is indispensable, as also the methods most suited for defining these parameters.

7.1 Parameters to be Defined

7.1.1 GEOMETRICAL PARAMETERS

Without doubt, it is imperative that the volume of the overburden and the workable material be determined as also their variations in thickness. This means, in practice, having at one's disposal:

a) Three detailed topographical maps:

— of the ground surface, which enables *inter alia* to properly direct the geophysical investigation and drilling,

     — of the top of the workable material,

     — of the top of the substratum, which will also serve as an aid for reclamation works, which accounts for its paramount importance in the contemporary context.

b) Two isopach maps:

— of the overburden thickness,

— of the thickness of the workable materials.

These last two maps, which serve as a basis for calculation of the cubature and the laying down of the plan of extraction, overburden removal and reclamation derive from the three topographical maps. Indeed, use of the detailed topography of the different horizons (ground surface, roof of the material, roof of the substratum) is inescapable, since exploitation of only the raw borehole data (that is, thicknesses) is often such as to result in erroneous compilation of documents.

Furthermore, it is imperative to know the piozometric surface of the groundwater since it has a direct bearing or the following:

— The very progress of exploitation (problem of flooding).

— Mode of overburden removal and even tolerable thickness of overburden, which can vary considerably depending on whether it is submerged or outside water.

— Choice of course of reclamation to be adopted and consequent movements in the ground (possibility of water level, for example).

— Mode of extraction of the gravels (dragline, grab, drag scraper, dredge etc.).

— Characteristics of the material extracted and hence its end-uses. In the case of a deposit out of contact with water, the material extracted will retain all its fines, whereas with submerged or partially submerged deposits, the material is invariably modified by extraction and loses a substantial part of its fines. In most cases this loss in fines is accompanied by a notable rise in the sand equivalent test (cf Chapter IV) (10 to 20 points or more depending on the mode of extraction) but can also be characterised by a somewhat too high loss in useful fines (between 100 and 300 µm, such as to warrant a correction by the addition of fine elements).

Furthermore, it is worthwhile to know the course and direction of groundwater flow since in some cases this may have a direct bearing on the manner of opening the extraction (problems of drainage or silting up; see Chapters II and VIII).

7.1.2 PARAMETERS DEFINING THE PROPERTIES OF THE MATERIAL

Though specifying the mineable material is the main consideration, it is equally important to know the nature of the overburden, since its extraction and re-use (forming part of the reclamation jobs and possible return to cultivation) can be quite different depending on whether it is plastic clay or slimy sand and/or its degree of submergence. The main properties of the minable material to be taken into account are:

a) Grading

Matching the processing plant to the deposit is possible only if the grain size of the whole material as also the inner grading of the sand are known. This is true for all deposits, including those that are homogeneous. In the case of heterogeneous deposits, which unfortunately are the most widespread, grading varies in both the vertical as well as lateral planes (Figure III. 34; also see Figure VIII.38 of Chapter VIII). A poor understanding of the distribution of its trend inside the deposit entails at all stages (scalping, screening, crushing) phenomena of overfeeding or underfeeding the machinery, with all the attendant consequences, among which a diminution of the operational capacity of the plant and a fall in the properties of the finished products are the two most serious.

Fig. III. 34. Two layers of widely different gradings overlying one another

This means that in heterogeneous deposits, to mitigate variations in grading, one would have to simultaneously open several points of extraction to achieve the maximum possible consistency in the feed material to the plant, which obviously implies a very thorough knowledge of the deposit (see Chapter VIII).

b) Contaminations and heterogeneities

The term 'contamination' usually means the presence of clay and/or silty clay loam in the deposit. These clays and silts can be uniformly distributed and affect the entire alluvial material: in this case the deposit will be totally contaminated and the sand equivalent will be low or very low but fairly consistent. Contrarily, contamination may be in the form of pockets or lenses of variable extent and interstratified in an apparently random manner throughout the alluvial deposit. This type of contamination is a major constraint in exploitation and marking out these lenses, when possible, helps avoid many rude jolts to the processing plant. The clayey contaminations after all represent a surface weathering of the alluvial deposits; as a result of phenomena of pedological nature the alluvia are transformed into decalcified reddish beds, rich in clays and iron oxides and with low sand equivalent, over highly variable thicknesses. The phenomena of repeated frost-thaw cycles (termed cryoturbation under periglacial climate) sometimes also alter the surface zones, the clayey materials then collecting into pockets of variable magnitude, but generally shallow under the surface of the deposit (Figure III.35).

If these contaminations due to fine and ultrafine materials remain the main heterogeneities of the alluvial deposits, certain other phenomena can also bring

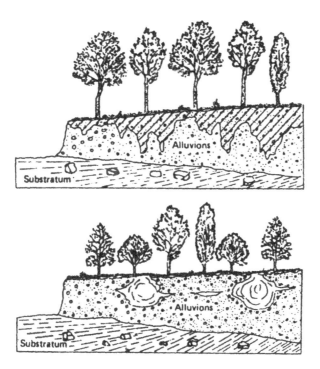

Fig. III.35. Contamination by the phenomena of redbeds (top) and cryoturbation (bottom)

about major dislocations in the progress of extraction and should therefore be demarcated beforehand. These are the hardened beds and large-size blocks. The former, frequent in deposits of the Paris region, are found in the zone of fluctuation of the water table. It happens that the presence of such horizons compels the quarry operator to either use an explosive or abandon a portion of the deposit. Such horizons are mainly troublesome when they reach several cubic metres.

c) Petrographic nature

The broad lines of the mineralogical, petrographic and mechanical characteristics of the materials are already known, starting with the stage of regional syntheses. However, a number of points remain to be specified when it comes to be a study at the deposit level.

— The percentages of soft or weathered elements (chalk, soft limestone, schists, various weathered igneous rocks) in the different aggregate sizes of the raw material.

— The exact nature and the hardness of the predominant elements: estimated first through petrographic examination and then by mechanical tests (micro-Deval in the presence of water, friability of sands), which enable testing the alluvial materials in the same fashion as the aggregates coming from massive rocks and determining their mechanical strength in a better manner than possible with such tests as lime determination, which take no note of the hardness of the limestones.

— The nature of the fines (< 80 μm) and their grading curve. It is important to know through a few checks—without multiplying the tests whether in a material containing 50% fines, the latter are predominantly silty or clayey, and whether these silts are primarily siliceous or calcareous.

— The organic matter content and in particular its harmfulness.

7.2 Techniques of Study

7.2.1 PRELIMINARY DATA ON TOP FOR STUDY OF A DEPOSIT

a) Data extracted from regional studies

When a river has been studied as part of a unified basin plan or of a regional inventory of resources, the broad trends in each zone are already known. These points have been elaborated in Section 5 above and we revert to them only to list the 'trends' available:

— mean grading within the zone of values it forms part of;

— mean and extreme thicknesses of the overburden and gravels;

— mineralogical and petrographic composition of the different aggregate sizes;

— main heterogeneities: homogeneous and heterogeneous contaminations, frequency and approximate size of heterogeneities.

b) Data extracted from the follow-up of exploitations in the same region

An accurate follow-up over one or more years of extractions situated in the same zone as the site under study, makes it possible to direct new surveys in a much

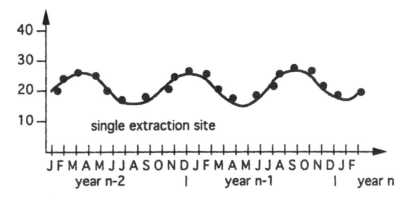

Fig. III.36. Monthly variation in sand contents at extraction. Broad valley

sand content (0/5 mm %)

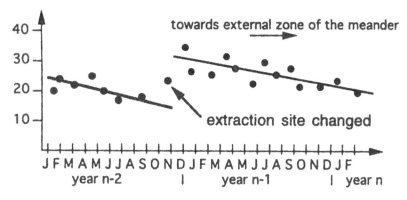

Fig. III.37. Monthly variation in sand contents at extraction. Meander deposit

more focussed manner than a simple regional synthesis. For instance, a few 'types of deposits' can be specified in the alluvial materials of the Paris region:

— Deposits which are homogeneous as a whole but heterogeneous in detail. Deposits of alluvial sands and gravels answering this description display low horizontal variations of grading. Generally they are the alluvially broad valley type and not the meandering form.

Whatever be the mode of quarry operation, the change in sand content at extraction versus time is in the form of an approximately sinusoidal wave of small amplitude and short or very short period (Figure III.36).

— Deposits which are heterogeneous as a whole and homogeneous in detail: Deposits of alluvial sands and gravels answering this description display pronounced horizontal variations in grading, which are often slow and progressive. Generally they are of the alluvially meandering valley type.

Whatever be the mode of quarry operation, the change in sand content at extraction versus time for any single point of extraction will follow an approximately sinusoidal shape of average to long period (Figure III.37).

7.2.2 METHODS OF STUDY

a) Use of aerial photographs: Topography

To start with, use of existing documents is highly profitable: aerial photographs to 1/25,000 to 1/30,000 scale (total coverage of France) and 1/25,000 and 1/30,000 topographical maps. Detailed analysis of these two documents (morphology, toponymy, photogeological interpretation) enables location of old channels, even totally abandoned, boundaries of earthworks etc. (Figure III.38).

A very precise levelling (to the centimetre) is also highly useful in the detailed study of a deposit.

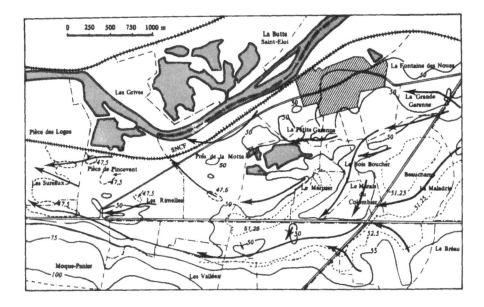

Fig. III.38. Lines of stream ways of river Seine

b) Geophysical resistivity line

The technique employed for prospecting alluvial deposits is most commonly that of resistivity with one or more lengths of line, depending on the deposit. The patterns to be used vary depending on the context.

Advantages of the method: fast and judicious (equipment not unwieldy and cost very low); when applied in the proper manner, it always yields a clear demarcation of the overburden from the deposit but requires calibration by means of mechanical drillings (Figure III.39).

Limitations of the method: when the sites are very complex (schemes with three or four zones poorly contrasted), one can hope to obtain the overburden only. Conclusions drawn from the data obtained from long line lengths (> 20 m) then remain circumspect, even after good prior calibration through drillings.

c) Geophysical survey: Refraction shooting

More rarely used, it is suitable only when the substratum is clearly contrasting (hard limestone for example) and the velocity of sound inside it higher than in the gravels. Furthermore, when the deposits are flooded, care should be taken to clearly recognise the level of the groundwater before the tests. In all cases this method can only give the total thickness of the loosely consolidated material and sometimes only the overburden-gravels boundary.

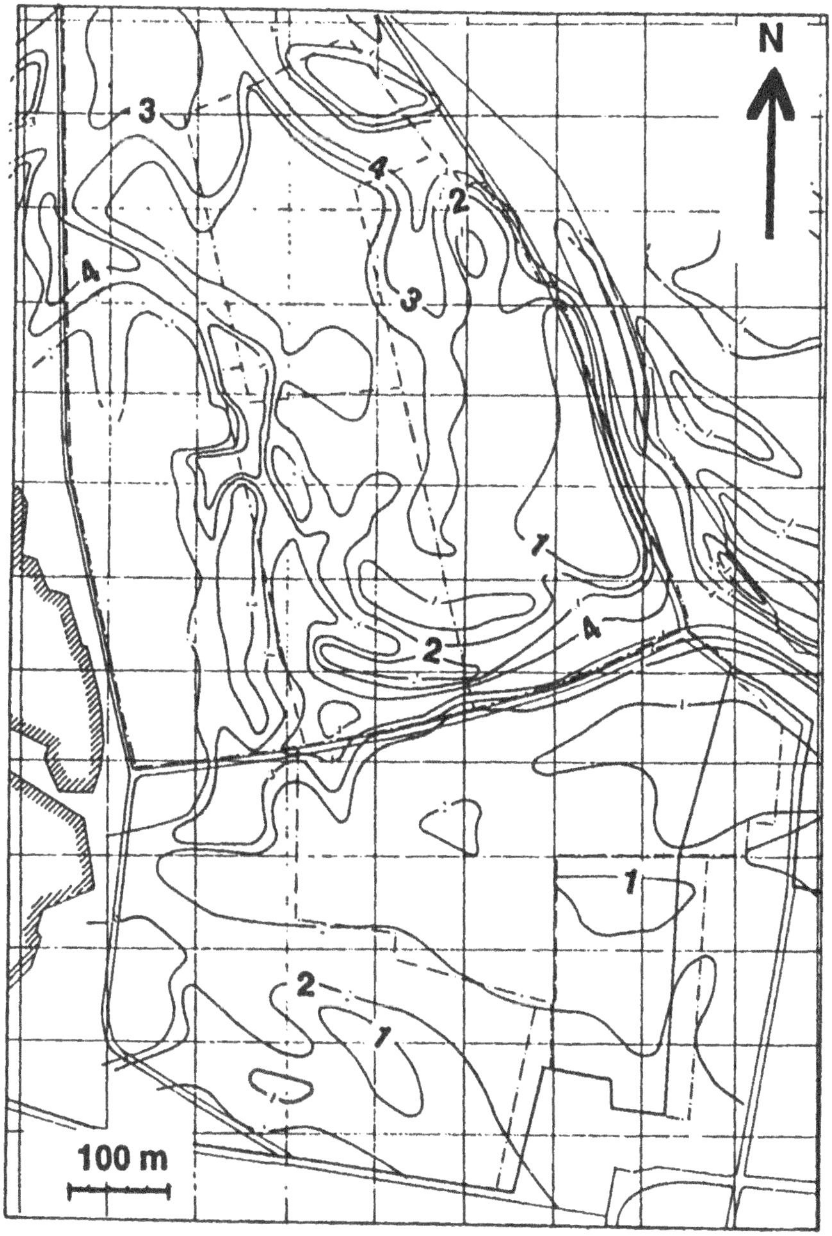

Fig. III.39. Thickness of overburden in a sand and gravel deposit

d) Geophysical survey: Artificial magneto-telluric method

This method has already been described. It is most effectively used in alluvial deposits for demarcation of the overburden.

e) *Boreholing*

Fast boreholing with non-representative sampling: These are generally relatively low-diameter boreholes (< 200 mm and often < 100 mm) carried out with screw augers working by simple rotation or rotation-percussion.

Advantages of the method: fast procedure (5 to 12 m/h) making it possible to correctly define the deposit's geometry (gravel-overburden and gravel-substratum boundaries at about 20 to 50 cm) and through this even calibrate the resistivity measurements. Generally, it is additionally possible to get an approximate idea of the material traversed, especially for fine beds (sands, clays).

For some years now, in conjunction with the methods of fast boring, use has also been made of recording the natural radioactivity in the borehole. In this case the fastest possible and hence less costly boring method (12–15 m/h) is chosen. The natural radioactivity probe is lowered directly into the string of hollow drill rods. The recordings enable precise determination, and at reduced cost, of the overburden-gravels and gravels-substratum boundaries, as also clay intercalations. A quantitative interpretation of the recordings should enable an approximate assessment of the clayey contamination of the deposit (Figure III.40).

Limitations of the method: though relatively economical, fast borings with or without log should, however, be used conscientiously, avoiding any systematic pattern *a priori*. Their use should be properly integrated into the total scheme of study. Furthermore, it should be mentioned that this method could be totally unusable in deposits containing very coarse elements (> 200 mm), which is the case, for example, with a number of glacial or fluvioglacial deposits.

'*Heavy borings*': We include under this term boreholes made essentially for obtaining representative grading curves. Their number could vary widely depending on the situation (that is the type of deposit) and previous study.

When there is no water, the means available are essentially as follows.

Hydraulic shovel: enables cutting trenches and hoisting quite adequate quantities of materials, even very large-size gravel. Further, trenches have the advantage of presenting a visible cross-section of the deposit. The sole drawback is a practical limit to the depth that can be cut, which is of the order of 8 m. This method is indispensable when processing tests (crushing-screening) are to be carried out for setting up plant machinery.

Single coil augers (ϕ 600 to 700 mm): the sampling increments are generally representative (D < 100 mm), the precision of the limits fairly good (20 to 50 cm) and the practical limit to depth of the order of 11–12 m.

Both methods are fast: 50 to 150 m/day.

Where the gravels are submerged, no method has yet proven entirely satisfactory. Some are relatively fast but lack precision, others more valid from the viewpoint of representativeness of the sample but very slow and hence quite costly.

Large diameter ($\simeq$ 400 mm) screw augers: driving is done normal to the deposit and the material is brought up by grabbing (without rotation). This method is suited to some types of gravel whose contour and cohesion are such that a sufficient quantity of material (including the large-size elements) is retained on the

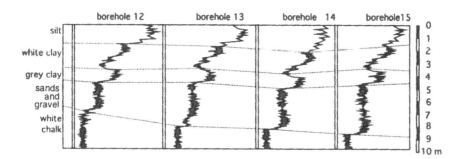

Fig. III.40. γ-ray log in an alluvial deposit.

Fig. III.41. Vibro-percussion-cum-rotation drilling rig

blades. Sampling is fast but limited to a 4 to 5 m depth for a 400-mm diameter (difficulties of grabbing) and to a maximum D of 100 mm for gravels.

This method can be employed for depths up to 6 to 7 m with a 200-mm auger but then is suitable only for gravels of D < 50 mm.

Beneto type drilling machines for big holes: this is the most commonly used drill. Sampling increments are generally good except for some loss of fines, even fine sand in both methods. However, it should be noted that this "sand loss" is not unlike that experienced during extraction proper. The diameters utilisable enable sampling almost all types of gravels regardless of depth.

The major and often very restrictive drawback of these methods is their slowness (6–10 m/day) and hence their cost; thus they are necessarily used very sparingly.

Vibro-percussion drilling machine (Figure III.41): this method enables sampling submerged gravels under conditions qualitatively similar to those of the Beneto type but at a decidedly faster pace (40 to 60 m/day).

The sequence of these methods and the patterns adopted obviously depend on the spread of the deposit, available earlier data, the stage of study (choice of deposit in a given zone or detailed study of a deposit for its exploitation).

REFERENCES*

Allard, P. and R. Campanac. 1976. The calcareous deposits of Provence-Côte d'Azur region. *Equipment Mécanique des Chantiers*. August–September.

Angot, D., J.H. Chezeaud and J. Pitot. 1973. The calcareous aggregates of Beauce. *Bull. de Liaison des Labratoires et des Ponts et Chaussées* (BL des LPC). Special issue, June.

Archimbaud, C. 1973. Evaluation of the properties of a rocky deposit. *BL des LPC*. Sept.–Oct.

Archimbaud, C., J.P. Joubert, A. Maldonado, A. Prax, L. Primel and J. Roy. 1973. The production of aggregates. *Review Travaux* 464a 'La Route', Nov.

Bernaix, J. 1974. Properties of rocks and rocky massifs. General report on theme I. *Proc. 3rd Cong. Internat. Soc. Rock Machanics, Denver (USA)*, Sept. vol. I, tA, pp. 39–68.

Berton, Y. and L. Primel. 1978. Methodology of aggregate resources surveys. Report on Aggregates Parafiscal Tax EG 45.

Blanc, J. 1977. Studies on the sand and aggregates deposit on the continental margin of Provence. *Equip. Méc. C. Mat.*, no. 155, March.

Blot, G. and D. Dufresne. 1974. A laboratory study of the clay content of alluvial materials and sedimentary rocks by measurement of their natural radioactivity. *BL des LPC*, no. 70, March–April.

Cauvin, J. and M. Lesauvage. 1971. Principal characteristics of the Marne alluvia. *BL des LPC*, no. 55, Oct.–Nov.

Champion, M., P. Maillard and P. Cario. 1971. Alluvia of the Loire in the central region. Inventory of production and deposits. *BL des LPC*, no. 56, Dec.

Chevassu, G. 1973. Surface geophysics applied to prospecting of massive rocks. *BL des LPC*, no. 67, Sept.–Oct.

Chevassu, G. 1976. Granitic rocks and their weathering. Geotechnical prospecting of the lines in Bretagne. LPC Research Report no. 61, Nov.

Chevassu, G., R. Lagabrielle and M. Rat. 1984. Detailed prospecting of deposits of aggregates. *Bull. AIGI*, no. 29.

---

*All entries all French except those marked with a cross (+)—General Editor.

⁺Corcer, A.J. and J.F. Gartner. 1984. Glacial geology: A key to aggregate investigations in Canada. *Bull. AIGI*, no. 29.

Cressard, A.P. and C.L. Augris, 1984. Marine aggregates in France. *Bull. AIGI*, no. 29.

Debelmas, J. 1974. *Geology of France*. Doin, Paris, 2 vols.

Demulder, E.F.J. 1984. A geological approach to traditional and alternative aggregates in the Netherlands. *Bull. AIGI*, no. 29.

Doridot, M. and M. Lesauvage. 1974. The study of alluvial materials in the valley of Yonne. *BL des LPC*, no. 69, Jan.–Feb.

Famechon, C. 1973. The sand of Landes. *BL des LPC*, no. 66, July–Aug.

Geoffray, J.M., A. Mishellany and J. Restituito. 1978. Basalts and pozzolanas of the Massif Central. *BL des LPC*, no. 92, Nov–Dec. 1977; no. 93, Jan.–Feb. 1978; no. 94, March–April.

Aggregates. 1975. LCPC 16-mm colour film, 32 min, made in collaboration with the Regional Laboratories of Angers, Clermont-Ferrand and Rouen.

Griveaux, B. 1974. The role of faults in the quarrying of massive rocks. *BL des LPC*. no. 73, Sept.–Oct.

Griveaux, B. 1976. Modified dynamic fragmentation test on chips from destructive drilling. *BL des LPC*, no. 82, March–April.

Griveaux, B. and A. Maldonado. 1972. Typical scheme of study of a quarry. *BL des LPC*, no. 57, Jan.–Feb.

Griveaux, B. and J. Peybernard. 1975. New methods of prospecting massive rock deposits. *Equipment Méchanique. Carrières et Matériaux*, no. 143, Nov.

⁺Hamilton, C.G. 1984. Exploration methods for aggregates. *Bull. AIGI*, no. 29.

Havard, H., N. Mignot, M. Potdevin and Sygfrid. 1976. Establishing a geotechnical catalogue of aggregates. *BL des LPC*, no. 84, July–Aug.

Heraud, H., J. Restituito, J. Terne and C. Tourenq. 1975. Reconnaissance of rock cuttings recovered by test-boring with hammer rock drill. *BL des LPC*, no. 77, May–June.

Horn, R.L. and J.L. Peragallo. 1984. Seismic reconnaissance of gravels. *Bull. AIGI*, no. 29.

Lagabrielle, R. and G. Chevassu. 1984. New geophysical methods for ground or aquatic deposits. *Bull. AIGI*, no. 29.

⁺Lenczewska, E. 1984. The application of photo-interpretation for recognizing and estimating alluvial soil beds in the area of flysch in the Polish Carlathian Mts. *Bull. AIGI*, no. 29.

Muraour, P., M. Frappa and J. Peragallo. 1984. Contribution of the technic of seismic reflection signals to the reconnaissance of aggregates under water. *Bull. AIGI*, no. 29.

Olivier, G. 1976. Petrographic counts in aggregates. *BL des LPC*, no. 83, May–June.

Panet, M. and L. Primel. 1975. Aggregates. Special issue no. X. Quality control and road construction. *BL des LPC*, March.

Panet, M., L. Primel and C. Archimbaud. 1972. Material deposits and temporary installations. *BL des LPC*, no. 60, July–Aug.

Primel, L. 1972. Regional syntheses of construction aggregate resources and roadworks. *BL des LPC*, no. 58, March–April.

Primel, L. 1976. Sand deposits in France. Information Day. 'Sand treatment for road subbases'. Bordeaux, Nov.

Primel, L. 1984. Multidisciplinary research on aggregates in the Bridge and Road Laboratories. *Bull. AIGI*, no. 29.

Rat, M. 1973. Geological reconnaissance of road alignments. Statement of prospecting methods and studies. *BL des LPC*, no. 68, Nov.–Dec.

Robert, J. 1973. Study of alluvial materials of the upper Garonne. *BL des LPC*, no. 68, Nov.–Dec.

Tourenq, C. 1974. Properties and classification of national construction materials. 2nd Internat. Cong. Engr. Geol., Sao Paulo, August.

Tourenq, C. and D. Fourmaintraux. 1971. Contribution of petrography to the physical and mechanical study of rocks. *BL des LPC*, no. 50, pp. 157–163.

Tourenq, C. and C. Archimbaud. 1974. Properties of limestones. *BL des LPC*, Sept.–Oct.

Villain, J. 1973. Materials of the Rhône valley between Switzerland and the confluence of the Ain. *BL des LPC*, no. 67, Sept.–Oct.

BL des LPC. 1977. Resources and prospecting of deposits. Special issue on aggregates, June 1977, 233 pp. This issue contains 21 articles devoted to the problems of inventorying and prospecting for aggregate deposits.

Bull. AIGI. 1984. International Symposium on Aggregates, *Bull. AIGI*, no. 29–30.

IV

Properties of Aggregates—Tests and Specifications

Claude Tourenq and Andre Denis

Since time immemorial the art of construction has been linked with the use of rocks. The remains of the walls of Jericho, 8000 B.C., constitute one of the first undeniable traces of their use in construction activities. Modern civil engineering continues to use these materials, mainly in the form of aggregates, the basic constituent for its technics. Thus, after water, aggregate is the product most used by man in the world.

Man has known throughout history how to choose from the available rocks those which suited him best: Our ancestors of the Palaeolithic age knew already how to use flint with very fine grains for making sharp tools, reserving the coarser sandstone for hammering tools.

Today, engineers impose on aggregates specifications relating to very diverse properties according to the type of use. It should be possible to determine each of these properties by conventional tests defined in specific standards. A compendium of the French Standard Institution: 'Association Française de Normalisation' (AFNOR) contains all the texts relating to the tests and specifications in force in France in the field of aggregates.

The properties of aggregates depend either on the nature of the original rock (mineralogy, physical and mechanical properties) or on the conditions of mining and processing (geometrical characteristics and fines content). But this distinction, convenient for dealing with the different properties of aggregates, allows a few intermediate cases: grading of alluvial aggregates influenced by the original grading in the alluvium, shape influenced by anisotropy or the nature of the original rock etc. Let us say that the geometrical characteristics and the fines content can be considerably improved by adopting suitable methods of extraction, crushing and sieving and that the mechanical characteristics can be influenced only to a very small extent by processing.

This distinction is useful, especially when viewed either in the context of the study of resources and deposits or for the purpose of controlling supplies.

1. CHARACTERISTICS OF AGGREGATES ASSOCIATED WITH THE PROPERTIES OF ROCKS

1.1 Identification of Rocks

Rocks are solid media, polycrystalline, heterogeneous and very complex, consisting of grains which belong to some ten mineral species and which differ in size and shape. The properties of rocks depend on the nature and state of weathering of the minerals.

Rocks are also discontinuous media, the assemblage of grains is never perfect and in between them lie roughly equidimensional voids termed pores and inter- or intracrystalline defects termed cracks, wherein one dimension is very small compared to the others. These discontinuities are mostly responsible for the mechanical behaviour of rocks.

Therefore, the identification of rocks necessarily includes several aspects, which are described in Standard* P18-557.

Petrographic identification enables determination of the mineralogical composition of the rock. This aspect is important if we want to detect those minerals that are harmful, say, with respect to binders. It also enables assessment of the size and shape of the crystals, their spatial distribution, the anisotropy of the rock and the state of weathering of the minerals.

If we want a more accurate estimate of weathering wherein clays play an important role, the methylene blue test described in Standard P18-592 gives a good estimate of their total specific surface area (see Section 2.4).

A study of the discontinuities is important because the mechanical strength and sensitivity of rocks to frost depend on them.

Pores are rather easily detected by measuring the volume of water V_w contained in the rock after complete saturation; if V_r is the volume of the sample tested, porosity n is denoted by the ratio

$$\frac{100\,V_w}{V_r}.$$

Cracks, which occupy a very small volume (n < 0.3%) are revealed because of their strong influence on the velocity of propagation of longitudinal waves V_l

Figure IV.1 illustrates very clearly the highly different effect of pores and cracks on V_l.

Since the velocities for different minerals could be fairly different, in order to compare all the rocks, it is advisable to use the index of continuity (IC) defined in Standard P18-556. IC is the ratio between the velocity measured on the rock and its calculated theoretical velocity V_l^0, assuming that it has neither pores nor cracks.

---

*All standards referred to in the text are French standards unless otherwise specified —Technical Editor.

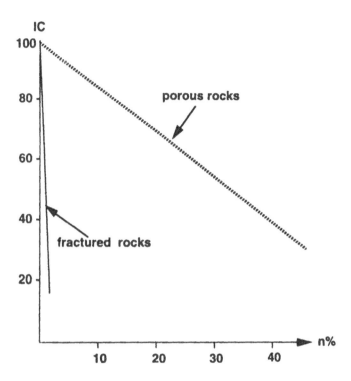

Fig. IV.1. Variation in the index of continuity (IC) as a function of porosity n for porous and cracked rocks

If C_i is the mineral i content of the rock in which the propagation velocity of the waves is V_i, V_i^o is given by the weighted harmonic mean of the propagation velocities in the minerals constituting the rock:

$$\frac{1}{V_i^o} = \sum \frac{C_i}{V_{l_i}}.$$

The index of continuity IC is then given by:

$$IC = \frac{V_l}{V_i^o} \, 100.$$

If we take into account the fact that the volume of the cracks is negligible compared to that of the pores, it is possible to calculate a degree of cracking

$$DF = 100 - 1.4\,n - IC$$

(1.4 being the slope of the regression line of the porous media of Fig. IV.1).

1.2 Characteristics of Aggregates Associated with the Mechanical Strength of Rocks

1.2.1 BRITTLE FAILURE OF ROCKS

In the field of low confining stresses, which corresponds to the conditions of use of aggregates in civil engineering, rocks have a brittle behaviour. We know from Griffith's works that brittle failure results from the uncontrolled propagation of cracks in a solid medium.

The compressive strength test (RC) is a good indicator of the mechanical properties of rocks. However, compared to it, preference should be given to the much less used tensile strength test (Brazilian test, Rtb), which involves crushing a cylinder compressed along two diametrically opposite generatrices (Fig. IV.2). It is to be noted that RC is approximately equal to 10 Rtb.

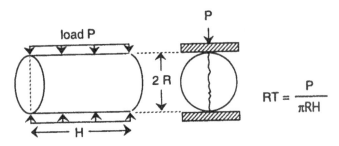

Fig. IV.2. Brazilian test

1.2.2 MECHANICAL STRENGTH TESTS OF AGGREGATES

The most commonly used tests on aggregates involve processes of rupture by fragmentation, similar to Rtb, and wear by friction.

The Los Angeles (LA) test causes fragmentation of aggregates in a ball mill: 5 kg coarse aggregates (4/6, 6/10, 10/14 or 25/50 mm) and, according to grain size, 7 to 12 steel balls of 430 g are introduced into a metallic testing cylinder of 711 mm diameter. After 500 rotations (1000 for the 25/50) at 33 rpm, the undersize at 1.6 mm related to the original mass, expresses the LA coefficient (Figure IV.3, Standard P18-573) in per cent. It should be noted that this test is carried out on a dry basis, water exerting very little influence on the result, and that a portion (20%) of the elements produced results from the wear by friction in the cylinder.

To expedite control tests, the French so-called "dynamic fragmentation test" (FD) borrowed from the British impact test is used. In this test a 14 kg weight falls from a height of 40 cm on 350 g of 4/6, 6/10 or 10/14 mm, 16, 22 or 28 times, the result being calculated as for the LA and calibrated on it (Fig. IV.4, Standard P18-574).

For the smallest aggregate sizes (fine aggregates or sand), the micro-Deval cylinder is used as a mill with steel balls (21 of 18 mm, 9 of 30 mm and the rest in 10 mm balls to make up 2.5 kg) acting on 500 g of 0.1/2 mm sand with 2.5 l

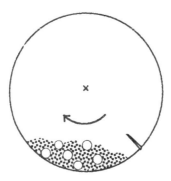

Fig. IV.3. Los Angeles cylinder

water. The coefficient of friability of the sands (FS) thus determined corresponds to the percentage of undersize in the 50 μm sieve (Standard P18-576).

The wearing away or attrition, i.e., production of fine elements by friction of aggregates against the metallic parts of the devices and against themselves, is probably the most frequent phenomenon. However, water plays an important role in the phenomena of wear and should not be neglected. Resistance to wear was measured right from 1870 by the Deval test designed for the macadam technique: 44 stones placed in a metallic cylinder rotating obliquely with respect to its axis, become worn by mutual friction. This test is still used in France for the ballasts of railway tracks either or a dry basis or in the presence of water.

The more recent micro-Deval test in the presence of water (MDE) consists of rotating in a cylinder of 200-mm diameter 500 g of 4/6, 6/10 or 10/14 mm fine aggregate with 2, 4 or 5 kg of 10 mm steel balls, in order to increase the friction surface and thereby reduce the duration of the test, and 2.5 l water. After 12,000 rotations in 2 hours, the percentage of elements passing through a 1.6 mm sieve represents the MDE coefficient of wear. The test can be carried out on a dry basis but the difference in results of sensitivity of the material to water is very significant (Fig. IV.5, Standard P18-572).

It is interesting to note the reverse effect, i.e., the wearing away of metallic parts caused by the aggregates or abrasion. The latter is readily measured with an abrasimeter (Standard P18-579) in which a standard steel plate is rotated at a given speed in a sample of 4/6 mm aggregate. The loss of mass of the steel plate yields an index of abrasion and the granulometric change in the aggregates enables assessment concomitantly of their resistance to fragmentation (Fig. IV.6).

When the particles which contribute towards wear are very fine, dusts on the surface or pavements for example, the surfaces of fine gravels are apt to smoothen under the action of the tyres and the resultant polishing may be unfavourable to vehicular braking on wet pavements.

The accelerated polishing test (Standards P18-575 and P18-578), inspired from the British, consists in subjecting a test slab of resin mortar in which 35 to 50

Fig. IV.4. Dynamic fragmentation machine (Impact test)

chippings of 6/10 mm aggregate have been embedded to two cycles of wear. Under the action of a tyre wheel with interposition first of a coarse emery, then a fine one and in the presence of water, the state of polishing is ascertained by means of a friction pendulum fitted with a rubber shoe (Fig. IV.7.).

It is possible to determine the "polishability" of rocks from their mineralogical composition. In fact, this characteristic depends on the rock's resistance to wear which is linked with the hardness of the minerals and the difference in hardness amongst these minerals, this difference being responsible for the residual roughness after polishing.

The polished stone value (PSV) obtained through the test described above varies from 0.30 for pure limestones to 0.65 for porous or weathered silicate rocks. There is often a conflict between the search for rocks with high mechanical strength

Fig. IV.5. Micro-Deval machine

Fig. IV.6. Abrasimeter

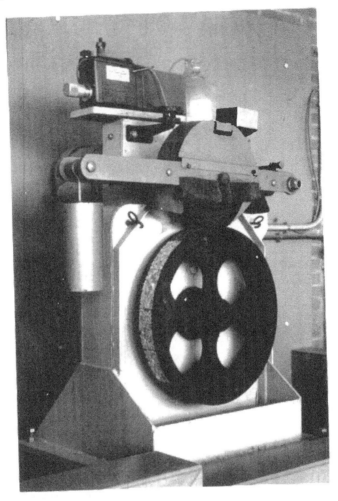

Fig. IV.7. Machine for polishing the chippings

generally associated with a very fine and polishable structure, and a high PSV assisted by granular structure and higher porosity.

1.2.3. SPECIFICATIONS RELATING TO MECHANICAL STRENGTH (Table IV.1)

According to the conditions of use of the aggregates, depending on whether they are incorporated or not in a composite material with hydraulic or bituminous binder, aggregates are subjected to very diverse static or dynamic stresses. This variability in conditions of use of aggregates justifies wide ranges of specifications relating to their mechanical strength.

It is also important to take into account the handling conditions (stocking etc.) and processing conditions (mixing, compaction etc.). During the work phase

Table IV.1: Specifications for roads and concrete

Techniques	Range traffic		
	Moderate	High	Very high
Surface courses	LA ≤ 25 MDE ≤ 20 PSV ≥ 0.50 SE ≥ 60 MBV ≤ 1	LA ≤ 20 MDE ≤ 15 PSV ≥ 0.50 SE ≥ 60 MBV ≤ 1	LA ≤ 15 MDE ≤ 10 PSV ≥ 0.55 SE ≥ 60 MBV ≤ 1
Unbound base courses	LA ≤ 30 MDE ≤ 25 SE ≥ 50 MBV ≤ 1.5	LA ≤ 25 MDE ≤ 20 SE ≥ 50 MBV ≤ 1.5	LA ≤ 20 MDE ≤ 15 SE ≥ 50 MBV ≤ 1.5
Bound base courses & unbound sub-base courses	LA ≤ 40 MDE ≤ 40 SE ≥ 50 MBV ≤ 1.5	LA ≤ 30 MDE ≤ 25 SE ≥ 50 MBV ≤ 1.5	LA ≤ 25 MDE ≤ 20 SE ≥ 50 MBV ≤ 1.5
Bound sub-base courses	LA ≤ 40 MDE ≤ 40 SE ≥ 50 MBV ≤ 1.5	LA ≤ 40 MDE ≤ 40 SE ≥ 50 MBV ≤ 1.5	LA ≤ 30 MDE ≤ 25 SE ≥ 50 MBV ≤ 1.5
Concrete for road courses	LA ≤ 25 MDE ≤ 20, PSV > 0.50 SE ≥ 60 MBV ≤ 1		
Other concretes: strength < 36 MPa	LA ≤ 40 SE ≥ 60 MBV ≤ 1		
Other concretes: strength ≥ 36 MPa	LA ≤ 30 SE ≥ 60 MBV ≤ 1		

LA: Los Angeles, MDE: Micro-Deval, PSV: Polished stone value, SE: Sand equivalent, MBV: Methylene blue value.

aggregates often undergo severe stresses which may lead to evolution of grading in very weak rocks with enrichment of fine particles due to fragmentation and attrition. These provisional conditions during work sometimes call for specifications which may appear severe with respect to the stresses to which the aggregates will be subjected once the structure is completed.

• *Hydraulic concretes*

The specifications are given in Table IV.1; the essentials of Standard P18-541 are given in these specifications.

The stresses to which these structures are subjected are basically static or dynamic with low frequencies for permanent works; they do not generally undergo any impact. The strength of aggregates is always higher than that of concrete and in most cases it is observed that cracks preferably follow the mortar-aggregate interface which constitutes the zone or least resistance. Only a few calcareous aggregates, less resistance and with stronger epitaxic bonds between cement and calcite sometimes reveal intragranular ruptures. Recent results revealed no significant influence of the hardness of the limestones—RC varying from 20 to 200 MPa and LA from 80 to 20—on the strength of the concretes, and this in spite of a grading change due to handling in the case of the most recent ones.

• Pavement materials (Table IV.1, Standard P18-101*)

Several factors are involved in the determination of levels of specifications:

— The type of technology wherein the requirements are obviously more severe for untreated or slow-setting materials than for those in which the binder plays its role immediately.

— The position of the layer in the pavement since the surface is subjected to higher stresses.

— The level of traffic wherein the daily number of heavy vehicles plays an essential role (effective load < 5 t).

— The nature of the works, e.g. strengthening under traffic, which calls for harder conditions and leads thereby to more stringent specifications (Table IV.1).

• Road foundation (sub-base and base courses)

In non-stabilized layers granular elements are not bound amongst themselves and hence under repeated traffic stresses undergo attrition, which affects the entire mass of the layer. The more gape in grading and the lower the compactness, the more important this phenomenon. This attrition produces sands and fines and modifies the grading of the material. In some cases the fines may be plastic. This development always takes place to the detriment of the mechanical stability of the layer; it can also reduce its permeability.

In layers stabilized with bituminous binders a 'flexible' bond occurs between the aggregates and the attrition is much lower than in non-stabilized layers.

In the case of hydraulic binders, after setting the aggregates bound amongst themselves no longer move with respect to one another. But for various reasons, of which shrinkage is the main one, these layers crack, giving rise to paths for water to infiltrate into the lower layers. The result may be a fall in the bearing capacity or at least an increase in deformability, which leads to accelerated fatigue of the entire structure of the pavement and considerable attrition of the aggregates on either side of the cracks. Attrition is accelerated in every case in the presence of water.

Specifications concerning the mechanical strength of aggregates for pavement layers therefore depend not only on the micro-Deval coefficient in the presence of water (MDE), but also the Los Angeles coefficient in order to take into account implementation conditions.

• Wearing courses

Coarse aggregates on the pavement surface are subjected to heavy impacts. The wheel of a truck which moves at 72 km/h is in contact with a particle of 10 mm for 1/200th of a second. Under these repeated impacts the particles may be fragmented. This fragmentation is particularly significant in the case of studded tyres; the less the particle is set in the binder and the less it is bound by the binder, the more readily fragmentation will take place. This explains the less stringent specifications for aggregates for surface dressings, bituminous concretes or hydraulic concretes.

The fast and repeated passage of vehicles on surface gravels has another effect. Slips take place during tyre-pavement contact in the presence of fine elements, some of which may be abrasive. Wear and polishing of the aggregates are in-

---

*replacd by XP 18-540 since 1997.

evitable. These phenomena reduce the coefficient of friction of the pavement and increase skidding risks and braking distances of vehicles. Maintenance of a good macrorugosity calls for good resistance to wear, an MDE close to 15; that of good microrugosity calls for good resistance to polishing and therefore a high polished stone value (PSV).

The compensation between these various coefficients (LA, MDE, PSV) is admitted within the limit of 5 units, except for macadam chippings. It appears that for rocks which resist wear very well (very low MDE), this limit could be increased, the maintenance of macrorugosity compensating for the loss of microrugosity by polishing.

• *Railway track ballasts*

The stability of the ballast layer, consisting of 25/50 mm aggregates, obtained by simple friction between these elements. The mechanical stresses borne by the ballast during execution (mechanical packing) and in service call for aggregates resisting fragmentation and wear.

The French specifications lay down a hardness coefficient DR_i resulting from a certain compensation between LA and DH. Depending on the type of sleeper and train speed, LA should be less than 22 or 25 and DH more than 10 or 8.

1.3 Resistance of Rocks to Weathering

Rocks have already undergone in their deposits the effects of hydrothermal internal actions, which will not recur, and meteorological external actions, which are continuous. The result is a state of equilibrium in the deposit acquired over thousands of years, which may be disturbed during extraction. By weatherability is meant the sensitivity of a rock to changes in the surrounding medium. The proclivity of rocks to cracking is one of the important phenomena of weatherability. Two essential factors are involved.

— Structure of the porous medium (extent of porosity and size of the pores) and state of cracking;

— Water content of the rock when frost takes place.

The sensitivity to frost of aggregates (Standard P18-593) is the relative decrease in the Los Angeles value measured on two samples of coarse aggregates (4/6, 6/10 or 10/14 mm). One test is carried out on undisturbed material and the other on a sample that has undergone, after saturation with water, 25 frost-thaw cycles between +25 and –25°C.

The physicochemical processes involved in other cases of weatherability are often complex. The changes recorded result most often from a chemical change in the fluids which impregnate the rock or which flow through it, or from a change in confinement conditions. They are due to:

— Swelling and expansion so the hydrophilic constituents (phyllitic minerals, clays or hydroxides) fixing the water or crystallizations in the cracks.

— Splitting through openings of pre-existing planary discontinuities (schistosity).

— Resumption of the process of dissolution and leaching of the already weathered sensitive zones.

Tests enabling revelation of the sensitivity of rocks to these developments have been devised but their interpretation continues to be delicate.

In certain concretes aggregates react with cement and form an expansive gel which leads to cracking of the concrete. This phenomenon is known as alkali-slica reaction and seems to be related to the use of cements having a significant alkaline content and the presence of minerals that readily liberate alkalies and silica.

1.4 Affinity

The affinity of aggregates is obviously not an intrinsic property of rocks; it can be defined only by considering simultaneously the binder and the aggregate. The surface finish of the aggregates, hence the cleanliness, is also involved.

1.4.1 HYDRAULIC BINDER-AGGREGATE AFFINITY

The bond between the cement paste and the aggregate plays an important role in the mechanical behaviour of concrete. This bond, established during hydration, depends on the nature of the two phases present, the proportion of cement and the curing conditions. Around the aggregates in the hardened concrete, a special zone of hydrated paste is observed whose texture is very fine at the interface, becoming progressively porous thereafter. It is this second zone which constitutes the weak spot in this bond.

In the case of calcareous materials a particularly effective epitaxic bond is developed which brings about a notable increase in strength.

1.4.2 BITUMINOUS BINDER-AGGREGATE AFFINITY

As a general rule, the wetting of a dry aggregate by a black binder presents no difficulty provided the binder is sufficiently fluid at the temperatures of application. On the contrary, if the aggregate is wetted without adhesive dope, wetting and displacement of water are not possible. The films of undoped bituminous binders on the surface of the aggregates are generally displaced by water thermodynamically. The rates of coating removal vary in a pronounced manner with the nature of the aggregates and the binder as also the latter's viscosity. For basalts and quartzites the rates of coating removal are positive whatever the binder; with calcareous aggregates the coating removal rate is exceptionally negative; this is also the case for a flint-tar combination.

The affinity of aggregates with bituminous binders is measured by the Vialit plate test: one hundred fine gravel pieces are affixed to a metallic plate coated with a binder and dislodgement then attempted by striking the plate on the back. One hundred minus the number of aggregate pieces dislodged and freed from the binder layer represents the affinity value. This test applies to surface coatings and the result considered satisfactory if equal to or greater than 80.

Passive affinity is also evaluated by trying to remove the earlier applied binder coating in water at 20 or 60°C, depending on the binder's viscosity (Fig. IV.8).

The resistance to water of a coated aggregate is also assessed by the Duriez test. The ratio of the compressive strength of a coated aggregate after seven days of immersion and its strength in the dry state, termed the immersion-compression

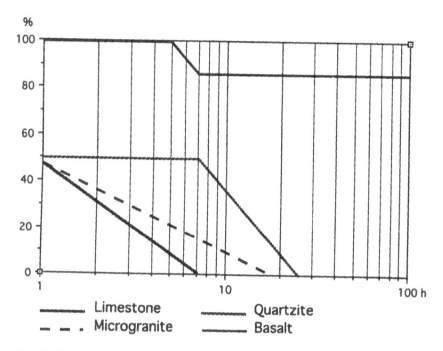

Fig. IV.8. Rate of coating removal versus duration of immersion (after R. Sauterey and colleagues)

ratio, gives a good idea of its sensitivity to water. This ratio should be greater than 0.75 for a good bituminous concrete and 0.65 for a gravel stabilized with bitumen.

2. CHARACTERISTICS OF AGGREGATES INFLUENCED BY MANUFACTURING CONDITIONS

2.1 Particle Size

The dimension of a particle is given by the opening size in millimetres of the square mesh of the sieve through which it squeezes.

The dimensions of the sides of the square mesh of sieves are the terms of a geometric progression of common ratio $\sqrt[10]{10} = 1.259$ (Table IV.2, except for 14 mm).

Table IV.3 gives the designation of various types of aggregates with reference to their particle size.

The terms filler, often used for fines, should be reserved for fines specially manufactured for particular uses.

It should be noted that between two successive dimensions in this progression, for example 1 and 1.25, the particle volume and hence its mass doubles.

Grading is characterised by the particle size distribution of the aggregates. It is determined by sieve analysis wherein the quantities of material separated between successive sieves in a series are determined.

Table IV.2. Mesh sizes used for grading of aggregates. Opening sizes in millimetres of square mesh of sieves

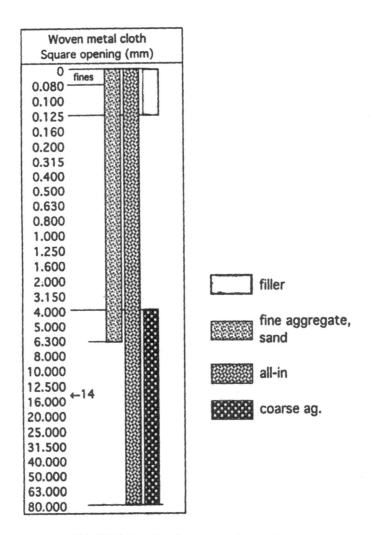

Table IV.3. Designation of main types of aggregates

Fines	0/D where D ≤ 0.080 mm
Fine aggregates (sands)	0/D where D ≤ 6.3 mm
Coarse aggregates	d/D where d ≥ 1 mm and D ≤ 31.5 mm
Pebbles	d/D where d ≥ 20 mm and D ≤ 80 mm
All-in	0/D where 6.3 < D ≤ 80 mm
Ballasts	d/D crushed where d ≥ 25 and D ≤ 50 mm

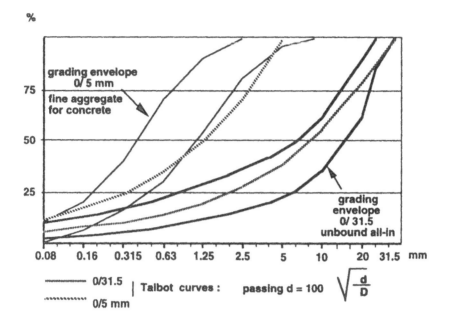

Fig. IV.9. Grading curves, envelopes of specifications and Talbot's curves

Sieving is done under such conditions wherein the probability of passing of a grain smaller in size than the mesh of the sieve in question is very close to 1. The results are depicted in Fig. IV.10 taking the dimensions (generally following a logarithmic scale) on the x axis and the percentage in mass passing through a given mesh (Standard P18-560) on the y axis.

For sands meant for use in concrete, a more generalised manner of expressing the grading is commonly used. It is the fineness modulus which, expressed in the form given by Abrams, corresponds to the hundredth of the sum of all that is retained on the 80, 40, 20, 10, 2.5, 1.25, 0.6, 0.3 and 0.16 mm sieves (Fig. IV.10).

Granulometric analysis on sieves presents few problems. Generally, sieving can be done from D to 0.0 mm following certain principles given in Table IV.4.

Table IV.4. Principle of sieving

Product to be analysed	Operations	
0/80 μm	See Section below	
0/D mm	Sieving under water or washing on 80 μm then dry sieving	Recover the fines if necessary
d/D	Dry sieving	

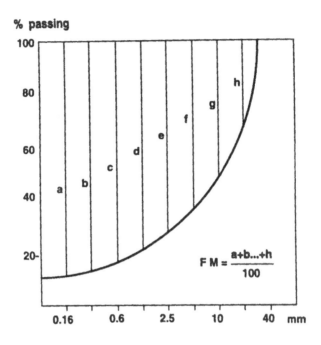

Fig. IV.10. Calculation of the fineness modulus (FM)

With regard to fines, since their dimensions can be less than μm, recourse has to be had to special techniques.

Whatever the procedure employed (Stokes law, laser diffraction, X-ray, densimetry etc.), the results obtained are highly satisfactory provided a thorough dispersion of the particles is ensured.

Mention should be made of an indirect method linked with the particle size of fines, the Rigden test. The porosity of compacted dry fines is measured in an impact apparatus specified in Standard P18-566 (Fig. IV.11). The volume of these voids should be proportional to the quantity of bitumen necessary for making a coated aggregate.

Automated particle size analysis of the fines is well resolved by methods based on X-ray or laser.

LCPC has developed an interesting device for quick determination of fines content of a sand without recourse to weighing or drying. The method simply involves introducing a densimeter at a given time and the mass of the sample is obtained by degauging. The precision obtained in ± 1 point for 10% fines is entirely adequate (Fig. IV.12).

In the fine gravels zone the videogranulometer developed in LCPC France using interception of light rays by particles and automatic processing of the results enables going beyond mere particle size determination to analysis of shape of the particle (Fig. IV.13).

Fig. IV.11. Impact device for the Rigden test

Table IV.5: Grading specifications

Specifications	Range of the zone of uniformity
Passing 1.58 D : 100% (1.25 D for ballast and chipping)	
Passing D : 85 to 99% (80 to 99% if D < 1.58 d)	10%
Passing (D + d)/2 if D < 2.5 d: 33 to 66%	25%
Passing d: 1 to 15%	10%
Passing 0.63 d: < 3% (5% if D < 5 mm)	
Sands:	
Between 0.5 mm and D	15%
at 0.5 mm	10%
at 0.08 mm	{ 4% if < 12% filler 6% if ≥ 12% filler

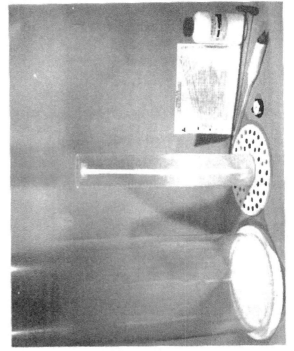

Fig. IV.12. Equipment for rapid determination of fines

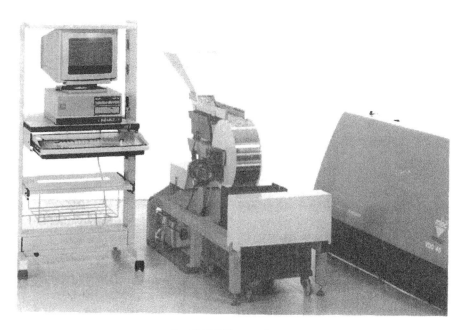

Fig. IV.13. Videogranulometer

A satisfactory method has been developed for 0/2 mm sands by laser.

Specifications related to techniques and traffic are given in Table IV.5.

Internal grading of an aggregate 0/D plays an important role in all its uses. This is defined by the envelopes of specification: limiting curves between which the granulometric curves obtained with the sample should lie. These particle-size envelopes, specific to each technique, are fairly wide and the search for a good uniformity makes it necessary to lay down the breadth of the envelope within which 95% of the curves obtained while testing one lot should fall. This envelope determines the zone of uniformity, which should lie within the specification envelope.

The quest for the greatest compactness has always been the guiding spirit in grading specifications. Commonsense indicated that the more compact a granular mixture, the less deformable and the higher its strength; the voids that the coarse gravels in contact leave between themselves can be filled by smaller ones and the voids still remaining can be filled by sand and so forth up to the finest elements. However, this logic should not be extended to the smallest particle sizes since the presence of very fine elements with large specific surface is fraught* with difficulties (which are dealt with later).

It is now well recognised, however, that an adequate percentage of fines improves the fatigue resistance of materials meant for pavement courses and the strength of cement concretes.

The pursuit of the best compactness at the theoretical level does not necessarily lead to the best grading at the practical level, however. Indeed, it is also advisable to ensure that the material be placed properly. The presence of reinforcement formworks for cement concrete and the means of vibration require sufficient workability properties of the mix placed. The workability of concretes is measured by the LCL workability meter. In the case of pavement courses it is equally important that the granular mix be capable of effective compaction by road rollers.

A number of theoretical approaches or practical methods have been proposed for defining the optimal grading to give the best compaction in practice. For example, mention can be made of Talbot's curves obtained from the formula

$$p = 100 \sqrt{\frac{d}{D}}$$

where p represents the percentage passing mesh d of an 0/D aggregate.

Experience has likewise shown that a relatively narrow envelope should be adopted for pavement courses too. In the field of pavements reference is often made in France to Robin's envelope (see Fig. IV.9), proposed in the Roads Directorate Programme of Study for 1955. For mixes in which the binder is in granular form, it is obviously appropriate to take this into account while defining the specification zone of the aggregates. In fact, while laying down these zones one should also take into account the change in particle size distribution arising from placement, especially in the case of the softest aggregates. An example of envelopes is presented in Fig. IV.9.

The drawbacks of defective grading are summarised in Fig. IV.14 which calls for some explanation.

— If the actual granulometric curve shows a depression in the zone of arrow 1, the material will be less workable, stiffer and more sensitive to segregation. Compaction would have to be very vigorous unless the rock is soft and likely to undergo change during placement. The same drawback will result from a lack of fine elements (arrow 3), leading to a loss in compactness.

— If the curve shows a hump in the zone of arrow 2, the material will be workable, can be placed with ease, but will contrarily lack stability and deform readily, especially if the surplus fine elements are spherical. Too great an amount of fine particles also leads to loss of compactness.

— Lastly, a hump following arrow 4 results in a lack of skeleton frame.

It is evident that adapting compaction to the grading used and the hardness of the material is essential, as much as proper utilisation of the means of compaction.

2.2 Angularity

Alluvial materials which have not undergone crushing have rounded shapes while crushed aggregates have sharp edges and possess a pronounced angularity. This should be slightly varied in the case of the softest materials where few, if any,

% passing

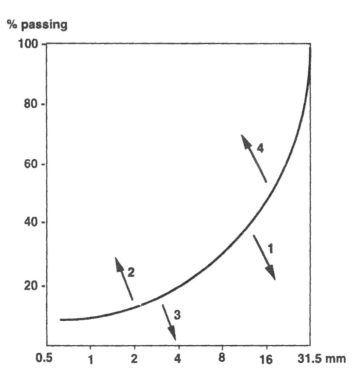

Fig. IV.14. Diagram of defective grading and its consequences

sharp edges persist. For the same compaction or void index, a good angularity confers a better mechanical stability on the granular mix: the angle of internal friction increases by about 7° in a triaxial test on changing over from a wholly rounded material to crushed massive rock.

On the other hand, the presence of rounded particles ensures better workability, which at the time of placement facilitates tight packing and results in good compactness. One faces the same dilemma referred to in the case of grading: whether to choose criteria which result in the best strength, with a risk of dislocation if an adequate compactness cannot be obtained, or improving placement conditions to obtain good compactness by admitting lower mechanical performance of the granular mix? The resolution to this problem presently hinges on the techniques and facilities of placement.

In the case of cement concretes with a high binder content, the choice is in favour of good workability through the use preferably of rounded materials. On the contrary, for pavement courses the accent is rather on obtaining better stability of the granular mix, either because there is no binder (non-stabilized mix) or because the binder sets slowly and hence good stability should be ensured for the course (gravel-slag), or because the stiffness of the binder is low (gravel-bitumen).

In the case of 0/D mix and d/D fine mixes, angularity indirectly increases with the index of crushing (IC) which is the percentage in mass of elements larger than D of the gravel of alluvial origin subjected to crushing. In the event of recombining, IC is given by the weighted average of the indexes of crushing of the various components.

Counts of the uncrushed faces give the results depicted in Fig. IV.15.

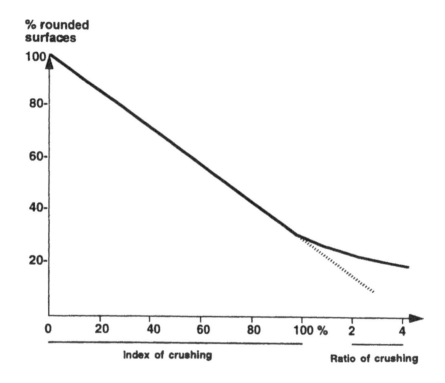

Fig. IV.15. Index of crushing (IC) is responsible for the disappearance of 70% of the rounded faces. The principles of ratio of crushing (RC) hardly improves the angularity

To preclude skidding on pavements, fine gravels with rounded faces are avoided. The specifications in this case are more stringent and refer to the ratio of crushing (RC), which is the ratio of the smallest size of the original material subjected to primary crushing, to D of the fine gravel obtained.

The method of estimating IC or RC is not very satisfactory: if the processing is complex or the manner of processing is not known, IC or RC cannot be measured; further, this amounts to considering as identical a 0/10 arising from the crushing of a 10/30 or a 10/100 for example. It would be better to consider the ratio of the D 50 of the original material to that of the end-product.

Efforts have been made to replace these definitions by a workability test which consists in measuring the time taken by a quantity of material under well-defined conditions. Standards P18-563 and P18-564 specify the time taken by a known quantity of sand to flow through the calibrated orifice of a funnel (Fig. IV.16) or

Fig. IV.16. Angularity meter for sand (Rouen regional laboratory)

by fine gravels through a trap. In the case of the latter, it is necessary to vibrate the entire set-up to ensure the flow (Fig. IV.17).

These methods based on the friction of the particles address all the geometrical properties of the aggregates: surface condition, shape, angularity and granularity.

A good correlation exists between the flow coefficient (EC) of the fine gravels, the crushing index and the number of rounded faces remaining in the material (see Fig. IV.15).

Mention should be made of the concept of surface condition.

The surface of a particle is invariably the outcome of breakage; it can change subsequently if the particle is carried along, for example, by a fluvial process, which tends to polish or roughen it.

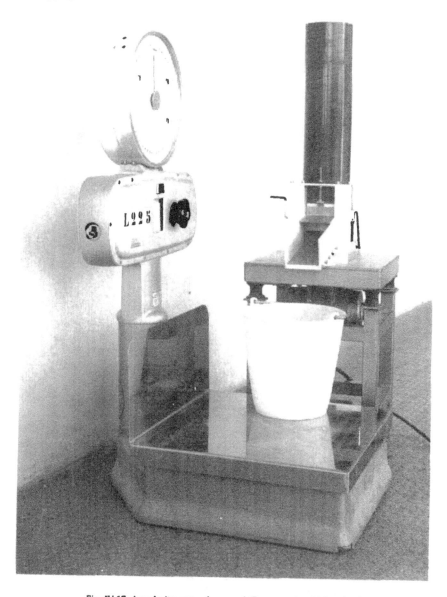

Fig. IV.17. Angularity meter for gravel (Rouen regional laboratory)

The roughness of a broken surface depends on the dimension of the rock minerals—the finer the grain, the smoother the surface—as for instance with flint.

It is obvious that this property plays a no mean role vis-à-vis the angle of internal friction of the granular media in the same manner as angularity. There is no simple method as yet for measuring this property but integrated with the measurement of flow, this approach can be deemed satisfactory.

2.3 Shape

At its simplest, shape can be characterised by the measurement of the three commonly recognised elementary dimensions:

— Length L and thickness T, which respectively are the largest and the smallest distance between two parallel planes tangential to the grain.

— Size S which is the minimal diameter of the circular hole (or square for sieves) through which the grain can pass.

The flakiness coefficient adopted by AFNOR in principle takes into account only flat particles (NF P18-561). It comes from a double sieving on sieves followed by bar sieves (Fig. IV.18).

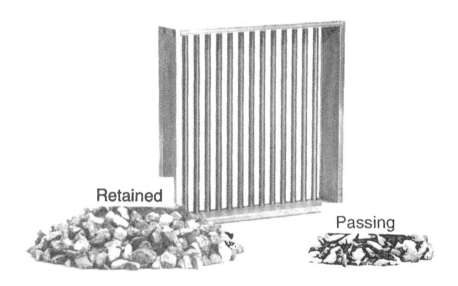

Fig. IV.18. Bar sieves used for the measurement of the flakiness coefficient

This test is quite fast and takes needles into account to a certain extent; these particles, due to the problematic shape of their grains, are invariably retained upon sieves larger than their actual S and are hence discharged onto grids of a wider spacing, which partly amounts to including them in the flat particles.

In addition, Standard P18-562 adopts determination of the mean thickness of fine gravels; this concept is highly significant for surface dressings because it conditions the pavement thickness. Difficulty arises from the fact that for equal mass a flat fine gravel occupies a larger surface area than one of proper shape, which explains the introduction of harmonic mean in the calculation of thicknesses in the Standard.

Hence these flat particles of equal mass have a smaller thickness for 'wearing' and most often entail an excess of binder (bleeding in the case of surface dressings).

For other usages, shape has a lower significance and its role is allied to that of angularity.

The presence of flat particles entails a slight decrease in the mechanical strength of fine gravels. For instance, the result of the Los Angeles test increases by 1 point when the percentage of flat particles increases by 10.

2.4 Contamination of Sands or Cleanliness

The term contamination here covers the presence of any substance present in the aggregate and capable of adversely affecting its technological properties.

The illustrative list given in Table IV.6 includes substances, each of which has a specific effect on binders. The clays invariably present in aggregate deposits constitute the most frequent risk (poorly washed alluvia, bad stripping), inadequate elimination, and in addition exert an effect at the placement stage regardless of whether a binder is used or not.

Other substances are more accidental or typical to particular deposits.

Table IV.6: Substances polluting aggregates

Type	Characteristics	Occurrence	Effect
Clays	Large specific surfaces, hydrophilic	All natural rocks	Acts as a screen vis-à-vis all binders, rehydration, deformation of cement concretes
Weathered minerals, feldspar	Sources of clay	Magmatic rocks	Acts as a screen vis-à-vis all binders, rehydration, deformation of cement concretes
Micas	Fairly large, specific surfaces	All rocks except limestones	Acts as a screen vis-à-vis binders
Iron hydroxides	Hydrophilic	Laterites	Expansion of binders
Pyrites, marcasite	Oxidisable	All rocks	Expansion (ettringite)
Gypsum, anhydrite	Soluble, reactive	Alluvia in dry climates	Expansion
Chlorides	Reactive	Marine sands and gravel	Corrosion of reinforcement in hydraulic concretes
Opal, microcrystalline silica	Reactive	All rocks	Expansion, fracture: hydraulic concretes
Organic acids	Reactive	Alluvia	Retardation or absence of stetting: hydraulic concretes

2.4.1 INVESTIGATION FOR CLAYS IN SANDS

• *Sand equivalent value (SE)*

The sand equivalent test (Fig. IV.19), borrowed from the Americans around 1950, has come in handy in solving the problem of clays in hydraulic concretes since their fines content remains around 1 to 2%.

But beyond 2%, the test has some drawbacks.

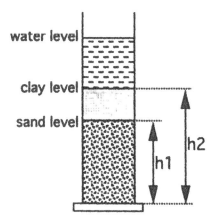

Fig. IV.19. Sand equivalent test

1. Too high a sensitivity to inert fines if they are even slightly on the finer side. For instance, 4% siliceous fines free from clays and less than 20 µm in size give a sand equivalent value of about 40.

2. Trapping part of the inert minerals (calcite, quartz, feldspars), which likens them to clays. These trapped particles can reach up to 200 µm.

3. Significance which the flocculate assumes with the least harmful clay, i.e., kaolinite, even though all studies on composite materials have clearly shown that 2 or 3 times more kaolinite over other clay minerals can be tolerated. But more kaolinite also means a very low sand equivalent value, which is incompatible with the specifications.

4. Lastly, dispersion of the clays, which increases with water content and hence often with duration of unprotected storage and multiple handling operations including transport. This in turn leads to the formation of a larger volume of flocculates, trapping even more fines. Thus the all-too-frequent and well-known drop in sand equivalent value between the moment of leaving the quarry and that of arrival at the site.

Nevertheless, it should not be concluded from these four points that the SE value has nothing but drawbacks. One should credit to it two advantages which are by no means negligible:

— When the fines content is low (below 2%), the test is very significant; a good SE value invariably means a good sand (so far as clays are concerned).

— Limiting the fines content to 10% in the SE test (which is then termed the SE value for pavement courses) somewhat mitigates drawbacks 1 and 2 given above (NF P18-597).

But defects 1, 3 and especially 4 always persist and can be partly resolved only through a test which exclusively takes clays into account. This test was developed by Tran Ngoc Lan in 1977, using methylene blue.

• *Methylene blue test*

The organic molecule methylene blue is absorbed by all minerals wettable by water; the quantity of blue absorbed by non-clay inert fines is negligible compared to the amount absorbed by clays.

The test consists of adding successive doses of blue to a suspension of water + fines and after each addition placing a drop of the mixture on a white filter paper. A dark blue central spot is observed surrounded by a colourless halo if the material is still capable of fixing the blue, or a spot of light blue if a surplus of blue has been used (Fig. IV.20). It is then easy to calculate the methylene blue value (MBV) of these fines proportionate to the total surface (outer + inner) of the clays present.

water halo   methylene blue halo

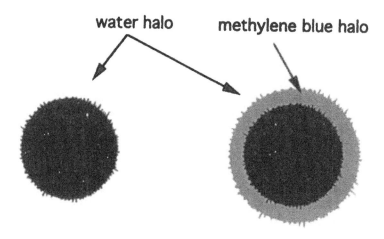

Fig. IV.20. Methylene blue test spots without and with halo

• *SE value-MBV relationship*

The SE value-MBV relationship adopted in Standard P18-101 is based on the following premise: the SE value should be retained for its advantages and VB should be used only if the SE value is doubtful, that is, to mitigate its drawbacks.

Therefore, for the purposes of control:

— If the SE value is good, MBV is not required, the sand is accepted;

— If the SE value is below the specification:

1) because there are inert fines smaller than 10 µm, MBV will be good;

2) because for too many inert fines are trapped in the flocculate, MBV will be good;

3) because there is more than 2% kaolinite in the sand, MBV will be good;

4) because the water content or transport has dispersed the clays, MBV will be good.

• Limitations of the spot test

But then there are minerals other than kaolinite in sands. Illite, montmorillonite, binary or ternary mixtures occur more often and fibrous clays less so.

Given that these clays, due to their internal surface, absorb much more blue than kaolinite, 4 to 15 times more, we greatly risk no longer being in a position to count upon MBV in cases, 2, 3 and 4 above, since any clay mineral other than kaolinite will lead to higher values of blue. For example, 2% montmorillonite in a sand lead to a MBV of 6.

Within the concept of cleanliness of sands as defined by Standard P18-101, there is thus a very real risk for the supplier whose sand, while having no impurity other than kaolinite, could give values of blue exceeding the specifications.

This further explains why the value of blue alone cannot be used; it should be related to the quantity of clays.

Fig. IV.21 shows the current interest of the expression 'SE value and MBV if ...

— SE value alone passes the materials situated in zones A, B and C.
— MBV alone passes the materials situated in zones A and D.
— SE value together with MBV is ... pass the materials in zones A, B, C and D.

Zones E and perhaps F include materials with low montmorillonite contamination, which could be accepted by adapting the tests.

The present specifications based on SE value are listed in Table IV.1

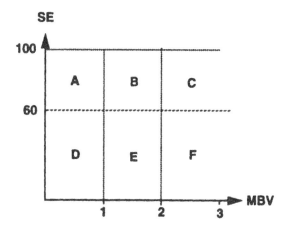

Fig. 4.21. Interest behind interrelating MBV and SE value

• Permissible clay contents

It may be noted that the present specification of 1 g blue for the stain test corresponds to about:

3% montmorillonite,
20% illite,

60% kaolinite in fines < 80 μm.

MBV goes to 1.5 for materials stabilized with hydraulic binders and to 2 for non-stabilized aggregates.

In view of the results known for bituminous concretes, it appears that it is not 15 times less montmorillonite, but 4 to 5 times which can be accepted.

With respect to aggregate-bitumen it should be noted that the permissible value for montmorillonite is lower (MBV 2), which has to do probably with the lower binder content and explains a higher sensitivity to water.

In the case of aggregate-cement mixtures, it is observed that the tensile strength and modulus are closely related to the total surface area of clays. This is clearly seen in Fig. IV.22 in which the tensile strength of a aggregate-cement mix correlates very well with MBV. this should permit an altogether different interpretation of the notion of cleanliness. One might consider that it is possible to utilise a more contaminated material provided: a thicker course is laid, contamination level is limited to that which would not lead to a plastic material (MBV < 5) and hence problem in placement, and the economic balance sheet remains positive.

In the case of hydraulic concretes the presence of clay necessitates an overdose of water or use of a plasticiser for obtaining good workability, which is not always favourable to their strength.

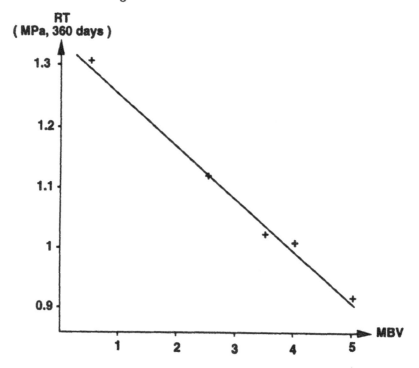

Fig. IV.22. Variation in tensile strength of aggregate cement mixture versus methylene blue value.
Rtb—Brazilian tensile splitting test; MBV: methylene blue value

Unikowski's studies underscore the same tendency to overestimate the role of swelling clays as for bitumen-coated aggregates. But the quantities of fines often being very low here, the blue spot test is the least precise.

Only the adoption of a more sensitive measurement of the blue value would pave the way for progress; the turbidimetric test holds promise in this regard.

• *Turbidimetric test*

A light beam of incident intensity I_i passing through a liquid medium is attenuated at its emergence from the medium by the colouring matter (methylene blue in this case) and by the floating particles (inert fines and clays) in the medium to give an intensity I. In the first case this attenuation is called *absorbance*, in the second *turbidity*.

These two principles are expressed by the same formula:

$$A = \log 10 \frac{I_i}{I}$$

Generally the coefficient varies linearly with the concentration of either colouring matter or solid matter as the case may be.

Hence if a suspension is prepared at first of fine inert solid particles and clays to have a given turbidity, the addition of methylene blue will change the value of A which will serve as means of comparison in the test.

For instance, the addition of a given quantity of blue to the water results in an absorbance of 0.7. Should clays be present in the water, they fix part of the blue and the absorbance would be less than 0.7 (Fig. IV.23).

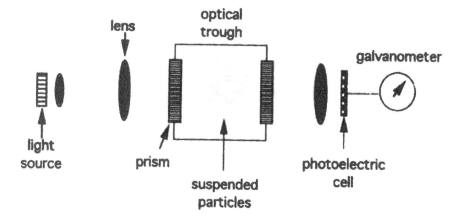

Fig. IV.23. Principle of turbidimetric measurement

The difference between these two values of A is a measure of the internal or total surface area of the clays accordingly, as the test is stopped at a larger or lower absorbance A.

Fig. IV.24. Alkali-aggregate reaction

2.4.2 ANOTHER TYPE OF CLAY CONTAMINATION

Clays can either cover the surface of fine gravels and form a screen vis-à-vis binders, or coalesce into small cohesive balls which can be found in the fine gravels.

A mere 0.5% clays in fine gravels suffices to cover all their surfaces. Therefore it should be ensured that the fines contaminating the fine gravels are not argillaceous. Thus in test for determining the surface cleanliness of fine gravels (NF P18-591), one is concerned with the quantitative aspect, complemented in the specifications by a limit to the value of blue for these elements.

The small clay balls are detected by the homogeneity test (NF P18-571) which separates the lightest elements of dense liquor: clays, organic fragments, weathered fine gravels etc.

2.4.3 OTHER SUBSTANCES CONTAMINATING AGGREGATES

All the other substances mentioned in Table IV.6 are identified through mineralogical (microscopy, X-ray) or chemical analysis.

Table IV.7. Repeatability and reproducibility of the main tests

Tests	Aggregate size, mm	Numerical value	Repeatability, r	Reproducibility, R
Particle density	10/14	2.62	0.009	0.01
		2.84 Tm3	0.006	0.02
Porosity	10/14	3	0.11	0.25
		4%	0.21	0,66
Water absorption	10/14	1	0.03	0.27
		1.4%	0.09	0.22
Grading	0/2	x	0.2 + 0.006 x	0.02 x
(% passing)	4/6 10/14	d D	1.25	2.75
	5 12.5	Intermediate sieve	2.85	5.70
Flakiness	4/14	8 to 20	1	1.8
Rigden's void index	0/0.08	40	1.1	1.8
Micro-Deval in presence of water	4/14	x = 2 to 30	0.35 + 0.04 x	0.4 + 0.09 x
Los Angeles	4/14	x = 10 to 30	0.47	0.58 + 0.04 x
Impact test	4/14	x = 10 to 30	0.70	1.15 + 0.04 x
Polished stone value	6/10	0.51	0.013	
Friability of sands	0.1/2	16	0.7	1.5
		38	2.2	3
Surficial cleanliness	4/14	x = 0.2 to 3	0.14 + 0.03 x	0.1 + 0.16 x
Methylene blue value (spot)	0/2	w = 0.07 to 4.25	0.08 x	0.16 x
Sand equivalent	0/5	60 to 80	1.2	2.5

Two types of contamination still remain difficult to control:

— Organic matter hindering the setting of hydraulic binders, because difficult to analyse, little known, quite diverse and harmful in very small concentration: < 0.1%.

As yet there is only the test on the composite (gravels or hydraulic concretes) which enables understanding the influence of these products on the setting of binders.

— Alkali-aggregate reactions, which favour the formation of swelling products inside concretes, pose an even more difficult problem. These products actually form several years after the concrete has been laid and call for a long-term prediction of a reaction whose presence can only be inferred from the products (Fig. IV.24) and their harmful effect. As with any long-term prediction, exiting methods lack certitude.

3. PROPER USE OF AGGREGATE TESTS

The moment one plans to implement specifications, it is essential to know the reliability of the tests. A survey of the network of the 18 French laboratories of Bridges and Roads was done and the results are given in Table IV.7. These values of typical distributions obtained in tests should enable a more balanced appraisal of the aggregates coming from our quarrying operations, by dissociating the scatter peculiar to the heterogeneity of the test from that due to the material.

REFERENCES*

AFNOR. 1984. Collection of French Standards. Aggregates. Paris, vol. 3.

Campanac, R. 1981. Adverse effects of argillaceous fines on the performance characteristics of a hot mix. *BLPC*, no. 111.

Coquillat, G., M. Delfau and R. Lesage, 1984. Possible use of medium to low hardness limestones in hydraulic concretes. *Bull. AIGI*, no. 30. Internat. Symp. Aggregates, Nice.

Denis, A., C. Tourenq and N.L. Tran. 1980. Water absorption capacity of soils and rocks. 26th Internat. Geol. Cong., Paris.

+Jones. 1964. New fast accurate test measures bentonite in drilling mud. *Oil Gas*, pp. 62–122.

Kergoet, M. and C. Cimpelli. 1980. Evaluation of behaviour of contaminated fine sands by the methylene blue test. *BLPC*, no. 108.

Leroux, A. and Z. Unikowski. 1980. Detection of the influence of clay fines in concrete aggregates. *BLPC*, no. 110. Internat. Symp. Aggregates, 1984 Nice. *Bull. AIGI*, no. 29–30.

Tourenq, C. and A. Denis. 1982. Tests for aggregates. *RR. LCPC*, no. 114, July.

Tran, Ngoc Lan. 1977. A new test for identification of soils: The methylene blue test. *BLPC*, no. 88.

Tran, Ngoc Lan. 1981. The use of methylene blue test in roadmaking earthworks. *BLPC*, no. 111.

+Unikowski, Z. 1982. Influence of clays on the properties of cement mortars. *RR. LPC*, no. 110, Paris.

+Yariv, S. and D. Lurie. 1971. Metachromasia in clay minerals, sorption of methylene blue by montmorillonite. *Israel Jour. Chem.*, Vol. 9.

*All entries in French except those marked with a cross (+)—General Editor.

V

Non-conventional Aggregates

Claude Tourenq, Pierre Dupont and Louis Primel

Use of non-conventional aggregates may be justified by the exhaustion of certain sources of conventional aggregates, especially alluvial ones, increased environmental requirements and the incidence of changes in transport cost. These factors have led to concern about material resources for the short and medium term.

The search for economy in realising various projects makes it necessary moreover to use locally available materials insofar as possible, resorting, if necessary, to materials still scarcely used in traditional practice because they do not always meet all the specifications in force.

But resorting to such materials should in no case result in loss of quality of the works to be carried out. This is possible only by developing specific methods suited to the geotechnical characteristics of the materials available (for example, either new methods of manufacture of aggregates, of manufacture and/or utilisation of the processed materials, or of new structures). It moreover implies a complete mastery of the materials used as well as a perfect knowledge of their limitations.

From this point of view, regional inventories of new aggregate resources were recently prepared by several institutions at the request of the advisory committee on road innovation of the directorate of roads.

In the field of pavements the regions which may really pose problems of supply are essentially the Paris basin and the Aquitaine basin; the identified alternative sources are represented by more or less soft limestones and by sands.

For hydraulic concretes evaluation of crushed products is a solution which should not be neglected in future.

Brief inventories of the potentialities of the regions which will in due course pose problems of resource substitution have already been prepared (Table V.1).

Some of the materials mentioned have never been used and are therefore almost not known; hence their evaluation will sometimes call for comprehensive studies and that, too, for several years. Others, on the contrary, are already well known to some engineers who have had the opportunity to use them at their sites. On the other hand, some engineers have shown reluctance to use aggregates whose characteristics may not comply with the Standards. In such cases, when a

Table V.1

Regions	Alternative materials	Soft limestone	Sands	Marine sands	Silt	Flint clay	Mining and metallurgical by-products
Aquitaine	•	•		•	•		
Bretagne				•			
Centre	•	•			•	•	
Champagne-Ardennes	•	•			•		
Ile de France	•	•			•		•
Limousin		•					•
Lorraine	•	•					•
Midi-Pyrénées	•						•
Nord Pas-de-Calais		•		•	•		•
Basse Normandie		•			•		
Haute Normandie	•	•		•	•	•	
Pays de Loire	•	•		•			
Picardie	•	•		•	•		
Poitou-Charentes	•	•		•			

parameter does not meet the specifications, it is always advisable to carry out a laboratory test on the binder-aggregate mixture.

For instance, for an aggregate-cement mixture, if one wants to use a clayey sand with a methylene blue value (MBV) of 2 instead of 1.5, a study under repetitive shear loading can reveal the difficulties of use, if any, as well as those of the couple tensile resistance Rt and strain modulus E and lead to a fairly good estimate of the possibilities of using the material.

It is more difficult to carry out a general study on the influence of one characteristic on a particular type of composite material. It is already under research but will take several years as it must be carried out systematically to ensure a better knowledge of the validity of the limits presently specified, not all of which have been studied in a comprehensive manner.

At the research level, things get singularly complicated as soon as one wants to involve only two parameters: Is a "cleaner" material more likely to be softer or have a defective grading? It is probably here that the most interesting potential for marginal materials lies but concomitantly the most difficult at the level of generalisation to all materials.

A beginning in this direction is the preparation of technical guides specific to each material to be promoted and in which a variable geographic zone may be involved—part of a region or several regions.

The purpose of these guides is to define the possibilities of using a material according to its geotechnical characteristics while simultaneously specifying the modalities as well as the limitations and precautions for using it on the basis of results already obtained. But for these technical notes to create a real movement of an industrial nature, it is essential that they address the main users, i.e., road as well as construction sectors.

In this context some guides for a few well-characterised materials, such as the limestones of Beauce and Charente-Maritime and sands of the Paris region, have been prepared, in order to popularise their use. However, only the road sector was taken into account in the first instance.

We have classified these materials into two groups:

1) Natural materials. These are either products not conforming to the specifications, the most frequent case, with special gradings or insufficient strength for example, with utilisation oriented towards pavements; or materials conforming to the specifications but not in conventional use, such as crushed magmatic rocks in hydraulic concretes.

2) By-products and wastes from industry: blast furnace slags, mine wastes and demolition materials, use of which enables reducing the volume of these wastes.

1. NATURAL MATERIALS

1.1 Non-alluvial Sands

Apart from the alluvial valleys, there are in France a number of geological horizons consisting essentially of sands, generally of marine origin, sometimes recently disturbed by wind, which represent reserves of several billions of tons: for example, the sands of Perche, Orléanais, northern Picardie, Fontainebleau and Beauchamp, Landes etc. (Fig. V.1).

The aforesaid sands extend geologically from the Secondary era (especially since the Cretaceous period) to recent Quaternary. Their uses are presently limited (essentially fills and subbases, sometimes aggregates for bases) or very specialised (sands for glassworks and foundry works, cellular type concretes). To develop their utilisation, especially that of fine sands which are very abundant particularly in the Paris basin, a technical guide defines their conditions of use. These sands of less than 0.5 mm are corrected by blanding with 0/6 calcareous or alluvial sands.

Various types of stabilization with binders (bitumen, cement, fly ash, granulated slags etc.) make possible in certain cases a performance compatible with 'nobler' uses, such as subbases for heavy traffic pavements or bases for pavements with less heavy traffic.

One can similarly study the development of utilisation of these sands in concretes, for instance to correct the grading of conventional concrete sands.

These technical studies should simultaneously take into account the economic conditions, for example: up to what percentage of binder (and which binder) is the utilisation of a fine sand from the Paris region of interest compared to that of an aggregate from a more distant place and consequently costlier, but which requires a lower consumption of binder.

1.2 Silts

Silts, eluvial or run-off sediments of detrital origin, occur in almost all regions but have developed over large areas and to significant thicknesses only in part of the Aquitaine and Paris basins. They are fine soils, intermediate between sands and clays, generally plastic and therefore sensitive to water. It should be noted, however, that loesses, well distributed in the Parisian region, Picardie and Nord, have

Fig. V.1. Fine sands of Fontainebleau, south of Paris

the characteristic of adhering well in vertical cuts. Silts generally meet the conditions of treatment with lime or cement. However, their use in pavement layers remains dicey. Their applications are presently few and limited to some test sites. Nevertheless the first results are encouraging and enable envisioning development of their utilisation in road technology, in particular for low and medium traffic pavements.

1.3 Soft Limestones

Beyond the strengths specified, Los Angeles < 40 and Micro-Deval with water < 35, limestones are often heterogeneous and freeze readily. As alternative materials, they can be classified into two categories, posing slightly different problems:

— Limestones corresponding to heterogeneous and discontinuous deposits, often with intercalations which are marly or difficult to quarry. They comprise in particular the limestones of the Jurassic border of the Paris basin (limestones from Barrois, Berry, Charentes, for example) and certain Tertiary limestones from the centre of the Paris basin (Beauce, Champigny, Château Landon) (Fig. V.2).

Fig. V.2. Tertiary lacustrine limestones of the Paris basin

A study of these limestones should first relate to the deposits, the purpose being to ascertain areas where the heterogeneities are so limited as to make it possible to continuously feed the processing plants.

Evidently limestones of this type are found also in the south-east, the Alps or the Jura but their study is of less interest because limestones of better quality or eruptive rocks of good quality are available in these same regions and, as mentioned earlier, also those which constitute the best massive rocks.

Technical guides are already available on the limestone of Beauce, on the soft limestones of Charente-Maritime and the limestones of Champigny.

— Very soft limestones with Los Angeles > 50 and Micro-Deval with water > 45. These rocks, which generally freeze readily but which are often homogeneous and clean, correspond in particular to the hardest facies of chalk (border of the

Fig. V.3. Chalk deposit

Paris basin, Fig. V.3) and to numerous other formations used most often as dimension stones.

As of now these limestones, occurring in almost all the sedimentary basins of France, are never quarried for the manufacture of aggregates. Here the order of urgency is reversed: One should first carry out studies about their possible uses, giving priority to limestones located in areas very poor in other sources of aggregates, followed by studies of those deposits in the second phase whose utilisation would prove economically viable (such deposits would pose more or less the same problems as encountered in the preceding type).

For the manufacture of concretes, all these limestones show disadvantages peculiar to crushed materials, i.e., an inferior workability, even though they are less angular than crushed eruptive rocks and generally have a high fines content. In fact, it is almost impossible to produce with these limestones sands poor in fines without resorting to washing, which may in certain sites pose technical problems, and in any case considerably increase costs.

These limestones represent very huge reserves. Yet one should not lose sight of the fact that some can only be quarried over relatively small thicknesses and will therefore be big consumers of space (less on average, however, than loose deposits, but more than the usual quarries of massive rocks).

1.4 Aggregates of Massive Rocks for Hydraulic Concretes

As for limestones, barring regions (Nord and south-east France) which have long used hard limestones, there is no equivalent utilisation elsewhere even though recent studies have shown that such is possible even with softer limestones.

Use of magmatic rocks is rarer still and research work has also shown that the current production from these quarries could certainly be used without having to modify the fines content.

2. BY-PRODUCTS AND WASTES FROM INDUSTRY

2.1 Blast Furnace Slag

Blast furnace slag is the main by-product of the iron and steel industry; it consists of all the mineral parts contained in the blast furnace charge (ore and additives) which remain after extraction of iron (Fig. V.4).

Fig. V.4. Discharging slag into a pit.

The slag can be slowly cooled in an open-air pit; a crystallised material having the appearance and properties of magmatic rock is then obtained. The product occurs in the form of a more or less fractured mass and may be more or less porous according to the conditions of cooling: the slower the cooling and in thin layers, the more crystallised and compact the slag. As in the case of all materials, the physical and mechanical properties of slag depend essentially on porosity.

After crushing and screening, the slag can be used in the road sector, mainly for the construction of pavement layers. However, if sufficiently compact, it can be used for making wearing courses. In fact, its microrugosity is maintained because of its porous texture.

The rapid cooling of slag in water leads to a hydraulic-setting binder (granulated slag), the largest particles of which play the role of correcting sand grain size for the stabilization of aggregates.

Production of slag depends entirely on the development of the iron and steel industry.

2.2 Steel Plant Slag

The steel plant slag is a by-product of the conversion of haematite cast-iron into steel. Treated as blast furnace slag, this material is handicapped by an insufficient removal of iron, which leads to a high particle density of about 3.3.

On the other hand, frequent residues of quicklime should be eliminated in order to guarantee sufficient stability of the product (risk of hydration). Immersion for one month in lukewarm water ensures stable aggregates for all pavement layers. The capacity of production is about 1.4 million tonnes per annum.

2.3 Coal Schists

The mining industry in general and coal mines in particular produce large quantities of sterile material. Part of this material is used in underground filling but the greater part has to be stocked in the open air, forming the spoil-heaps so characteristic of mining countries (Fig. V.5).

Fig. V.5. Spoil-heap of coal schists

In the spoil-heaps, especially in the oldest ones which contain schists with high coal content, the phenomenon of coal combustion may take palace. The result is an evolution of the intrinsic characteristics of schists that have undergone more or less high temperatures at different times.

According to the changes undergone, coal schists can be classified in increasing order of burn into black, orange, red or violet.

In building construction, red schists are used as aggregates for construction either as bricks or elements of construction such as chimney-flue tiles. Their good fire-resistance and insulation characteristics are particularly appreciated.

In the field of civil engineering 'all-in' schists are used as filling materials. Unburnt schists are generally preferred because they have undergone no agglomeration by combustion, are thus easier to extract from the spoil-heap and hence can be transported to the site at lower cost.

Red and black 'all-in' schists or the 'all-in' schists screened into 0/80 mm are used as subbase materials. Red schists, in the form of 0/20 mm gravel, are already in use for surfacing sidewalks, courtyards and by-lanes.

Studies have been carried out in road technology with red schists from Nord to consider their possible use as 0/20 or 0/31.5 mm all-in stabilized with hydraulic binders. Laboratory results are comparable to those recorded with traditional aggregates. Field tests have shown an acceptable behaviour provided that the thickness is about 10% greater than that of conventional designs.

2.4 Demolition Materials

Pavements (fatigued concretes, non-recyclable courses), apartment buildings and public buildings are demolished in all countries of the world. This activity results in the production of materials that can be characterised as follows:

— Heterogeneity of the overall deposit but possibly great homogeneity locally or in time;

— Scattering of raw product but concentration ensured by demolition workers;

— Dumping of rubble and consequently an almost total loss of the deposit whatever its quality may have been.

The greater the urbanisation, the more important the deposit. This ties up with the fact that conventional aggregates have become locally very scarce and transport very costly.

To wit, the potentialities offered by Paris and its environs are estimated at nearly one million tonnes per annum; in the Nord-Pas-de-Calais region more than 10,000 hectares of industrial fallow lands are to be rehabilitated in about ten years. This market initially got underway during the eighties in Paris following the destruction of slaughterhouses in Villette. Since then, more installations have come up in Ile-de-France and in the Lille region (Fig. V.6).

Manufacturing aggregates from demolition concrete is now operational. The quality of aggregates from crushed concretes has considerably improved since 1982 thanks to the technical progress achieved in the installations whose general design is now quite efficient. However, the highly special source of these deposits does entail more frequent checking of certain points, such as the presence of sulphates or light and crumbly particles. Generally speaking, the homogeneity of the characteristics of aggregates coming from a given installation is cause for additional

Fig. V.6. Demolition reinforced concrete

concern compared to natural aggregates, given the changes in the nature of the demolition products effected by treatment.

These few reservations notwithstanding, the possible uses for this type of material are important in the road sector, more particularly in the field of low and medium traffic pavements. It is in the latter field that they should be preferably used and it is this usage which should be encouraged.

As for recycling old bitumen-coated aggregates, several years of research turned this into a successful operational technique. However, the tonnage of recycled old coated aggregates appears to have reached a ceiling—for the time being about 500,000 t per year—which represents less than 2% of the national production.

Several reasons can be advanced to explain this situation:

— Absence of strong constraints: the need to save bitumen is less acute at present, aggregate resources are abundant and environmental problems are slightly less acute now than in the past;

— The number of suitable plants represents hardly 10% of the total number.

— Road designers show a certain reluctance towards a technology which is still new and are handicapped by the difficulties of transfer of old materials from the public area to another.

— Competing technologies, particularly those using modified binders, have been simultaneously developed.

2.5 *In-situ* Retreatment of Pavements

This technology seeks to recreate from a degraded pavement unsuitable for the traffic to be carried, a homogeneous and stable structure. It consists of on-the-spot treatment of existent materials, with the addition of binder and, if necessary, of additional materials, with a view to obtaining a new subbase. Retreatment improves bearing capacity and the profile of the pavement while concomitantly limiting the addition of 'fresh' aggregates and modifying as little as possible the geometrical characteristics.

The binder generally used is cement, which best withstands the presence of clay, frequent in old pavements, and confers on the treated material a much higher rigidity and resistance to rutting.

This technology is hardly new; it made its appearance in France during the fifties. Subsequently it fell into disuse but rising costs in road technology in the past few years have made it relevant once again. Its field of application concerns more particularly low or moderate traffic conventional pavements. To date, some million square metres have been treated by this process.

REFERENCES*

AIGI. 1964. International symposium on aggregates, Nice. May, no. 30, theme VI: Artificial aggregates and special specifications.

Andrieux, P. and J.H. Colombel. 1976. Use of fly ash in road technology. *BLPC*, no. 83, May.

ANNABA. 1986. First international seminar on evaluation of industrial wastes. October.

Aubert, J. and J.M. Lorrain. 1977. Lacustrine limestones in the Paris region. *BLPC*, Special issue IV: Aggregates. June.

CETE BORDEAUX. 1976. Treatment of sands for pavement layers. *Journ. Inf. LCPC*, November.

Fr. DOC. 1974. Solid wastes, proposals for a policy. French Documentation, Paris.

Famechon, C. 1977. Sand from Landes. *BLPC*, Special issue IV: Aggregates. June.

+Gutt, W., P.J. Nixon, M.A. Smith, W.H. Harrison and A.D. Russel. 1974. A survey of the locations, disposal and prospective uses of the major industrial by-products and waste materials. Building Research Establishment, Dept. of Environment, GB, February.

+Harrison, W.H. 1974. Synthetic aggregate source and resources. BRE, Dept. Envir., GB, December.

Krass, K. and W. Fix. 1977. LD slags. A new aggregate for road construction. *Strasse und Autobahn*, no. 8, p. 326.

OCDE. 1981. Use of marginal aggregates in road construction. OCDE-LCPC Report, November.

Primel, L. and C. Tourenq. 1976. Alternative aggregates. *Ann. Mines*, December.

Resource Recovery and Conservation. Elsevier Scientific Publ. Co., vol. I (1975).

Sauterey, R. 1976. Use of industrial wastes in building construction and public works. *Ass. Int. Env.*, December.

---

*All entries in French except those marked with a cross (+)—General Editor.

VI

Presentation of a Quarrying Operation

Georges Arquié, Christian Guizol and Jacques Lassartesse

If it is obviously not necessary to stress the fact that economic life requires aggregates for constructing transport infrastructures or buildings of any kind, and if it suffices for that to refer to the first chapter of this book, it is necessary to mention an essential feature of the aggregate market: the transport cost of these materials is very often of the same order of magnitude and sometimes much higher than the initial cost, so much so that efforts were made to locate deposits close to the places of consumption.

But places of consumption are mostly places with a large concentration of population, i.e., towns, and since these are almost always situated in the valleys, it is in the vicinity of these valleys that efforts were made to locate deposits.*

It is not surprising that these quarrying operations were visible and their nuisance raised popular outcry resulting in the laying down of rules and regulations, already presented in Chapter II. The latest one arising from these various constraints and imposed by legislation is that of rehabilitation, which influences the manner of working a deposit, and can almost render mining impossible because not profitable.

The aggregates produced should have characteristics which, over time, become more and more specific. Chapter IV explained why and clearly highlighted the new requirement of users who want materials which are constant in time.

To produce aggregates, starting from a natural deposit, involves crushing, washing, scraping, recrushing, removing dust, stocking, screening, resorting...in order to finally achieve materials that are constant in time and conform to very accurate specifications..., and then to sell them after weighing, reconstituting, loading, dispatching, transporting etc.

---

*This led professionals to operate alluvial deposits, which explains the fact that this type of installation presently produces about half of the aggregates sold in France (it is, moreover, proper to note that the costs of production are generally lower than for deposits of massive rocks).

And it means doing all this without too much alteration of the environment.

1. FROM A NATURAL DEPOSIT

The supplier of aggregates has no control over his raw material because the deposit which he is quarrying is the result of geological forces and cannot be changed (hard rocks or sand and gravel).

A good geological knowledge of the site enables overcoming this difficulty by correct selection of the deposit. The quarry owner can, moreover, with a given deposit, direct, at least to some extent, his operation towards a particular bed or a particular zone. We shall see that he can also, through intelligent mining, obtain a selective and homogenised raw material.

1.1 Selection of Deposit

For the preliminary study of a deposit, see Chapter III. Here we simply emphasise the criteria for selection.

The market study imposes certain constraints on the future operator:
— of distance. He will try to minimise the distance between his future plant and the centre(s) of gravity of the expected market;
— of compliance with the specifications relating to the mechanical characteristics of the material.

If the future producer intends to deliver products for concrete, the qualities of the rock will differ from those for aggregates for a pavement wearing course (and in particular pavements with heavy traffic).

But quite often the future producer is faced with a non-specialised market, i.e., he will have to deliver products for concrete as well as road aggregates or ballast for rail tracks. The selection of deposits then becomes more delicate but the possibilities of operation become wider (he can reserve one portion of the deposit for a particular use and another portion for another use).

Furthermore, the producer is often installed on a deposit for a long time and meanwhile the market changes. It is then that a joint study of the deposit and the market enables the producer to move towards a particular segment of the market rather than towards another.

Taking into account this market study, the strength characteristics of the stone will have a great influence on the decision of whether to quarry a deposit or not. This point is too obvious for elaboration.

But it appears to us that one particular characteristic is too often neglected: the homogeneity of the deposit.

It is certainly possible to take advantage of a heterogeneous deposit (see Chapter VII for massive rock deposits and Chapter VIII for sand and gravel deposits). Moreover, a perfectly homogeneous deposit is rare, if not completely non-existent. But it is clear that the more homogeneous the deposit, the easier its quarrying.

1.2 Orienting the Quarrying of a Given Deposit

This orientation is possible only if the quarry operator has carefully studied his deposit (i.e., not only its working face, but also behind the face) as well as the market.

2. OBTAINING MATERIALS WHICH ARE CONSTANT IN TIME

How does one deliver materials which are constant in time, given such a varied raw material as a natural deposit?

By operating in the same manner as cement manufacturers do, whose plants manufacture an artificial deposit in the form of a prehomogenised stock.

More and more efforts are being made to constitute in the quarry (be it a deposit of massive rocks or one of sand and gravel) a primary stock (or stock of precrushed material or a prestock) which serves as a reserve of raw material for the processing plant; this primary stock should constitute as homogeneous a reserve as possible.

We shall revert to this constitution of a homogeneous reserve, which has very different characteristics depending on whether the deposit is one of massive rocks or alluvial materials. In both cases this primary stock constitutes the link between two main parts of the quarry installations.

3. PRINCIPAL FUNCTIONS IN A QUARRYING OPERATION

These are mainly four:

— Working the deposit, which ends with discharging the materials into the primary receiving hopper.

— The primary station which starts from this hopper and takes the material up to the primary stock mentioned in the preceding section.

— The processing plant which takes its raw materials from this same primary stock and leads to the various finished products to be conditioned and marketed.

— Development after working. It is in fact impossible today (and the rules mentioned in Chapter II do not permit such a practice) to abandon a deposit without restoring the site to a proper appearance or even studying and preparing the displaced soil for future use.

Then we have the following different functions for an aggregate quarry (Table VI.1).

Table VI.1: Functions to be carried out in mining aggregates

Deposits of massive rocks		Group of functions	Deposits of sands and gravels	
Functions	Corresponding chapters		Coressponding chapters	Functions
Stripping (1)	7	Quarrying the deposit	8	Stripping (1)
Stoping	7		8	Extraction (2)
Fragmentation of large blocks	7			
Loading	7		8	Loading (2)
Transport to primary crushing section	7		13	Transport to primary crushing section (3)
Primary crushing section including extra॰ ᴐn	11	Primary stock	10	Primary crushing section
Pre-elimination	10 and 11		10	Scalping
Primary crushing (9)	9		9	Primary crushing (4)
Post-elimination (10)	10 and 11		12	Scrubbing
Stock of precrushed materials (5)	14	Primary stock	14	Primary stock
Reduction: crushing and grinding (6)	9	Processing plant	9	Reduction: crushing and grinding (6)
Classification: screening and other processes (6)	10		10	Classification: screening and other processes (6)
Washing (6)	12		12	Washing (6)
Removing the dust (6)	13			(7)
Connections between equipment (6)	14		14	Connections between equipment (6)
Stocking	14		14	Stocking
Recomposition	14		14	Recomposition
Weighing	14		14	Weighing
Dispatch	14		14	Dispatch
Development after quarrying (8)	2 and 12	Development after quarrying	2 and 7	Development after quarrying

Remarks:
(1) Stripping must often be done to prepare the future management of the quarry.
(2) In sand and gravel deposits the functions of extraction and loading are often done with the same equipment.
(3) It often happens that transport to primary section is combined with extraction (suction dredges).
(4) In most cases of quarrying deposits of sand and gravel, the boulders eliminated by scalping are abandoned. But it may happen that they are crushed to obtain more angular materials. In this case the processing plant is divided into two sections: one for the rounded material and the other for the crushed material.
(5) Often called prestock for the sake of brevity.
(6) The order of these functions is not fixed and often some are involved several times in the processing chain. There are also many "recycling procedures".
(7) There is generally no removal of the dust in sand and gravel plants.
(8) We do not like the use of the term 'rehabilitation of the site' (which would mean the site is restored to its original condition), nor even of the term 'redevelopment' as it is obvious that development after mining need not necessarily restore the soils to their earlier use.
(9) There are sometimes two levels of crushing so as to reduce the grain size of precrushed material.
(10) These operations are often termed prescreening.

4. HOMOGENISATION AND STOCKING FUNCTIONS IN DEPOSITS OF MASSIVE ROCKS

It is in these deposits that the prestock best fulfils the functions indicated in Section 2 above. Nevertheless, we do come across in certain cases some major divergences from this scheme.

It may be recalled that the formation of deposits of massive rocks used for the production of aggregates (limestone or eruptive or metamorphic deposits) has already been described in Chapter III.

4.1 Stripping

The nature of the overburden, the methods for stripping it out and commercial utilisation of the resultant materials are dealt with in Chapter VII while the use of these materials in development after mining has been studied in Chapter II.

4.2 Elimination of Impurities Contained in the Deposit

We shall not deal here with the elimination of the top portion of the deposit, often weathered, because this elimination is achieved during stripping (see Chapter VII).

On the other hand, we feel it would be useful to say a few words about the impurities contained in the deposit itself, which come from cracks being filled with earth or clay or from highly localised alterations.

These impurities can only be eliminated by separating upstream of the prestock the fine material likely to contain this harmful matter (e.g., clay) from the coarser materials. This operation can take place at different stages:

— Before the primary crusher. Feeding then does the function of screening. This is what we shall term pre-elimination.

— Immediately after the primary crusher. In this case along with the fines contained in the stoping products, those which adhere to the coarsest elements after this first operation and which are released through a passage in the primary crusher are eliminated, but rock fines resulting from the attrition of blocks in the primary crusher are also eliminated. We shall term this operation post-elimination.

The grading of the all-in resulting from elimination varies according to the degree of pollution of the deposit. It can range from 0/10 to 0/60 mm. This 'prescreened all-in' can be sold as such and can have the same uses as the 0/D from the weathered parts of the deposit and eliminated by stripping. However, as this material resulting from scalping (scalpings) contains a large amount of 'unweathered' material along with a relatively small amount of 'clayey' fines, it may be interesting to evaluate this product by elimination of the fines under water. The various hydraulic methods of washing and classification are described in Chapter XII. They enable in a product 0/D, with D ranging from 10 to 16 mm, elimination of the fraction 0/d, with d ranging from 40 to 60 μm.

4.3 Stock of Precrushed Materials (see Chapter XIV)

For the sake of brevity, this is often called prestock. It constitutes the reserve of raw material for the processing plant.

Its main function is thus as a reserve stock to avoid breakdown of the plant upstream. These breakdowns can be accidental—maintenance operations due to wearing out of the equipment—or systematic, the plant downstream having different and, generally, longer working hours than those of the plant upstream.

It is often technical means which enable proceeding from a rhythm of discontinuous operation to one of continuous operation. In fact the upstream of the prestock—the primary crusher—is fed in a discontinuous manner, with the trucks or loaders bringing the blocks. The processing plant should, on the contrary, work in a continuous manner, which is rendered possible by prestocking.

Finally, it is in our interest to take advantage of this need for a stock in order to homogenise the material. Obviously this function assumes elimination of impurities mentioned in Sections 4.1 and 4.2.

But it also assumes that efforts are made to carry out the mining in such a way as to eliminate or reduce certain disparities which may exist in the deposit. It is particularly from the manner in which stoping and loading are carried out that this result is obtained, as shown in Chapter VII. The principle consists in carrying out extraction in such a way as to bring to the primary section a suitable mixture of the various sound beds of the quarry.

The manner of constituting the stock of precrushed material is explained in Chapter XV.

5. HOMOGENISATION AND STOCKPILING FUNCTIONS IN SAND AND GRAVEL DEPOSITS

This practice of an extraction skilfully carried out to give to the primitive rock a suitable cocktail of the various "sound" beds of the quarry, which we have just now emphasised in the quarrying of deposits of massive rocks, we shall find again in a pure state, so to speak, in the quarrying of deposits of sands and gravels. In fact, this is because the corresponding deposits present themselves in a particular manner, which is analysed in Chapter VIII.

Exactly in the same way as for a deposit of massive rocks, the future operator should, before taking any decision about quarrying, carry out a double study of the deposit on the one hand, and the market on the other.

The deposits are distinguished by:
— thickness of the overburden,
— thickness of the sand and gravel layers and their heterogeneity,
— position of the water table.

The method of quarrying the deposit depends on the foregoing and efforts should be made to reconstitute a supply of all-in material, close to the average composition that can be expected.

5.1 Washing or Scrubbing If Need Be

We shall emphasise just one point here.

It may happen that the deposit is polluted by clayey fines and that these are distributed in such a manner that it is impossible to abandon the mining of certain areas containing them.

We are then compelled to carry out strong washing or scrubbing either on the all-in material (upstream of the plant) or on the sands (downstream of the plant), when such methods would enable obtaining profitable commercial material.

5.2 Primary Stock

The operator is generally able to bring close to the feeding hopper of his processing plant an all-in material with mean characteristics similar to those of his deposit. He then contents himself with a small safety stock ranging from a few hours to a few days of feed material.

When the heterogeneity of the deposit is too distinct, a systematic reconstitution of the all-in should be carried out in order to maintain a constant supply to the plant. the latter is always calculated for given gradings and the rather reduced possibilities of variations in supply. A skilfully constituted mixture of material from areas of the deposit in which the gradings are complemantary is therefore brought to the feeding hopper. Homogenising the all-in material implies having a well-regulated plant and hence finished products of constant quality, which is the final basic objective.

Few gravel pits have plants as sophisticated as those of quarries of massive rocks. One contents one's self with having different piles on the ground with proportional recoveries by loaders.

Of the two functions of the prestock, homogenisation and buffer stock, only the first still exists.

5.3 Processing Plant

We find in the processing plant the same functions as those for the mining of deposits of massive rocks, as shown in Table VI.1.

However, a special point is worth mentioning. If the round material normally obtained from the mining of sand and gravel deposits is generally very suitable for the manufacture of hydraulic concretes, it is much less suitable for road layers which require (see Chapter IV) a minimum of angularity. We are then forced to crush the material to obtain sharp edges.

Consequently, we see in some processing plants a duplication of installations. We may say in such cases that there are two separate processing plants, one for round materials and the other for crushed material with, however, some connections between the two installations.

This case is distinct from that of the installations which produce round material enriched in certain aggregate sizes (in deficit in the deposit compared to the market requirements) by a suitable limited crushing.

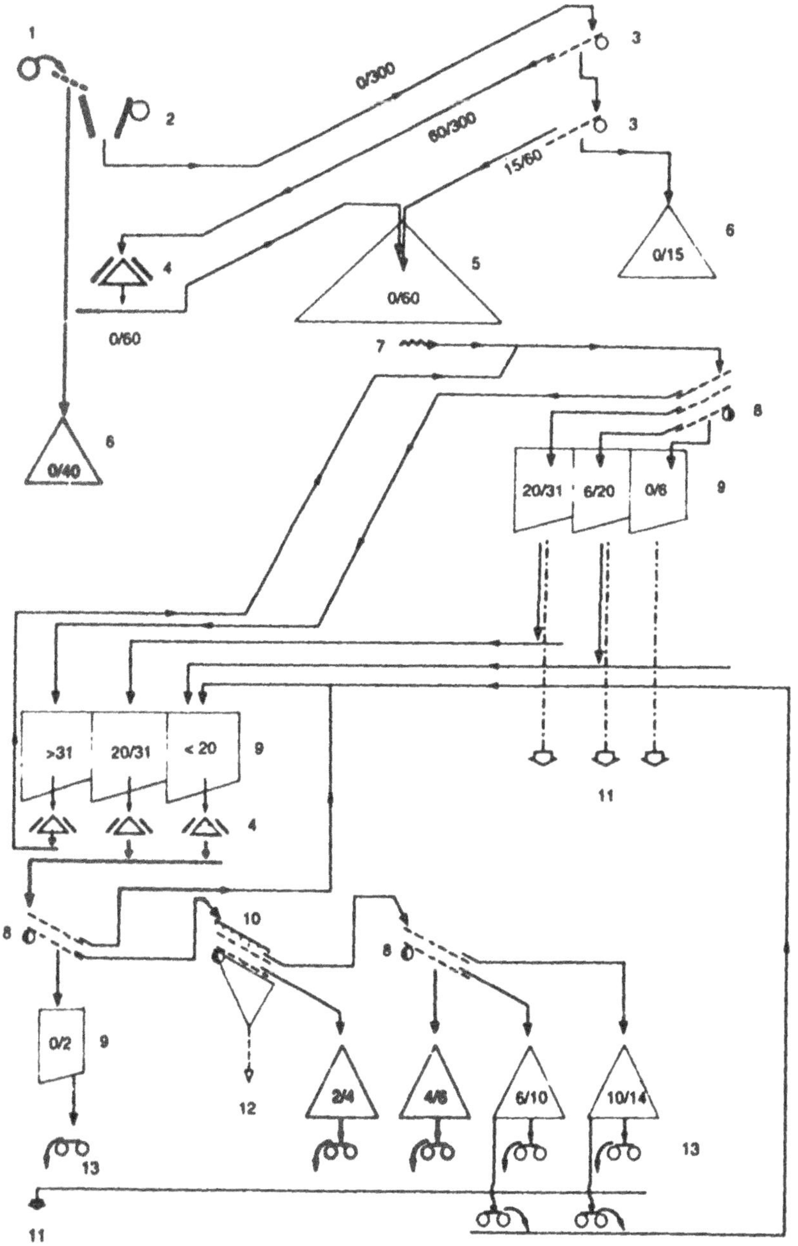

1 : Plate feeder, 2 : Primary jaw crusher, 3 : Eccentric screen, 4 : Gyratory crusher, 5 : Stockpile, 6 : Trash, 7 : Vibrating feeder, 8 : Counterweighted screen, 9 : Bin storage, 10 : Rinse screen, 11 : Loading, 12 : Treatment of water, 13 : Belt batcher

Fig. VI.1. Flow sheet in a massive rock quarry

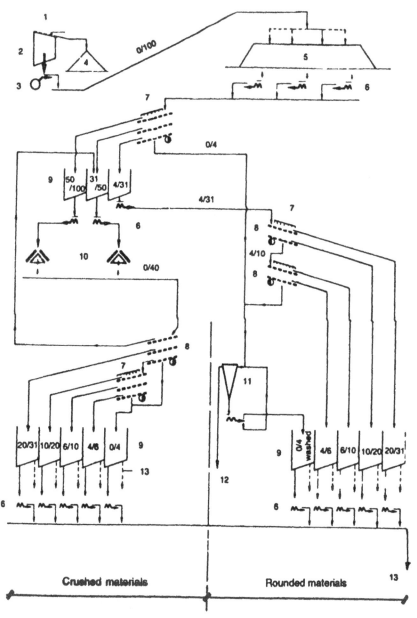

1: Scalping screen, 2 : Hopper, 3 : Plate feeder, 4 : Very coarse aggregates,
5 : Stockpile, 6 : Vibrating feeder, 7 : Rince screen, 8 : Counter weighted screen,
9 : Bin storage, 10 : Gyratory crusher, 11 : Hydrocyclone, 12 : Sump, 13 : Loading.

Fig. VI.2. Flowsheet in a sand and gravel pit

6. FLOW SHEETS

Figures VI.1 and VI.2 give, by way of illustration, flow sheets which show how the various functions mentioned in Table VI.1 can be organised. They should not be considered standard flow sheets; more comprehensive and more accurate details are given in Chapter XV.

7. AUTOMATION

A new function has appeared in aggregate producing operations, that of automation of the installations. Chapter XVI is devoted to this innovation. Automation is more and more utilised.

8. CONCLUSION

We have seen how an installation can be split into major parts and then into functions, which will be dealt with in the following chapters of this book.

But one should not content one's self with this anatomical description of the quarrying of aggregates, this type of dissection. The entire installation constitutes a whole which should remain homogeneous and it is absolutely necessary to effect a synthesis. Chapter XV is devoted to that.

VII

Stripping, Extraction, Loading and Transport in Quarrying Massive Rocks

Jean Lot and Georges de la Rupelle

Many are the preoccupations of an operator exploiting a massive rock quarry before the aggregate proper becomes a reality. They comprise uncovering the aggregate material in his deposit, dividing it into parts, extracting it and then transporting it—all at the most favourable cost price, one acceptable to the processing plants, which will then produce the aggregate proper through many stages of size reduction and grading.

The whole series of operations preceding the processing, denoted by the overall term *extraction*, represents the total cost price for a ton of the material delivered to the processing plant. The overall level of this cost price should reflect the relative magnitude of each of these constituent phases: quarrying + loading + transport and how each in turn influences the rest.

The quarry operator will also have to appraise the site on which he will bring to bear his expertise as an entrepreneur. Hence he carefully considers:

— the nature of the rock, type of massif,
— the method of exploitation, environment.

He then decides his plan of action, encompassing

— stripping,
— quarrying,
— loading,
— transport (haulage) (Fig. VII.1).

1. DEPOSIT AND METHODS OF EXPLOITATION

1.1 Nature of Rock and Type of Massif

Under the designation 'massive rocks' there are materials of three origins: igneous, metamorphic and sedimentary (see Chapter III).

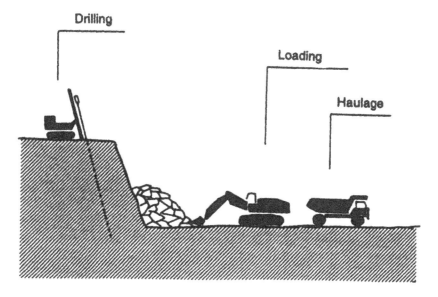

Fig. VII.1. Extraction: Elementary phases.

The soundness of any preliminary study of a deposit resides in making available through a complete and minute prospecting of the deposit, data on its quality, homogeneity, surficial extent as well as at depth, to formulate a master plan for exploitation:

— magnitude of stripping,
— extraction method to be applied,
— direction of cutting the working faces,
— objective of homogenisation,
— maintaining stability of the massif,
— controlling hydrology.

1.2 Methods of Exploitation

Before taking up the extraction techniques proper to massive rocks, it is desirable to present the methods which should cover almost the whole run of exploitation.

• *By Successive Horizontal Benches in Full Width*

A working face is made at one end of the deposit over its full width, preferably in cross-cut, which will advance parallel to itself so as to ensure extraction of a complete bench of a height determined by considerations of the fragmentation. This is the commenest method and is applied to most massive rock deposits (Fig. VII.2, A followed by B).

In the case of a deposit which is thin in depth, a single bench will exhaust the site. If the deposit is thick, a second bench will be extracted following the same

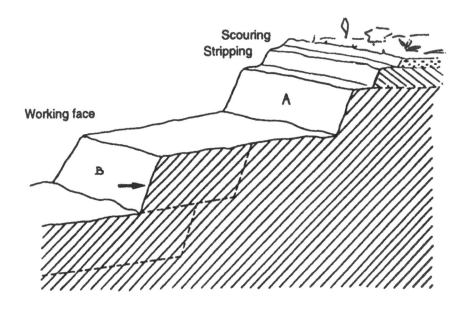

Fig. VII.2. Method of exploitation by successive horizontal benches in full width

principle, then a third and so on. For the purpose of homogenisation at the opening of the quarry, two benches may be worked simultaneously but the aim of the method remains, namely, extraction of the entire area before taking up a new level.

In this method overburden stripping works are often undertaking simultaneously over the entire quarrying site. Each bench constituting one level of traffic, the method is favourable to the shifting of the primary receiving station corresponding to each level. On the other hand, reclamation is possible only at the last stage and the boundary slopes alone can be considered the ultimate limit.

• By Simultaneous Horizontal Benches

A first working face is opened, followed immediately by a second one at a lower level so that both the working faces come under operation almost simultaneously. More faces can thus be opened depending on the deposit thickness (Fig. VII.2, A and B).

This method is frequently resorted to for deposits of medium and low thickness compared to their area (for example, basalt deposit, limestone deposit). The works of overburden stripping and extraction proper can be carried on concurrently, which spreads out the stripping operation vis-à-vis time. When applicable, this method affords the possibility of a more immediate reclamation on the heels of extraction operations (case of restoring cultivation).

Contrarily, simultaneous exploitation at different levels makes location of the primary receiving station practically unchangeable.

Depending on the topography of the site, two basic configurations can be distinguished, both of which will contend for the choice of method to be adopted.

— Extraction by terracing: safety considerations related to the stability of the massif tend to favour extraction from the top downwards by an attack at the summit.

— Extraction by pitting: perfect illustration of an excavation created by successive horizontal benches.

Leaving aside their visual impact from the point of view of the environment, it is evident that the problem of water flow determines these two configurations: the first along a more direct flow of surface waters while the second could influence the level of the water table by catchment in the pit. The problem of dewatering than becomes an additional headache for the quarry manager.

There can, of course, be mixed configurations, with extraction done in the flank in the first phase and then in the pit.

1.3 Environment

To solve his problem of extraction, the quarry manager should take into account several environmental factors which, broadly speaking, comprise:

1. External constraints related to the situation of the site, with which he cannot argue:

— Geographic, topographic and hydrological constraints inherent to the emplacement of the quarry, necessitating care during exploitation and restoration thereafter, in addition to safeguarding the permanence of groundwater by restitution and without pollution;

— Sociological constraints and constraints of proximity related to the quarry's immediate surroundings in terms of its historical, archaeological, tourist and even agricultural aspect;

— Climatic constraints incidental to the regime of winds and rain, which may aggravate possible pollution even though the mineral (dusts, muds) threat has been contained;

— Geological constraints directly dependent on the structure of the deposit, nature of the rock, overburden, dip and depth are features peculiar to every exploitation;

— Regulatory constraints that require compliance albeit some are flexible.

2. Inherent constraints, which can be partly overcome by choice of mining methods and equipment:

— Constraints of deposit associated with its quality and homogeneity, which are overcome by the expertise of the quarry manager;

— Equipment constraints: adaptation of available resources: drilling equipment, collecting machines, hauling machines, characteristics of the primary crushing station;

— Constraints in quarrying, arising quite often from taking up additional commitments vis-à-vis the external environment, to which safety precautions must be added;

— Economic constraints stemming from the local market and transport facilities.

These various sensitive issues and the obligation to find responses and remedies for them are repeatedly mentioned in the impact assessment report which accompanies any application for authorisation for mining, up to reclamation of the mined area. All the details concerning this subject have been given in Chapter II.

2. UNCOVERING

After the preliminary exploration studies of the site to be exploited, isolating the massif from its geological surroundings and laying bare its surface call for preparatory works that literally 'uncover' the rock to be quarried.

There are two essential phases.

2.1 Scraping the Soil Cover

This refers to clearing the first layer of overburden and essentially concerns the top soil. This soil is to be clearly differentiated from other mine wastes in the matter of utilisation owing to its relative chemical, physical and biological properties, which make it suitable for revegetation of the site.

The soil materials scraped are therefore carefully stored to preserve their fertile properties.

While exploitation is in progress, the scraped soil materials can aid in revegetation of the earthworks and fills. Once exploitation is over, they will form part of the reclamation of the site: revegetation, remodelling and landscaping.

If the site permits and there is no outcrop, scraping can be effected by a power scraper. However, it is more often done with a bulldozer and the scraped material left as cordons at the boundary or transported to the storage site.

2.2 Stripping

The depth at which a deposit is reached may vary widely and consequently surficial alternation may be greater or less: position of groundwater, seasonal fluctuation in the flow regime, impregnation by water in contact at the fissures.

Subjected thus to physicochemical actions, the rock deteriorates and its properties are diminished. It is imperative that the deposit be purged of its contact layers to preserve the maximum homogeneity of the good material. The more massive and compact the deposit, the less deep the weathering; only the areas of contact would need to be purged. A simple stripping done by conventional earthmoving machinery enables expurgation of the wastes, which can be used in fills.

Beyond the superficial contact areas, sometimes progressive weathering is noticed with depth along the fissures even though the rock may be intact in the interstices. In this case it is desirable to resort to a premitting operation for a few metres depth, a kind of shake-up blasting that enables the machines to uncover beyond the simple rocky contact (see Section 3). In that case the materials thus obtained furnish a quality fill material for reinstatement works.

Production downstream of the quarry which would conform to the desired quality levels for cleanliness of the end products and a consistent hardness of the aggregates would be largely guaranteed if the stripping is carried out with care despite the costs in total exploitation.

Depending on the area of the site, the quality of the rock and the strategic position of the deposit, overburden stripping may represent 10 to 15% of the volume to be extracted. Furthermore, the economic and commercial context should be such as to permit accepting the inevitable increment.

2.2.1 THE MACHINERY

Removal of the overburden calls for the usual earthmoving equipment. This operation is done with the quarry operator's own machinery if it meets the needs or is subcontracted to firms that undertake earthmoving jobs, taking advantage of their slack seasons and their machinery, which often is better suited for this kind of work.

— The scraper, saldom used except when the deposit is spread over a wide area, can handle a large thickness of overburden and the wastes are relatively homogeneous. The smallest outcrop hinders the progress of this machine.

— The bulldozer is better adapted to any kind of quarry terrain, especially when equipped with multiple scraping teeth. However, the area should be relatively limited since the magnitude of the volume scraped and the place for its piling would require movement of too huge a volume of material for this single machine. Scraping constitutes an additional plus point but cannot substitute for premitting in the matter of cost as well as fragmentation of the wastes.

— The hydraulic shovel, certainly the best suited machine whatever be the area and depth to be purged. Recourse to a back-acting bucket offers real possibilities of scraping the contact zones and enables sloping the boundaries of exploitation, which is clean, natural and safe.

Of course, the shovel requires transport vehicles up to the point of piling the wastes. Depending on the bearing capacity of the terrains, a quarry truck is used or, better still, an earthworking tipper and trailer machines, which are better able to adapt themselves to all types of terrain.

It may be mentioned here by way of a reminder that more powerful stripping machines, such as the dragline and the bucket wheel excavator, used in the field of mining, are not at all justified for quarrying work.

2.2.2 UTILISATION

Scraping and stripping are necessary for guaranteeing the quality and cleanliness of the product, despite the eliminations necessary in the processing of the material, which will be performed on value-added products.

Wastes from the overburden have no commercial outlet except possibly for fills.

The main concern of the quarry manager is to manage the storage of these materials. If a major portion can be used in making cordons demarcating different areas in the quarry site, protective mounds, landscape earthworks, the surplus should be stocked at the nearest point without, however, taking up surfaces which are to be quarried sooner or later since rehandling of wastes could mean an additional expenditure, something that must always be avoided. Depending on the quarrying method adopted, reclamation is done as the quarrying progresses or upon its completion. This option decides how far or near to the subsequent place of reutilisation the waste should be piled.

3. QUARRYING

After stripping, the rock has to be fragmented so that the material can be easily collected by the machinery and conveyed to the processing plant. This operation is carried out with the help of explosives.

Quarrying by deep vertical mines seems already to be the mode most favoured by quarry managers. Quarrying by benches of low height, not exceeding six metres, no longer gives the desired output in volume to be competitive. Surface mining methods are very well known to be dangerous and are forbidden by the Mines service, barring very special exemptions.

Widespread adoption of the deep vertical mines method of quarrying is not only a sequel to the experience gained at a number of sites by quarriers who have long professional standing and are apt to have a good understanding of their deposit, but to the training imparted by the profession of aggregate producers during preparation of candidates for the certificate of blasting superintendent, a training characterised by the adoption of Swedish principles as developed by Langefors.

3.1 Structure of the Massif

The structure of the rocky massif is an important parameter in the study of fragmentation by explosives. Depending on the state of discontinuity (faults, diaclasis, strata, ...), the job of drilling and the explosive effect engender different results in terms of yield. The state of discontinuity is determined in advance by geological investigation, either through direct measurement by extraction of cores or indirect measurement by the seismic or microseismic method; the latter method enables recording the speed of the compression waves, a criterion which depends on the type of rock.

The discontinuities thus revealed are taken into consideration:

— For selecting and placing the explosive so as to take advantage of the planes of discontinuities: for example, with a quarry head regulated upon a plane of sub-horizontal fracture; or

— For distribution of the explosive charge so as to avoid fly rock caused by the zones of least resistance. But the best benefit yet resides in utilising these natural discontinuities by orienting the quarrying faces with reference to the dip of the stratification planes:

Cross-cut working face, along the dip: The explosive performs well, is favourable to a well-scattered pile of broken rocks and a uniform quarry head, but entails the risk of unstable blocks sliding at the top as a result of the rear effect.

Cross-cut working face, against the dip: Always favourable to explosive performance, without fear of instability at the top. But the tendency is against scattering of the pile of broken rocks, the quarry head is less uniform and the risk of undercutting present.

Working face in the direction of the dip: Orientation which assures utmost safety but the juxtaposition of discontinuities of different strengths engenders risk of an irregular profile with a 'jagged' toe in the working face in the case of highly inclined beds.

3.2 Blasting Plan

Adaptation of the blasting plan to the needs of the quarry manager is drawn either from previous blasting experience in the case of a quarry worked earlier, or from the result of specimen blastings for the quarrying carried out for opening new working face.

The following objectives are to be borne in mind:
— Magnitude of production and mode of quarrying adopted.
— Geometric requirements:

Such a toe as will aid a good planeness of the mine head and permit the proper advance of the loading equipment;

Straightness of the working face to secure control of the advance of the working faces with due regard to the structure of the massif.
— Dimensional requirements:

Block-size distribution of fragmented rocks which should be compatible with the primary crusher's opening sizes;

Grading curve of the run-of-mine fragments; too many fine elements is often an inconvenience since the operations of primary sorting will eliminate a 0/30 or 0/40 zone, the quantity of which decidedly affects the overall yield.
— Safety imperatives:

At the working face proper: All conditions which permit preparatory drill work as also loading operations to be carried on in absolute safety: neat rupture of the face, absence of overhang, little rear effect, good spread of the rock pile;

In the neighbourhood: Absence of fly rock, minimum ground vibration and noise abatement are among the factors valued by environmentalists and which quite often govern the grant of a quarrying permit.

The blasting plan should therefore take into account the above objectives and adopt the blasting parameters necessary to fulfil them. This, in brief, influences the choice of the following:
— height of the working face,
— inclination of the hole drilled,
— borehole pattern or the ratio bench width/distance between holes,
— number of rows of holes,
— hole diameter,
— explosive used.

3.2.1 WORKING FACE HEIGHT

The height adopted is 12 to 15 m in order to remain within the frame of the relevant regulation currently in force in France. But exemptions are often granted for working faces up to a maximum of 20 to 25 m if the deposit is known to be solid and stable, but subject to particular conditions of quarrying. Technical, economic or safety considerations as listed below are to be borne in mind.

— The drilling work loses its efficacy when the hole is too deep.

— With an abrasive material, the wear and service life of the drill will depend on the duration of drilling for the same volume of rock broken.

— Extension of the working faces depends on the magnitude of the unit quarrying operation, which governs the advance of the loading machinery.

— The attack of the drill upon the surface at the start of each hole, up to its final centring, is one cause for loss in yield.

— The toe is influenced by the extra depth of drilling, the bottom charge and also the actual position of this charge with respect to a possible deviation in drilling.

— The effect of stemming (tamping) at the top of each hole quite obviously depends on the number of holes for the same volume of rock quarried.

— The stability of the massif, hence of the unbroken rock, despite an obligation to purge, may be at risk, which increases with the height of the working face.

— The drilling rig, its attachments, and consequently the diameter, place a limit on the height (see Table VII.1). A very empirical approach indicates a height of the order of 200 times the drilling diameter (d), say about 15 m for a 76 mm drilling diameter and about 20 m for a 101 mm diameter.

3.2.2 INCLINATION OF THE HOLE DRILLED

Though the technique is called 'deep vertical mines' it is favourable to inclination of the holes. The regulation limits the angle of the hole bored to 20° with respect to the vertical. This limit is fixed partly by the possibility of charging the cartridges by gravity. In practice, a 15° inclination is more often employed (Fig. VII.3).

The resultant advantages are as follows:

— Better rupture at the foot with less crater effect;

— Extension of the hole, hence of the explosive charges which are more favourable to the output;

— Fewer blocks at the top, with less rear effect at the edge of the bench;

— Greater safety because of minimal overhang.

The constraint of the inclination lies in the relative precision of positioning the drilling machine; the guiding support-slide of the hammer should be in a vertical plane orthogonal to the working face, then finely adjusted at the angle adopted. These precautions help to contain the risks of deviation and shift in positioning of the explosive bottom charge. The calculated thickness of the bench to be quarried (V) is often corrected by a value equal to 2 to 3% of the working face height (H): thickness $= V - 0.02$ to 0.03 H.

3.2.3 BOREHOLE PATTERN

The borehole mesh size is the unit bench-width area between two adjacent holes, or the area between four adjacent holes if there are several rows of holes drilled.

It may be emphasised here that in vertical mines the bench-width corresponds to the thickness of the bench-slice to be quarried obtained from the formulae. On the contrary, the real bench width, in the case of inclined mines, is the one measurable on the surface, which is why a correction needs to be carried out in terms of cosine of the inclination for connecting the thickness and the bench width.

Earlier, the relation between bench width and drilling diameter was defined as per the Langefors method. In first approximation and depending on the rock type, the quarry operator can adopt a bench width value of 40 d to 45 d by keeping the hole spacing 1.2 to 1.3 times the bench width.

For example, a drilling diameter of 76 mm is compatible with a bench width of 3 m and a spacing of about 4 m, while for a diameter of 102 mm the bench width and the spacing are respectively about 4 m and 5 m.

A close spacing of the holes vis-à-vis the magnitude of the bench width tends to produce a larger number of blocks, from which it could be said:

borehole pattern $< V^2$ = stone-block (riprap) quarry,
borehole pattern $> 1.2 V^2$ = aggregate quarry.

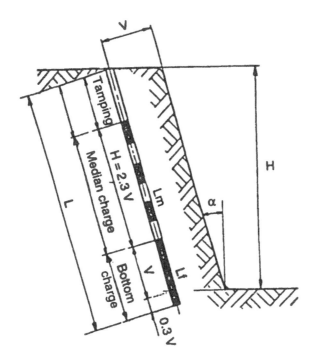

$$\text{Drilling length} = L = \frac{H}{\cos \alpha} + 0.3 \, V$$

Linear bottom charge = Lf
Linear median charge = Lm
Bottom charge = Lf x 1.3 V
Median charge = Lm (L - 2.3 V)

Fig. VII.3

3.2.4 NUMBER OF ROWS

When the site does not permit blasting over a large width of working face, thougl the production might be high and the machines powerful enough, it becomes neces sary to provide for several rows of holes, generally two to three. This possibilit is contingent upon the use of blasting with microdelays so that the bench-slic ahead is loosened before blasting the second and so on.

In practice, it is advisable to retain the same thickness to be broken up for th first and second row, for a given scheme of explosive charge, but reduce the thir row by 0.20 to 0.30 m so as to increase the specific charge and facilitate spreadinj the rock pile from the previous rows.

3.2.5 BLAST-HOLE DIAMETER

The diameter of a blast hole is for the most part that of the drill bit and thi parameter is fixed depending on the drilling rig available to the quarry manageı The type of drilling rig again depends on the kind of material to be quarried, it hardness and internal discontinuities.

Let us recall that when the fundamental formulae are applied, the diamete determines the maximum bench width and that in turn the bottom charge.

Until the end of the 1970s, it was common to drill in 76 mm diameter. Witl the advent of more powerful hydraulic drilling rigs, the diameters have presentl increased to 89 to 102 mm, with a tendency to rise to 110 mm and even 115 mn and above in the case of high production quarries of upwards of 1 million tonne per annum (Table VII.1).

Table VII.1

Diameter of drilling	Lf (kg/m)	Lm (kg/m)	Maximum bench width (m)	Bottom charge (kg)
63	5.0	1.6	3.0	15
76	5.6	2.3	3.5	26
89	7.7	3.1	4.1	41
101	10.0	4.0	4.7	60
127	15.5	6.0	5.9	120

3.3 Drilling

Depending on the blasting scheme adopted, application of the explosive is don after demarcating on the ground the location of the set of holes for the kind o quarrying in question and their actual boring.

The drilling technique for blast holes comprises:

— a rotary movement,
— a downward movement, most often with percussion (except for 'tricone machines),

— blowing of compressed air to expel the cuttings and to cool the drill bit.

These movements are transmitted to the drill bit through a string of extension rods (percussive drilling with hammer outside the hole) or directly by the hammer itself (drilling with down-the-hole hammer) connected to the drilling machine by a string of extension rods.

The drilling machine should be related to the production envisaged; the accessories for it have recently undergone change with the advent of hydraulic rigs. Notable amongst those used in massive rock quarries are rotary percussive drilling through an 'open-hole' or 'down-the-hole' hammer.

The drilling rig comprises a drill carriage and a compressor. The capacity of the compressor is chosen according to whether it is pneumatic or hydraulic since in the latter case it only assures supply of the compressed air needed for blowing during drilling. The latest machines incorporate a compressor unit as part of the vehicle.

Accordingly, the quarry operator has a choice of three solutions:

Pneumatic open-hole percussive drilling with a service pressure of 7 bars, drill bit worked by a string of extension rods of 32 or 38 mm diameter, which limits the drilling diameter to a maximum of 76 mm for maximum depths of 18 m, that is, the length of the hole at the inclination desired for a working face of the order of 15 m and an excess depth generally fixed at 30% the breadth of the bench-slice to be quarried.

The instantaneous speed varies from 15 to 25 m/h depending on the rock type and diameter.

Pneumatic down-the-hole percussive drilling with a service pressure up to 17 and 20 bars; the machine with extension rods of 89 mm diameter gives a better straightness of hole, limiting the deviations and risks of jamming by the presence of the hammer immediately behind the drill bit. The instantaneous speed goes up to 30 and 40 m/h but at a high cost of energy consumption and direct wear on the hammer.

Hydraulic open-hole drilling with hammer employing a pressure up to 250 bars, using a small quantity of compressed air for blowing only, the air being supplied from a compressor that is increasingly being integrated with the drilling by machine manufacturers. The energy economy, the greatly improved penetration performance, the flexibility in adjustment, and the very low noise level have enabled an ever wider development and adoption of these new machines despite their greater sophistication, which demands training and adaptation of the personnel.

Depending on the rock, the drilling speed can go up to 50 m/h. With this type of drilling machine, the drilling accessories also have progressed in respect of increase in diameter, viz., extension rods of 45 and 51 mm (Table VII.2). Nevertheless, the current trend in development of hydraulic drilling is towards limiting the height of the working faces, preferably to 15 m, so as to minimise the fears of deviation with regard to performance.

Table VII.2

Drilling diameter (mm)	Diameter of extension rods (mm)	Diameter of air blowing (mm)	Maximum depth (m)
64-76	38	15.0	18
76-89-102	45	18.0	24
89-102-115	51	21.5	30

The advancements made in machinery development are reflected in working conditions:

— Control cabin which enables the driller to be more efficient regardless of climatic conditions.

— Rod changer which relieves the operator of the increasingly heavy handling operations.

— Dust collector, for an agreeable environment in the immediate vicinity and the surrounding space.

— Control console with adjustment of pressure and rotation, anti-jamming device etc.

3.3.1 DRILLING ACCESSORIES CONSTITUTING THE OPERATION 'CONSUMABLE'

These can amount to 50% of the drilled meterage cost.

— Drill bits are of the 'X' or 'button' type according to whether the drill head houses tungsten carbide discs or inserts. Essentially depending on the rock (hardness and SiO_2 content), the service life can range from a few hundred metres to several thousands. Grinding of the bit is a factor which determines longevity, which in the case of buttons includes the geometry, which should be preserved.

— Extension rods, connected by couplings or, lately, by a male-female junction with highly specific threads for absorbing the fatigue of transmitting the striking energy. Adoption of automatic extension-rod changers (becoming more frequent) permits drilling with heavy extension rods of 45 mm and 51 mm diameter which is favourable to better straightness of the hole. The service life can touch 3000 to 5000 m (Table VII.3).

3.3.2 EXECUTION OF DRILLING

Whatever the attention paid to determining the blasting plan, the results turn out to be correct only to the extent that the blasting plan is rigorously executed.

In this sense drilling forms the basis for correct blasting. Therefore attention is paid to:

— Good demarcation of the borehole pattern.

— Vertical positioning to ensure the length to be drilled; the holes will then be checked before charging the explosive to adjust the extra depth.

— Inclination of the holes which, in the event of bad positioning of the slide, gives rise to deviations (Fig. VII.4).

The drillman is directly responsible for the positioning of the drilling machine, alignment of the slide in the vertical plane, its inclination and 'priming' the hole with the first extension rod.

His experience is equally valuable in sensing the repercussions due to the discontinuities and heterogeneities of the deposit at the time of drilling. His intervention in the running of the machine, careful monitoring of the drilling and relaying his findings to the geologist confer upon the drillman a decisive role that should be recognised as a real speciality.

Table VII.3: Drilling equipment
Extract from Mineral Industry 'Blasting'

Hole diameter (mm)	Mode	Type	Drill bit	Power (hp) Energy P/H*	Productivity per day (m³)
≤ 41		Hand hammer	Monoblock drill	12/25 P/H	< 100
≤ 41		Hand hammer with drill advancer	Drill or small extension rod of 1.2 m	12/25 P/H	< 100
≤ 64		Wagon drill	Handle, coupling, extension rod, drill bit	25/50 P/H	< 500
≤ 76		Automatic drill-pneumatic tyred	same	60/100 P/H	< 1000
64–102		Automatic drill-crawler mounted (light)	same	100/250 P/H	1500–2000
76–110		same (heavy)	same	200/300 P/H	2000–3000
76–89 105–170(3)		Light down-the-hole drill	Handle, tube drill bit	200/300 H-P	3000–4000
> 170		Heavy tricone drill	Tube (weight rod) drill bit	300 H	> 4000

*P : Pneumatic, H : Hydraulic

3.4 Blasting

3.4.1 THE EXPLOSIVE

The method directly inspired by Langefors introduces a coefficient of the explosive power in determination of the blasting parameters. The nature of the material (Table VII.4), the discontinuity of the deposit and the possible presence of water in the mine holes are to be taken into account in the choice of explosive: dynamite slurry, nitrates or nitrate fuel, in cartridge or bulk.

Table VII.4

Specific consumptions	
Hard and homogeneous rocks: diorite, quartzite, andesite, gneiss, granite	80 to 130 g/t
Hard but highly fractured rocks	100 to 150 g/t
Soft rocks: limestone	50 to 80 g/t
Very soft rocks: gypsum	30 to 50 g/t

It is advisable to match elements of impedance in order to judge the appropriateness of the explosive chosen for the type of rock. The impedance of the rock is the product of the seismic velocity (speed of compression waves) in the rock and the density.

$$IR = Vl \times Tr \quad (Vl \text{ in m/s, dr in g/cm}^3)$$

The impedance of the explosive is the product of the speed of detonation and the density of charge.

$$IE = Vd \times Tc \quad (Vd \text{ in m/s, dc in g/cm}^3)$$

The yield passes through a maximum when:

$$0.4 < \frac{IE}{IR} < 0.7$$

Earlier blasting experience and the characteristics of the deposit are also helpful in choosing the type of explosive, without forgetting the economic considerations (Table VII.5).

Table VII.5

Explosives	Detonation speed	Density	Strength	Priming	Shock sensitivity	Inflammability	Fly rock	Sensitivity to moisture	Sensitivity to frost
Dynamites	4000 to 8000	1.10 to 1.53	+	0	0	+	–	–	0
Nitrates	4400 to 4700	1.10 to 1.15	0	0	0	0	0	0	–
Bulk powder	3250 to 4000	0.78 to 0.95	–	–	–	0	0	+	–
Gels or slurries	3300 to 5600	1.20 to 1.50	+	–	–	–	+	–	0

+ high or substantial
0 moderate
– feeble or ineffective

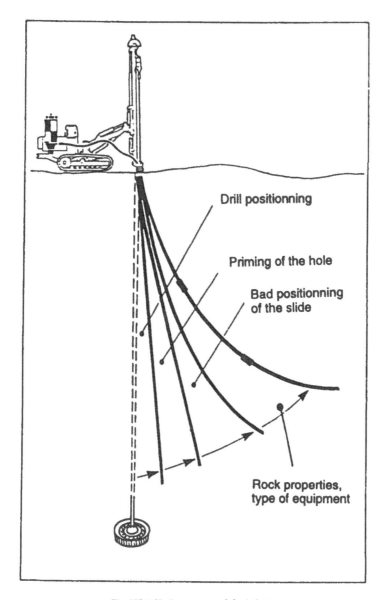

Fig. VII.4. Various causes of deviation

— The bottom charge (Lf) often consists of dynamits cartridges or equivalent gels;
— The median or column charge (Lm) is continuous with a less powerful explosive, preferably in bulk, such as nitrate-fuel, if absence of water permits, or discontinuous, stepped charges, which then necessitates explosives in cartridges (Table VII.6).

Table VII.6

	Examples of blasting schemes				
	Diorite	Trap rock	Quartzite	Soft limestone	Gypsum
Borehole pattern E × V	3.5 × 3	5 × 4	6 × 5	3.7 × 4	4.5 × 4.5
Rows (no.)	2	2	1	1	1
Height (m)	22	25	25	18	18.50
Inclination	6°	5°	16°	15°	0
Diameter (mm)	89	102	110	89	rotary 125
Bottom charge	F15 Ø 70 25 kg	Sofranex Ø 85 48 kg	F15 Ø 90 Symagel 505 45 kg	Gelaurite 2000 4 kg	Anfo cartridge
Median charge	N 135 52 kg	F15 84 kg stages	D7, F15 75 kg, 54 kg	D7 72 kg	Anfo 32 kg
Stemming	3 m	4 m	5 m	3 m	3.30 m
Blasting scheme	30 holes	15 holes	10 holes	16 holes	10 holes

Despite their interest from the economic viewpoint and their excellent density of charge, pumpable slurries are hindered in adoption due to the insufficiency of unit tonne blasted in most aggregate quarries. Moreover, the presence of fractures or faults in the deposit obliges the quarry manager to employ cartridged products. These bulk explosives are liable to freely diffuse inside the massif and even create discontinuities which are difficult to prime, which entails a major drawback of limiting the blasting efficiency and retaining unexploded explosives in the blasted products.

All quarries follow manual gravity charging of blast holes, that is, by free fall of the cartridges inside the hole or pouring in bulk. Each hole charged is then stopped by a stemming for a height practically equal to the bench width. Quite often the chippings from drilling are used for this purpose but increasingly a more granular material, 4/6 fine gravel for example, is used to facilitate some pervious-ness to the gases produced from the explosion and also to prevent a violent break-ing loose of the stemming material, without fracturing zones at the head of the hole.

3.4.2 PRIMING

The explosives are fired by priming, which until recently excluded the presence of a blasting cap inside the hole in the case of all deep mines of length exceeding 6 m. In this case priming is ensured by a detonating fuse attached to the first cartridge positioned at the bottom of the hole, after checking the exact depth of the hole. The detonating fuse itself is lighted by a blasting cap on the surface.

Use of an electric blasting cap is fairly common. Aggregate quarries which resort to lighting with a slow fuse are rare. Regulatory constraints had earlier prohibited placing the cap inside the hole but permission has now been granted for deep mines. Earlier exemptions permitted applying this new principle: an electrical or non-electrical cap, housed inside the cartridge, primes at the bottom of the hole, then the charge in the entire column, with two alternative safety provisions—either a detonating fuse along the hole primed at the top by a second

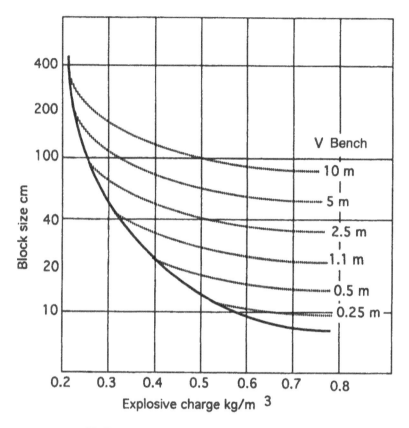

Fig. VII.5. Rock size distribution versus explosive charge

delay blasting cap or a second cap in the charge at the top before stemming, which also is of the delay type.

It appears that this mode of priming is currently coming into wider use in several quarries which have reported satisfaction in several respects:

— Better blasting efficiency: better control on rock size distribution and granulometry (Fig. VII.5).

— Cleaner fracture at the foot.

— No fly rock, less noise.

— Less ground vibration effects.

In order to facilitate the sequential break-up of each column to be blasted, a detonator with microdelay is quite often used. It minimises the nuisance from ground vibrations since each unit charge detonates separately. Depending on the position of the zone to be blasted on the working face, the lighting sequence is distributed around the central mine (5-4-3-2-1-0-1-2-3-4-5) or the bench to be blasted is swept across (0-1-2-3-4-5 ...) by breaking up towards the free end.

3.5 Result of Blasting

The fragmented materials collected after the blast for loading into trucks or dumpers, or direct feeding to the crushing machinery, enables assessment of blasting efficiency.

3.5.1 APPEARANCE OF WORKING FACE

A more or less clean fracture, cutting along the line of blast holes and the possible effects behind the blasting give an idea of the fragmentation. The resul is considered good if the stability of the rock face does not seem dubious: block: unstable and possible overhangs which would necessitate purging. These ob servations should lead to correction of the pattern of blasting—which is alway: possible.

3.5.2 PATTERN OF BROKEN-ROCK HEAPS

The kind of equipment generally used for collecting broken rocks is adapted to : greater or lesser extent to the results of blasting.

— The front-end loading shovel: the power of this type of equipment i: adapted to the heap of rocks dislodged from the working face but neverthele: heaped for the most part at the foot of the face. Lack of mobility does not permi sorting the blocks, which makes the operation more cost-intensive.

— The back-acting shovel: to be effective, the machine should be positionec above the quarry head level, and hence calls for creation of a platform above the heap. A back-acting shovel is perfectly suited for a relatively high heap since the machine proceeds by undermining, always working towards the top of the heap and by knocking off the top portions of the face if possible. In this manner o working the problem of sorting becomes even more cost-intensive even though the capacity of the bucket is duly matched to the maximum permissible rock sizes fo: the crusher.

— Pneumatic-tyred loader: first of all, the machinery is highly mobile bu much less powerful than the shovel. It is then preferable to spread the heap and reduce its height, so as to avoid the 'wall' effect which might result in : much too compact heap. The rock size distribution should be wide enough when working with the loader. The blocks are easily sorted and stocked bu there is no formal sizing, which could lead to disruptions in the feed to the primary crusher.

Whatever the type of equipment, after collection of the broken rocks it is im perative that the floor be properly sectionalised so that the quarry head is plane and the geometry neat for laying out future blastings.

3.5.3 ROCK SIZE DISTRIBUTION (BLOCKOMETRY)

The idea behind sorting the resultant rock fragments has already been men tioned. It should be emphasised, however, that apart from providing a possible source of blocks suitable for stone-bedding, the presence of these large ele ments, which enables assessing how well the blasting pattern is suited to the deposit, will warrant complementary operations, which add to the primary blasting cost:

— Fragmentation by blasting: when not forbidden, fragmentation by blasting with a shallow drill hole and placing of a cartridge becomes a highly expensive duplication of blasting, with noise and risk of fly rock besides.

— Fall of mass or 'drop-ball' is a simple technique which implies mobilisation of equipment adapted for this use and a specially skilful operator.

— Rock-breaking hydraulic hammer: where fragmentation by blasting is prohibited, a rock-breaking hammer remains as yet the sole solution capable of measuring up to the task, with an output adequate for the secondary breakage, despite the investment cost of the vehicle + hammer and the cost of running the unit.

But let it be noted that thorough control of fragmented rock size at the stage of blasting remains the quarry operator's foremost objective.

3.5.4 TRENDS

The list of objectives and constraints to be taken into account in the exploitation of massive rock quarries for aggregates throws into relief the importance of adapting the parameters of the blasting pattern to the characteristics of the deposit and the quarrying equipment. The processing installation permits a primary crushing station, depending on its own daily output. The maximum feed size is determined by the intake and screen mesh characteristics of the primary crusher. This data in turn helps in determining the size of the loading and transport machinery at the extraction end. At this stage matching height, mesh size, diameter and explosive becomes clear. Everything is fixed—almost. Recourse to hydraulic power in drilling equipment offers possibilities of drilling larger diameters: 102 or 110 mm are adopted without difficulty. This increase in diameter offers a whole range of perspectives on the change in mesh size, with a specific charge adjusted for the effect of the diameter. The importance of geometric perfection of the drilling contour over the positioning at the bottom of the blast hole prompts limiting the risk of deviation by confining one's self to heights around 15 m.

The increasing use of cartridged slurries, nitrate fuel emulsions, with a more favourable density of explosive charge, is confirmed by results comparable to that of using dynamites, but with a heightened safety factor.

All these trends do not necessarily lead to greater volumes of broken rocks, since acquisition of the site and advance of the working faces depend as much on the quality of the deposit and the production target. On the other hand, increase in quantities of specific (explosive) charges is an unmistakable trend for improving fragmentation from blasting and enhancing efficiency in the subsequent operations of collection, loading and crushing.

Quarry managers are also turning their attention towards new priming techniques. A detonator at the bottom of the blast hole is an example. More than one detonator inside the hole is also to be envisaged to overcome the discontinuities in the bench-slice to be fragmented. 'Non-electric' systems are available and with a well-adapted formation sequential priming is poised to make inroads into quarrying with a whole range of possible blasting programmes which will overcome the risks of vibrations.

4. LOADING AND TRANSPORT

4.1 State of the Problem

The broken rocks lying at the foot of the working face should be collected and transported to the site of primary crushing.

The preferred domain for mobile loading and transport equipment powered by diesel engines is that situated between the deposit and the crushing site, concomitant with minimising in every way possible the distance between the different points of extraction and the primary processing site. The objective thereafter is to derive the maximum benefit of the advantages of transport by conveyor belts between the primary crushing site and the stationary plants of final processing, the transport downstream of primary crushing being the essential domain of the conveyor belt.

Therefore the task is one of designing a processing infrastructure which makes it possible to take advantage of the merits of both these systems of handling using them judiciously in their complementary domains.

A stationary crusher calls for major and costly civil engineering works and hence does not permit remaining in step with the advance in quarrying over period. This results in a costly lengthening of the transport distances for portable machinery over the passing years.

A portable crusher moving quite frequently, sometimes daily, is very attractive with a single working face of several hundred metres length advancing by successive parallel bench-slices. The platform should be plane and quite firm to facilitate movement of the crusher and slippable belt conveyors situated downstream.

The mobile conveyor helps maintain a short transport distance with the broken rock heap. The crusher can be fed directly by a shovel if the distance always remains very small or, more commonly, by a tyre-mounted loader to ensure collection of the material and its transport.

A single machine fulfilling two or three functions thus minimises investment and consequently brings down operational and maintenance expenses vis-à-vis alternatives involving separate loading and transport machinery.

But then this solution is applicable only so long as the transport distance does not exceed 150 m. Beyond this the loader becomes overstrained and its efficiency drops to a very low level.

With more than one rock face exploited simultaneously at different levels and generally at several points, a balanced solution in a great many cases consists in having a semi-portable crusher whose shifting at more or less spaced predetermined points of time enables remaining in step with the progress in exploitation One can thus shift it regularly depending on the problems to be resolved. Therefore it is a technical compromise affording the possibility of periodically stationing the primary processing plant under the best conditions. Tipping the extracted material from one face to another, whenever possible, needs to be noted in this case. In effect, this solution enables taking advantage of gravity for transferring the materials to the lower level where the primary crusher is stationed, thus avoiding long distances of transport by dumpers.

It is worthwhile seeking the simplest and the most balanced solution in each individual case. The problem is to effect handling whatever the instantaneous production with equipment that is limited but well adapted to the actual needs and the inevitable periodic fluctuations in production dictated by the market and outlets.

While the number of machines should be sufficient, it is advisable to hold no more than what is actually needed. This avoids proliferation of separate loading and transport equipment which is more or less well maintained and in good working condition.

This consideration eliminates the accumulation of undersized, heterogeneous and oversized machinery in excess of the actual needs of exploitation.

A balanced choice of sizes and types of handling equipment facilitates a phased renewal of the fleet and also avoids abnormal and hazardous ageing.

A proper balance of sizing of the equipment inter se and with respect to exploitation needs is one of the essential factors for productivity. It makes for good conditions of operation and safety, permitting in addition smooth coping with fluctuations in production.

4.2 Loading Machinery

4.2.1 PNEUMATIC-TYRED LOADERS

By virtue of its very high mobility and flexibility in use, the pneumatic-tyred loader has become one of the commonly used collecting machines in quarries. It is a self-propelled loading machine equipped with a hoistable bucket in the front and an articulated chassis whose pivot is situated between two axles, permitting a steering angle of the axle to the right or the left, which is generally 35° (and sometime 45°).

The horizontal effort of penetration into the heap of broken rocks is made by the forward movement of the loader. The strike at the heap should be made with the front axle not steered so as to avoid overstraining the machine.

The capacity of the machine's bucket should be selected with due regard to the density of the loose materials. Overstraining the machine due to overloading should be avoided on the one hand, and the risks of unbalancing on the other. For these reasons, the payload should always remain at 50% of the limiting load of static equilibrium of the loader steered to the maximum. This has led manufacturers to design different bucket sizes essentially for the general run of density 1.6 to 1.8 and for that of lighter products of the order of 1.1.

The shape of the bucket is important and its geometry should be such as to facilitate its filling. There are two main types of buckets:

— Rock bucket with V-shaped cutter with or without teeth, a model specially destined for collecting heaps of broken rocks in quarries of massive rocks.

— Straight-edged bucket, often designed under the name of sand bucket, suited to collecting fine materials, that is, of smaller particle size.

The bucket which suits quarries of massive rocks is the rock type one with a V-shaped cutter that facilitates penetration into the heap, and size for a density of 1.6 to 1.8 of the loose material. The presence of teeth improves penetration but in the last-mentioned case it is particularly advisable to guard against machine

operators using the loader for clearing rocks to avoid the inevitable serious an
rapid deterioration resulting from lever forces exerted on the machine.

Development in loaders has meant development and improvement in tubeles
pneumatic tyres (without chamber) for civil engineering applications, incorporatin
a radial frame provided with tread bands having a thick anti-abrasive rubber laye:
These pneumatic tyres have appropriate treads for imparting a satisfactory grip
depending on the nature of the ground, and ensuring a long service life of sever;
thousand hours (generally 4000 to 5000 h).

Correct size of the tyres with respect to the characteristics of the machine a
also a regular check on inflation pressures during cold weather are of paramour
importance. Good durability of the tyres is additionally dependent on corre(
operation of the machine, especially avoiding a backward skid while striking th
heap of broken rocks.

The preferred area for loaders is quarry platforms where the ground is firr
and dry. Indeed the presence of hard and sharp elements in water is the greate:
enemy of tyres. That is why it is recommended that working platforms b
thoroughly drained, with slopes for water removal and prevention of formatio:
of pools at the foot of the working face.

The width of the bucket should be slightly larger than the outer edge-to-edg
spacing of the tyres so as to make way for the wheels and to protect the flanks (
the tyres at the time of striking the heap.

In some special cases, owing to particular problems of grip on highly slipper
grounds or works carried out continually upon grounds carrying hard and shar]
elements in the water, quarry operators find it necessary either to encase the tyre
in chains or to make use of special tyres inlaid with steel spikes (studded tyre:
example Beadless, Caterpillar). Actually, this is just a makeshift solution sinc
chains or studs entail too strong a grip (on the ground) at the time of return t
strike the heaps again, and give rise to constant multiple and dangerous vibration
that exacerbate strain on the machine.

Earlier, when no special tyres existed for heavy-duty loaders, which otherwis
had satisfactory durability, it was necessary to protect them by using chains. Th
development of excellent civil engineering tyres incorporating a high-quality fram
to fulfil the service conditions of heavy-duty loaders has enabled elimination c
chains and consequently considerably improved the service life of these machine
while significantly reducing their maintenance costs.

The choice of loader size should be made with due regard to that of th
dumpers and tippers used. In principle, a good balance is secured by loading th
dumper or tipper in four buckets. The maximum hoisting height of the loader'
arms should permit discharging the bucket inclined at 45° into the dumper tul
without problem.

In loading 40 to 50 t dumpers one after the other at a uniform rhythm, a 550(
litre loader of 375 to 420 hp assures an industrial average handling rate of 380 t
400 t/h whereas in loading-cum-transport, it gives an industrial average handlin;
rate of 280 to 300 t/h for a 100 m haul on a proper haulage track.

In loading-cum-transport it is advisable to choose most particularly loader
which are highly rugged and reliable, sufficiently powerful and amply dimensione(

for the volume of production expected. One should avoid the pitfall of going in for undersized equipment and be on guard against loaders of brilliant performance but limited sturdiness, equipped with buckets which are overdimensioned in relation to the mechanical properties of the machinery.

Indeed to cope with recurring instances of non-availability of loaders resulting from their rapid deterioration, undertakings are constrained to increase the number of such loaders of inadequate characteristics in order to achieve their production targets. The outcome is an encumbering accumulation of an odd assortment of ill-suited machines, a situation which then becomes very difficult to rectify considering the magnitude of the investments involved.

The normal service life of a pneumatic-tyred loader that is well suited to the job, of rugged quality, and maintained normally and systematically, is 20,000 h in mainline service with an average availability of 90% excluding programmed preventive maintenance operation during its service life. After 20,000 h, it can still log 5000 to 6000 h as a relief and standby machine.

For 15,000 h working, the total expenses on mechanical maintenance (excluding tyres and lubricants) of a 4200 to 11,000 litre loader, of size well suited for its use, normally range between 80 and 90% of its value at acquisition when new; this, of course, is subject to proper operation of the machine and normal preventive maintenance to avoid mechanical breakages and abnormal deterioration.

Loaders in the capacity range between 5 to 5.5 m^3 (370 to 450 hp motor) and 9 to 10 m^3 (690 to 730 hp motor) are most common in quarries of massive rocks.

The smallest loaders with the minimal characteristics to normally cope with the operating conditions in quarries of massive rocks are those of 4 m^3 (270 to 300 hp motor).

When the hoisting height of the loader is found to be inadequate for loading some transport vehicles, recourse is occasionally had to extending the arms. In this case it is prudent to reduce the bucket capacity and thereby save the engine from overstrain. Actually, though, this is only a makeshift.

Without detailing all the accessories which are needed, it is relevant nevertheless to recapitulate some important points:
— Safety belts
— Suspended seat, adjustable to the driver's weight
— Instructions for help in emergencies
— Reversing signal (audio signal and reversing lights that come on automatically when reversing)
— Lighting and signalling devices
— Safety device to prevent starting up while in gear
— Breaking devices (manoeuvring, parking, emergency)
— External soundproofing
— Internal soundproofing within the cabin
— Front and rear windscreen wipers and washing attachments
— Heating-de-icing (and air-conditioning if need be)
Note: Great attention should be paid to the quality of the flexible shafts when replacing them, to prevent their breakage in service.

4.2.2 HYDRAULIC SHOVELS

A shovel is an excavating machine made up of a supporting chassis (mounted o tyres or caterpillars) and an orientable turntable for rotation on this chassis. Th turntable comprises an engine, driver's cabin and arms or boom with a bucke attached. A hydraulic shovel is a shovel with hydraulically controlled movemen

The hydraulic pumps used on the latest generations of shovels are of variabl flow with electronic regulation, which permits great flexibility in operation whil keeping the maximum operating pressure within 300 to 350 bars in most case: The idea behind limiting the pressure is mainly to obtain a good durability of th flexible shafts.

The hydraulic shovels used in quarries of massive rocks are generally mounte on caterpillars. The width of the tracks is selected depending on the nature of th ground. Given hard rock, it is advisable to go for narrow tracks since these enta less strain on the other elements of the bogie compared to broad tracks. Indee wide tracks are meant primarily to improve bearing on soft soil.

The translatory movement is obtained by an independent hydraulic drive c each caterpillar made up of a hydraulic motor and reducer.

As with pneumatic-tyred loaders, the increasing use of hydraulic shovels i open-cast mines and quarries is attributable to the appearance on the market c rugged and reliable machines of adequate size and power answering to the opera tional needs and conditions.

A caterpillar-mounted hydraulic shovel is attractive in situations entailing onl limited movements. Indeed, in jobs requiring sizeable and frequent translator movements, the caterpillar gear gets overworked, resulting in high maintenanc costs. In other words, loaders and shovels have their own preferred areas of ap plication, there being, of course, borderline areas wherein either could be adopte non-exclusively.

If the output of a pneumatic-tyred loader is governed by the size of its bucke since the machine has to manoeuvre to load the dumpers or trailer dumpers, thi is not true of a shovel. In effect, because of its turntable, the shovel pivots rapidl into place without having to manoeuvre like the pneumatic-tyred loader. So muc so that its output depends essentially on how rapidly the bucket can penetrate th heap of broken rocks. This penetration is facilitated by the fact that the shove works with narrow buckets and sometimes even very narrow ones, whereas th bucket of a pneumatic-tyred loader has to be very broad since it should be broade than the machine proper. This explains why a 700 hp front-end loading shove having a 5.50 m^3 bucket is found to give a decidedly higher output than a 6.5 m bucket. Likewise, a 525 hp back-acting shovel with a 4.3 m^3 bucket 1900 mm wid has a positively higher production than that obtained with a 5.5 m^3 bucket 230 mm wide.

Thus two types of machine exist: back-acting and front-end loading.

• *Back-acting equipment*

Loading the bucket is done by drawing it towards the supporting chassis. Thi technique is particularly attractive for extracting soft overburden materials an loading the dumpers meant for carrying away the wastes.

To collect the broken rocks, the shovel should be mounted above them in order to load the dumpers. This invariably poses a certain number of problems considering the risk of slopes sliding etc.

This technique arises from the fact that the initial kinematics of the first-generation hydraulic shovels was designed for back-acting work.

• *Front-end loading equipment*

Loading the bucket is done by pushing it towards the material.

Initially the front-end loading equipment derived directly from an imitation of that employed for draglines, incorporating technical compromises in the various features of the back-acting equipment. This resulted in fatigue arising from difficulties in penetration and also in mediocre efficiency.

To remedy these problems, manufacturers have now updated the kinematics. The main improvements consist in achieving a horizontal thrust of the bucket with its level being automatically maintained, and in increasing the forces of penetration and undermining, as also those of lifting of the boom.

In addition, the manufacturers have been striving to secure the bucket's angle at hoisting to be maintained constant to the utmost possible extent. To improve the output, several machines are nowadays equipped with devices to ensure rapid descent of the bucket.

The introduction on the market of powerful and rugged hydraulic shovels designed with kinematics well adapted to front-end loading has led to an expansion in their use for the collection of broken rocks. Thanks to the new designs, these machines enable maintaining a horizontal ground surface.

There are two major types of front-end loading buckets: the trap bucket discharging by the bottom (Fig. VII.6) and the front-discharging bucket. The trap bucket helps contain the size of the material within limits at loading by eliminating big blocks. It affords the possibility of greater precision in loading the dumpers. The trap bucket loads itself like the front-discharging bucket and can also be used as a front-discharging bucket. While it is naturally attractive in the case of sticky materials, it costs more and entails higher maintenance expenditure than a front-discharging one.

As in the front-end loaders, in the back-acting ones also the manufacturers offer a range of three bucket sizes referenced to the densities of loose materials:

— mineral density 2.6,
— common density 1.8,
— coal density 1.1.

In quarries of massive rocks the size normally adopted is 1.8 density.

The shovels currently on the market can be delivered either in the front-end loading version or the back-acting version (Fig. VII.7).

The most commonly used shovels for loading 50 to 55 t dumpers for massive rocks are:

— 500 to 550 hp shovels weighing 90 t,
— 710 to 750 hp shovels weighing 110 to 120 t.

The smallest shovels used in quarries of massive rocks are normally 400 to 450 hp weighing 70 to 80 t.

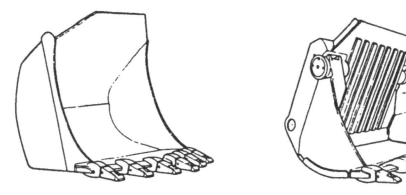

Fig. VII.6. Trap bucket discharging by the bottom (Left); Front-discharging bucket (Right)

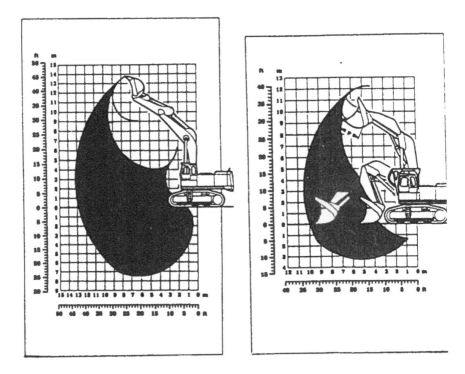

Fig. VII.7. Example of back-acting and front-end loading 525 hp hydraulic shovel machines weighing 90 t

Shovels of 1000 hp weighing 180 to 200 t are to be found in quarries using dumpers of 80 to 100 t payload.

The motors of shovels are generally diesel-fuelled except in certain special cases where it is possible to use electric motors.

Further, it should be noted that if the loading machine always works on ground covered with water, it is preferable to use a shovel on caterpillars rather than a pneumatic-tyred one.

4.2.3 DRAGLINES

These are mechanical shovels in which the movements of the bucket are controlled by ropes.

The machine works by front-end loading and comprises a supporting chassis mounted on caterpillars and a turntable which can be oriented by rotation on this chassis, which also comprises motor winches and the driver's cabin.

The kinematics of the bucket comprises (Fig. VII.8) a fixed boom whose inclination can be regulated, and which transfers to the chassis the forces of the head and a mobile arm supporting the bucket.

The bucket is pulled upwards in a curved path. Discharge is effected by opening the bottom.

Draglines were widely employed in quarries between 1950 and 1965. After 1965 they disappeared, yielding place to pneumatic-tyred loaders and hydraulic shovels which are better adapted to traditional exploitation.

The largest hydraulic shovels reach a weight of 250 to 280 t with 1200 to 1400 hp nominal power and are equipped with front-end loading buckets of 13 to 14 m^3 for a density of 1.8, or 18 to 20 m^3 for a density of 1.1 (coal).

But in these large sizes electric-powered draglines are back in favour everywhere and an entire series of models weighing 200, 250, 350, 500, 700 and 900 t is available, destined essentially for exploitation of huge open-cast mines.

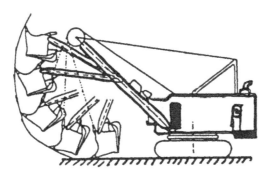

Fig. VII.8. Dragline. Kinematics of the bucket

4.3 Transport Equipment

4.3.1 DUMPER

These are special trucks with a rigid chassis particularly designed exclusively for the quarry, provided with a tipping bucket, and are extensively used in quarries of massive rocks.

The dumper comprises a robust chassis resting on the steering wheels in the front and on a single axle in the rear, which is the driving axle, with twin wheels A guard iron is fixed between each of the twin wheels.

Good durability of the tyres is directly attributable to the use of adequately dimensioned casings (example: 2400 × 35 for 50 t dumpers). The choice of rubber and tread depends on the nature of the ground, its condition and work conditions (distances, speeds and loads).

There are essentially two kinds of rubber: one meant to withstand abrasion cuts and tearing-off (case with most massive rock quarries), the other envisaged for solving the problems of heating up in the case of fast transport over long distances, especially on concrete and asphalt roads.

The kind of tread is one of the important aspects as it makes for a satisfactory grip, preventing skids.

The development and improvement in tyres with radial tubeless frame by leading world manufacturers has made available excellent casings. The latest advances made have resulted in new tread types in order to eliminate, for instance the phenomena of vibrations caused by some profiles of wearing courses on hard ground. These improvements relate also to the tyres of loaders working in loading-transport on hard ground where one occasionally meets with vibration phenomena resulting from the pattern of tyre treads.

The inflation pressures are decided for each size and type of tyre depending on the maximum axle load.

In the choice of type of tyres and monitoring their soundness and wear manufacturers mostly go by the principle of kilometre-tonne per hour in the case of dumpers.

As with pneumatic-tyred loaders, regular checking of inflation pressures during cold weather is crucial.

It needs to be noted that civil engineering treads with hard anti-abrasive rubber are specially designed for running on tracks (and not on asphalt or concrete roads).

Hard and sharp elements present in water being the greatest enemy of tyres it is important to carry out a proper maintenance of the tracks with profiles which enable water drainage.

The size of the dumpers to be adopted in each case depends on the hourly tonnages to be transported and the distances to cover. This involves a study of the cycles. One full cycle comprises:
— time for loading,
— time for the outward trip,
— time for manoeuvre and waiting at the tipping site,
— time for discharge,
— time for the return trip,
— time for manoeuvre and waiting at loading.

The turn-around of the dumpers should be smooth and without an idle period, which results especially from inordinate delays at loading or discharge.

In other words, the size of the dumper should be chosen to match that of the loading machine and appropriate to the capabilities of reception and tipping. The loading machines, transport machines and primary crusher should be well balanced as a group.

Properly laid out and maintained tracks make for uniform cycles.

The size of the dumpers most commonly used in massive rock quarries are:
— 25 to 28 t capacity dumpers of 350 hp,
— 35 to 38 t capacity dumpers of 450 hp,
— 50 to 55 t capacity dumpers of 650 hp,
— 75 to 80 t capacity dumpers of 850 hp.

It may be noted that the 50 to 55 t capacity dumber is the one most used in massive rock quarries of 500,000 to 2,000,000 t/year output.

Where massive handling is involved, as for the removal of several tens of millions tonnes of wastes for the stripping of overburden of mineral deposits, the undertakings in question find it necessary to employ 900 to 1300 hp dumpers of capacity 110 to 130 t, with 15 to 23 m^3 shovels, or very large 1100 hp loaders of capacity 16 m^3.

The dumpers are rugged and reliable machines capable of working 32,000 to 40,000 h as mainline equipment, subject of course to systematic and normal preventive maintenance.

The total expenses on mechanical maintenance (excluding tyres and lubricants) of dumpers for 30,000 h work are of the order of 90% of their value when new.

These machines are equally suited for climbing as well as descending with load; in the latter case it is imperative for the machine to incorporate an effective hydraulic speed governor on the gearbox.

Lastly, the total weight of a 50 t capacity dumper with load is of the order of 90 t and that of an 80 t dumper 135 t. This remark underscores the magnitude of the rolling mass to be managed while descending with load.

The ideal arrangement would be to equip the dumper with oil-immersed multidisc brakes and a powerful hydraulic speed governor in the gearbox so as to provide two totally independent braking devices to take care of any defect. The magnitude of the moving load makes it imperative to observe this basic safety criterion in "downward slopes" with load.

Manufacturers have been striving to improve the control and operation of gearboxes by means of electrohydraulic devices which combine electronic units of control with hydraulic components destined to carry out the necessary operations as sensed by these components.

Without detailing all the other accessories needed, it is nevertheless appropriate to mention a few points:
— Emergency instructions
— Lighting (headlight, dipped headlights, rear lights)
— Signalling change in direction (from and rear blinkers)
— Reverse lights coming on automatically when reversing the machine
— Audio reverse signalling
— Safety start, which prevents starting the engine while in gear
— Running brakes
— Emergency brakes
— Hand brakes
— External soundproofing
— Internal soundproofing inside cabin
— Strengthened safety guard for cabin
— Safety at lowering the tipping bucket (signalling)

— Safety at raising the tipping bucket to prevent accidental drawing out of the lifting jacks

— Rock tipping bucket which is strengthened and sheet-lined to prevent its damage by the fall of blocks at loading

— Seat with shock absorbers adjustable in position and adjustable to the driver's weight

— Passenger seat

— Safety belts.

Note: Dumpers up to 100 t carrying capacity are now equipped with mechanical transmission, following the development of excellent gearboxes answering the requirements in question. Electrical transmission becomes attractive beyond 100 t carrying capacity. A diesel engine rotating at a fixed regime has a dynamo coupled to it which powers the electric motors placed in the wheels. But mention should be made of the competition from mechanical transmissions witnessed these days since the availability of torque converters enables equipping dumpers of 130 t carrying capacity. Besides, if dumpers with mechanical transmission are well suited for running on level ground, they are often ill-adapted for climbing gradients with load.

4.3.2 TRAILER DUMPERS

The trailer dumper is a mobile transport machine having a coupling between the tractor and the tipper. Its favoured area is not the larger quarry of massive rocks but it can offer an attractive solution in special cases (stripping of overburden, small temporary quarries serving specific road, highway or bridge structures).

Trailer dumpers are versatile transport equipment specially designed to be able to run on bad ground which might be soft and pitted to a greater or lesser extent. Their turning circle is generally 10 to 20% smaller than that for dumper trucks of the same capacity.

A manufacturer designed a 4×4 trailer dumper of 22 t with a special provision of two transverse wheels which could be lifted and lowered by means of two hydraulic jacks to enable making a U-turn within a 10 m space.

An entire series of trailer dumpers is available on the market, especially between 22 and 32-t carrying capacity with 2 axles of 4×4 and 3 axles with either 6×6 or 6×4.

Trailer dumpers with 3 axles of 6×6 and 36 and 50 t carrying capacity have just recently been introduced on the market.

The high-tonnage trailer dumper is not meant to compete with dumper trucks of the same capacity running on good tracks but rather to tackle transport on bad ground. Actually, since trailer dumpers are designed with a low unit axle load and have low-pressure tyres, their wheels sink less into soft ground in view of the decrease in pressure on the ground.

Since a 3-axle trailer dumper consequently has a smaller load per axle for the same tonnage transported, it permits transport over roads whose maximum permissible load is limited because of bridges, which a 2-axle dumper truck would not be allowed to use.

Trailer dumpers being narrower as a general rule by 20% compared to dumper trucks of the same capacity and of lower height and about 20% longer, solved the problems of transport in cramped sites and also in a few underground galleries.

Hence the trailer dumper is essentially a machine suited to the needs of earthwork jobs. This explains why a count in France in 1988 put the number of equipment of this type in service at 2000 against 600 dumper trucks of 25 to 50 t capacity sold in the preceding 12 years.

Another count showed 900 dumper trucks of 32 to 80 t capacity and 450 trailer dumpers of 22 to 50-t capacity.

If the trailer dumper is attractive for a certain number of application, it is profitable only for relatively short transport distances not exceeding 1500 to 2000 metres on average.

Considering their design, trailer dumpers are less suited than dumper trucks for descending with load, especially on slippery roads or tracks. Indeed, in this case there is no small risk of disaligning the coupled machines to assume a V-shape under the push of the tipper's weight with its load on the tractor in front.

Likewise, considering the tipper's smaller width in relation to its lifting height and length, it is advisable, at the time of discharging, to ensure an adequate strength and planeness of ground to prevent the risk of lateral overturning of trailer dumpers.

In other words, dumper trucks as well as trailer dumpers have their own specific areas of use, not excluding a border zone where either of the two could be adopted.

4.3.3 OTHER MODES OF TRANSPORT

Three-axle rear-dump trucks (4 × 4 to 6 × 4 to 6 × 6) employed in some small-scale exploitations.

Special 8 × 8 vehicles of standard road width with 20 m^3 quarry tipper

Dumpers with lateral or bottom discharge hitched to roadway tractors.

5. OPERATION AND MAINTENANCE

5.1 Operating Conditions—Tracks—Platforms

A judicious choice of the resources deployed for ensuring the best possible conditions for handling material constitutes the starting point of prevention.

Organisation of material-handling jobs should be done with unceasing care to secure the best possible working conditions so as to minimise the physical fatigue and nervous tension of the operating personnel.

It is imperative that the jobs be performed at the most uniform rhythm possible, and flexibly besides.

The conditions of the platforms and tracks, as well as their layout, are important factors for obtaining good working conditions and hence safety.

To this end quarry platforms should be well maintained with gentle slopes and drains to evacuate the water and it is worthwhile to check the levels with a telescope.

Use of a service bulldozer and periodic rolling with a blade grader are particularly recommended in this regard.

The layout of the tracks for movement of the dumpers should be particularly well examined side by side with minimising the slopes. Naturally, there should never be a slope greater than 20% and as far as possible, it is desirable not to exceed 8 to 10% in a straight line and 5 to 6% in curves.

It is advisable to try for the largest possible radii of bend and these should be at least equal to the magnitude of the steering diameter of the dumpers and trucks. (A steering diameter of 20 m dictates a 20 m minimum radius of bend.)

The tracks should be as far away as possible from the foot of the walls and slopes. The order of the Ministry of Industry and Research, dated 13 February 1984, gives directions in this regard to French quarry managers which are highly profitable to observe.

The profile of the tracks should permit water drainage.

It is pertinent to provide for a gentle superelevation of the bends, which should not, however, exceed 3% so that the machines do not skid sideways in periods of freezing or slush.

It is recommended for safety that whenever possible a counterslope be constructed in stone opposite the descent before each bend to stop a vehicle in distress.

It is equally essential to install a prominent and reflecting signal reminder of the speeds to be observed, and the obligatory engagement of speed governors on loaded dumpers during descent along the tracks.

It is advisable to provide for easy cross-traffic of the material-handling machines and protection of pedestrians.

A judicious alignment of the tracks is important since one can have good tracks laid out and executed properly only as long as one does not have to constantly repair them.

Lastly, the track networks should be so planned as to minimise dust emissions to the utmost. Certainly, doing away with any visible dust emission whatsoever is improbable as quarrying could otherwise only be carried out in rainy countries. However, a balance between what is feasible and what is not is possible with periodic spraying of water on the tracks.

For working by night, it is imperative that the mobile loading and transport equipment be equipped with powerful lights whose characteristics are specified for French quarry operators in the order of the Ministry of Industry and Research, dated 13 February 1984.

At the points of discharge into the hoppers, such as those feeding the primary crushers, the movement area should be conscientiously planned and executed with effective lighting.

Special attention should be paid to comfort and sound-proofing of the cabins of loaders, shovels and dumpers. The seats must be of the suspension type, adjustable to the operator's weight. It should naturally not be forgotten to periodically replace the seats since after a certain period of service, their condition deteriorates.

Another important factor of comfort is heating and air-conditioning equipment for machine cabins.

With regard to material-handling machines such as loaders equipped with ROPS safety cabins for protection in the even of overturning, it is crucial that operators fasten their safety belts. It is difficult to ascertain whether this injunction is observed or not.

Wearing a safety helmet, though accepted without demur by pedestrians, is generally disregarded by operators of material-handling machines.

Other Points

• *Fire safety*

It is pertinent to recall the need for equipping mobile loading and transport equipment with versatile ABC powder industrial fire extinguishers capable of acting upon class A fires (fires of solids, such as rubber, plastics, textiles), class B fires (hydrocarbons) and class C fires (gas), in conformance with the standard specifications NF. S61915 of October 1966 and NF. S61900 of December 1984.

The extinguishers for use on board the machines should be of a reliable and very rugged model, resistant to vibrations, and should be designed for such use in job sites exposed to inclement weather conditions.

It is recommended that an extinguisher with 2 kilos ABC powder be installed inside the cabin and one or two extinguishers with 4 to 6 kilos ABC powder on the outside of the machine at readily accessible points.

A periodic check of the extinguishers (at least once a year) is advisable.

• *Wireless telephone link*

Wireless telephone devices that enable constant contact between the operators of machines and the quarry controller are increasing in use. Equipment of high quality, rugged and reliable, designed for fitting on board civil engineering machines and capable of withstanding vibrations is needed.

5.2 Maintenance—Training—Safety

A judicious choice of the material-handling machines facilitates maintenance operations.

Recruitment of personnel who have the physical fitness requisite for operating these machines is absolutely mandatory. To entrust the running to such machines to persons who lack the necessary physical attributes is foolhardy.

Furthermore, the operator should possess the basic knowledge indispensable for his job.

Thus, driving dumpers weighing 80 to 100 t when loaded and running at 50 km/h on tracks in quarries is not usually entrusted to persons who have not passed the driving test for heavy-duty road trucks of 3.5 t total weight.

That is why French legislation, by an order dated 13 February 1984 was obliged to define precise rules pertinent to this subject.

Public pressure also obliged mine managers in France to explicitly state and publicise at each site rules/instructions for the concerned personnel, to wit:

— Rules for monitoring maintenance and performance of material-handling machinery.

— Rules for maintenance of haulage tracks in quarries and mines.

— Rules concerning operation of the machines.

Obviously, drawing up a list of rules/regulations and handing it over to the concerned personnel would hardly achieve the objective. It is equally necessary to make the personnel learn them and subsequently to monitor their strict application.

Such monitoring calls for day-to-day, constant action on the part of those responsible to prevent omissions, errors etc. It is a thankless task of unending reminders but is the sole means of preventing accidents that could arise from carelessness and nonchalance.

It should also be recognised that the responsible personnel are subjected to pressures of every kind, especially when they are overly vigilant in this regard vis-à-vis their workforce.

Depending on individual mentalities, the operators are wont to bypass regulations. Automobile driving in some countries is proof thereof, if such is needed.

So to keep a watch over the observance of safety and prevention rules, the manager in open-cast mines and quarries has to be a competent person invested with the necessary powers of authority.

To clarify the situation regarding this subject, order 80-331 of 7 May 1980, supplemented by order 84-993 of 22 October 1984, has made it obligatory for every mine operator in France to designate officially and by name the physical person placed in charge of the technical direction of the works in quarries and mines.

In other words, it is absolutely indispensable for the personnel to be specially qualified and aware of the precautions to be taken under the safety rules, which must be imperatively respected.

The number and the qualifications of the officers-in-charge of the planning, direction, execution and supervision of the works should measure up to the nature and the extent of the works.

Likewise, maintenance personnel should possess the knowledge necessary to keep track of the condition of the machines under their charge and timely rectify any possible faults to avoid the risk of accidents.

They must therefore be qualified and know how to determine the limits of their intervention in preventing dangerous errors from being committed in repairing and setting operations. In other words, the officer-in-charge must exercise his judgement in sending for the manufacturer's after-sales service specialists when restoration of the operation of the machine lies beyond his means.

Pottering is always dangerous in this department.

What is more, the maintenance of loading and transport machinery should be done in a planned manner, with due regard to the timetable of actions indicated by the fabricator whose technical assistance should be sought for delicate checks.

For their part, the operators should be watchful in maintaining their machines in sound working condition and help in the routine daily and weekly checks as also in the regular cleaning of the equipment.

Each operator should be interested in having a machine that is clean and in sound working condition. In other words, if it is obviously indispensable for the operators to be versatile to ensure the progress of exploitation, contrarily commonality of use of the machines by several crews wherein no one feels responsible should be avoided.

Remark regarding diesel engines

According to the studies conducted by large petroleum companies, the diesel engine per se accounts for one-third of the maintenance and repair costs (excluding tyres) of loading and transport equipment.

Monitoring the condition of diesel engines, in particular by the methodology of spectrometric analyses of oils to follow the progressive presence of fine metallic particles, testing dilution, detecting the presence of water and confirming the silica content, has proven extremely useful. This method permits intervention neither too early nor too late in most cases of progressive wear and prevents unscheduled dismantlings, or premature or overdelayed general preventive overhauls.

Good machinery, a sound organisation, judicious choice of exploitation methods, well-laid-out haulage tracks, maintenance in accordance with the rules of the art, and personnel who are sincere, competent, qualified and conscientious, in addition to a sound environment—these are the essential factors for productivity, profitability, and prevention and safeguarding against accidents and injuries.

REFERENCES*

[+]Rock Blasting. Langefors and Kulstrom.
Practical Manual on Works with Explosive. CEFICEM.
Rock embankment—blasting with explosive. *Revue Générale des Routes*, no. 593.
'Blasting'. Mineral Industry Company pamphlet.

---

*All entries in French except those marked with a cross (+)—General Editor.

VIII

Stripping, Extraction, Loading and Transport in the Exploitation of Sands and Gravels Deposits

Alain Prax

After touching a peak figure of nearly 500 million tonnes in 1991, the output of quarry products in France stabilised around 450 Mt in 1994.

Aggregates accounted for a major portion of this output, more than 80%, say 370 Mt, *of which sands and gravels constituted 186 Mt.*

Numbering around 2400 today, and relatively widely dispersed throughout the national territory, sands and gravels deposits under exploitation in France for the production of aggregates display a very wide range of intrinsic properties aside from conditions of the deposits.

The methods of working employed on these sites cover almost every type likely to be encountered in open-cast exploitation of 'sands and gravels deposits'.

We propose to draw up a list of these methods and to describe the situations in which particular ones are preferred, after a rapid preliminary survey directed towards delineating the specific character of the workings of French sands and gravels deposits.

1. SPECIFIC CHARACTER OF EXPLOITATIONS

Sands and gravels accounted for nearly 50% of the total production of aggregates in 1994[1] at the national level. But this ratio is subject to considerable variations at the regional or departmental level, obviously depending on the local geological contexts, which are especially varied in our country.

---

[1]This ratio has slowly declined for some ten years now because the 'exploitable' resources are rapidly becoming scarce in some regions.

1.1 The Deposits

The most commonly exploited resources are sands and gravels formations (old Quaternary sediments), widely dispersed in the middle and lower valleys of the principal rivers (Rhine, Rhône, Loire, Garonne, Seine) and some of their tributaries (Moselle, Saône, Durance, Allier, Vienne, Marne, Eure). But these are also glacial or fluvioglacial formations, which are highly evolved in the Alps, Pyrenees and Vosges as well as the Lyons region.

The marine formations of the continental shelf, again of fluviatile origin for the most part, are even today still least in demand.[2]

The intrinsic characteristics and the situational conditions of the deposits of these different formations quite clearly exert a decisive influence on the operations of extraction.

But one should not forget to include in them the land and environmental constraints which, though practically unknown in the field of mines, are *often fraught with heavy consequences* in the field of quarries.

1.1.1 INTRINSIC CHARACTERISTICS

• *D_{max} (mean maximum diameter)*

This varies from 100 to 200 mm in most cases but exceeds these values (> 200 mm) in glacial or fluvioglacial formations and in the torrential sediments of mountain zones.

On the contrary, it does not exceed 20 mm in the 'limestone soils' (Seine upstream, Aube, Meuse, upper valley of the Marne). ᐧ

• *Grading*

This has a direct effect on the permeability of in-situ alluvia and the rapidity with which water drains off it during extraction (submerged formations).

The *water evacuation technique of working,*[3] inconceivable in clean and coarse formations, on the contrary has proven perfectly viable in other formations (workings in the lower valley of the Marne, Oise and Loing), though subject to increasing regulatory constraints (protection of groundwater table).

Use of large-scale water-evacuating machinery thus becomes inescapable, unless too onerous, in the case of submerged or partially submerged deposits, which

---

[2]Main marine aggregate quarries (extraction at depths of 20 to 30 m maximum):
—estuaries of the Seine, Loire and Gironde,
—rias on the coast of Brittany,
—estimated level of production in 1994: 3 Mt.
[3]Lowering the entire water table: requires maximum permeabilities of the order of 10^{-3} to 10^{-4} m/s.

are relatively rich in sands, when one has recourse to transport by belt conveyor for the extracted pit run (workings along the Lower Seine).

• Petrographic composition

The *free silica* content (quartz, flint, ...) of a formation, besides its in-situ compact ness, contributes directly to the *abrasiveness of the materials to be extracted*.

Most sands and gravels formations are polygenic (silico-calcareous) and hence behave in a particularly variable manner. However, there exist at the two extremes formations which are entirely siliceous (100% free silica: Eure Loire, ...) or entirely calcareous (Seine upstream, Aube, Tille, Ain, Drome, high valley of the Doubs ...).

1.1.2 CONDITIONS OF DEPOSIT

• Thickness

Most formations under exploitation scarcely ever exceed 10 to 12 m bed thickness (6 to 7 m on average).

Formation of the Rhine valley, with exploitable bed thicknesses of 30 to 60 m are an exception to this rule. So are the glacial and fluvioglacial formations (Lyon and Alps regions: thicknesses = 20 to 50 m or more).

• Groundwater table

Nearly 70% of the sands and gravels formations exploited in France are submerged at least in part, under the water table.

Depending on the relative positions on the water table, the top and bottom (bedrock) of the deposit, three cases can be delineated:
— 30%, *'Surface' deposit*: the level of the water table is below the bottom o the deposit (can be worked 'dry');
— 70%, *'submerged' deposit*: the level of the water table is above the top of the deposit (can be worked 'wet');
'partially submerged' deposit: the level of the water table is midway between those of the top and bottom of the deposit (generally workable in two distinc phases, 'dry,' then 'wet').

• Grain size heterogeneities

These are invariably conspicuous in the vertical plane, the base of the formation being coarser than the top.

On the contrary, lateral heterogeneities are not always sharp, *with the exception of meander deposits* (Lower Seine).

• Intercalated clays

This is a phenomenon frequently encountered in a number of formations (Marne Yonne, Seine, Rhine glacial moraines...). In most cases it constitutes a major nuisance in extraction operations.

• Hardened crusts and erratic blocks

Relatively compact, even hardened crusts (conglomerates calcretes) are frequent in the oldest sands and gravels formations (the Crau, upper terraces of the Rhône...). They constitute a major, even forbidding nuisance in extraction operations, which can sometimes be mitigated by a premining operation.

Nor are erratic blocks, frequent in glacial or fluvioglacial formations, free from nuisance problems.

• Morphology of the top and bottom of the deposits

These are essentially variable, even at the regional level, and often incompatible with some methods of extraction, in the same way as intercalated clays and hardened crusts.

Which is why *preliminary exploration reconnaissance* is mandatory for any working (see Chapter III).

1.2 Size of Exploitations

Having marked out the principal zones of production of alluvial aggregates (Fig. VIII.1), it is in order to proceed to the workings represented by the 2400 quarries under exploitation in 1994 over the national territory. What are their sizes and how are they distributed at the national and departmental levels?

Table VIII.1, based on data collected from the Technical Service of Quarries (Ministry of Industry) help answer these questions; the situation in 1994 was thus as follows:

— more than 60% of the national output came from less than 10% of the sand and gravel pits of capacity greater than 200,000 t/yr;

— about 50% of the pits produced less than 15,000 t/yr;

— about 70% of the pits (capacity < 50,000 t/yr) produced in a decidedly intermittent fashion;

— only 66 pits (2.7%) produced more than 500,000 t/yr.

Table VIII.1: Quarries of alluvial aggregates. Production by level-wise distribution. All of France, 1994 (as per Technical Service of Quarries, Ministry of Industry)

Production level 1000 t/yr	Theoretical hourly capacity*	Number of workings	Share in the total (%)
0 to 15	Intermittent exploitation	1047	43.6
15 to 50	Intermittent exploitation	535	22.3
50 to 150	35 to 100**	448	18.6
150 to 250	100 to 165	168	6.8
250 to 500	165 to 335	144	6.0
500 to 1000	335 to 665	56	2.3
> 1000	> 665	10	0.4
Total		2403	100

*Basis = 1500 h of effective production per annum
**Frequent intermittent exploitation

No more than some 50 departments had sands and gravels pits of a capacity > 250,000 t/yr and only some 10 had a significant number of pits of large capacity (> 500,000 t/yr), the break-up of which is as follows: Lower Rhine 13 Seine-and-Marne 5 and Eure 4; mouths of the Rhône, Upper Rhine and Yvelines 3; Upper Garonne, Isere, Rhône and Oise 2.

Again the pattern of production is highly diversified, marked by the presence of all modes, ranging from the artisanal to the industrial...

In such a situation recourse to large-scale production methods (> 500 to 600 t/h) remains rather the exception and, in almost all cases, the exigencies are met by machinery with a capacity between 100 and 400–500 t/h.

It should be noted that for small-size pits (< 150,000 tonnes) recourse to over sized but multifunctional extraction machinery is quite often the rule. In fact the machinery in this case fulfils multiple tasks: extraction, stripping, even collection or loading[4].

2. STRIPPING AND RECLAMATION

Stripping operations comprise scraping (extracting), transporting and dumping temporarily or permanently, the dead-ground (also called 'steriles' or 'overburden' overlying a deposit, to enable its exploitation.

The operations of reclamation, or more precisely of *'post-mining rehabilitation of the soils and the mine site'*, made obligatory since the 1970 and 1977 revisions of the Mining Code, are expressly bound in practice with the stripping operations unless there is a possibility of refilling the site itself with the sedimentary material (debris, various waste products). Whatever be the subsequent reuse of the site (agricultural, industrial, recreational, residential etc.), it effectively consists of creating a new landscape after working the deposit, with *the bottom of the deposit forming the framework of this landscape*, after permanent dumping of the excess steriles.

2.1 General Method

Accomplishing the stripping and reclamation operations in a cost-effective manner is more a question of the method than the means. In other words, it is necessary to ensure:

— first of all, that the reclamation scheme is technically and economically compatible with the geotechnical characteristics of the site in question;

— and then, simultaneously carry out the stripping and reclamation operations in the course of working the deposit.

These conditions presuppose a *prior detailed geotechnical investigation of the deposit* and this even in those zones where the sands and gravels formations have long been exploited.

Detailed geotechnical investigation of the deposit
— topography of the natural terrains,
— topography of the top of the deposit (= 'top of the sand'),
— topography of the bottom of the deposit (= 'substratum' = 'bedrock'),
— ISOPIEZES of the groundwater table (= level of the water table),

---

[4]Backacter hydraulic shovels, pneumatic-tyred loaders.

— overburden ISOPACHS (= 'overburden thickness'),

— sands and gravels ISOPACHS (= thickness of sands and gravels) (see Chapter II and III).

While on this subject, it may be recalled that under the present economic conditions, those *deposit zones* wherein that ratio of *overburden to usable deposit thickness* is equal *to or above 1* are deemed *unexploitable.*

2.1.1 TECHNICAL COMPATIBILITY OF SITE/RECLAMATION SCHEME

The technical compatibility of the reclamation scheme with the site rules out options such as the creation of a water body, when the groundwater level is below the midway to the deposit's bottom or very close to it. This also implies that the volume of the surface reliefs to be achieved in this case (surface deposit) shall not exceed the volume of available steriles.

Furthermore, these surface reliefs can only form a *superstructure,* wedged upon the *infrastructure* (or framework) constituted by the deposit's bottom (Fig. VIII.1).

In the contrary case, that is, partially submerged or 'fluvial' deposits, if one imagines, for instance, a site wherein the depth of the steriles is no more than one-fourth of the total alluvial overlap, the area of filling to the pre-existing natural ground level will be no more than one-fourth of the exploitable area, the water table occupying the rest (Fig. VIII.2).

Large complex alluvial deposits, which extend over wide areas and straddle several terraces (Fig. VIII.3), are the ones in respect of which compatibility of reclamation schemes with site is technically the most difficult to resolve.

In this case the bottom of the deposit comprises broad zones situated between levels which are clearly above the groundwater table and the others which are definitely below it, as well as the transition zones.

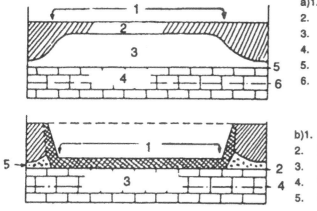

a)1. Exploitable zone
 2. Steriles
 3. Sands and gravels
 4. Bedrock
 5. Bottom of deposit
 6. Water table level

b)1. Permanently dumped steriles
 2. Bottom of deposit
 3. Bedrock
 4. Water table level
 5. Sands Gravels

Fig. VIII.1. Surface deposits.
a) before working; b) after working

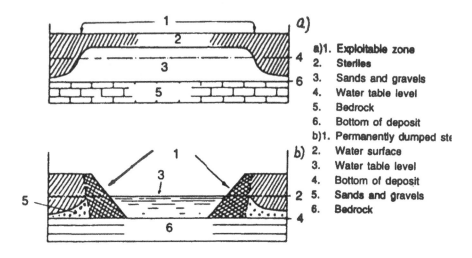

a)1. Exploitable zone
2. Steriles
3. Sands and gravels
4. Water table level
5. Bedrock
6. Bottom of deposit
b)1. Permanently dumped st€
b) 2. Water surface
3. Water table level
4. Bottom of deposit
5. Sands and gravels
6. Bedrock

Fig. VIII. 2. Partially submerged or fluvial deposits.
a) before working; b) after working

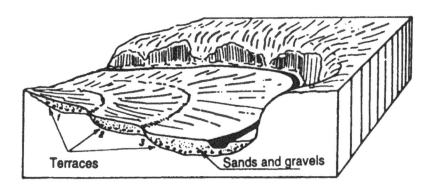

Fig. VIII.3. Meander deposit.

The problem in such cases is often compounded by the near absence of steriles at the ground level zones (positive levels of the bottom of the deposit vis-à-vis the water table) and an abundance of steriles at the 'partially submerged' or 'fluvial' zones (negative levels of the bottom of the deposit vis-à-vis the water table), which imply substantial transports of earth over long distances. Still it is possible to lay down a course of reclamation compatible with this type of site. But one must a priori be in a position to implement very intensive preliminary geotechnical investigations (Fig. VIII.4).

Let us lastly note that quite often the complex contour of the exploitable zone boundary, as well as the possible presence of islets of steriles enable easy achieve-

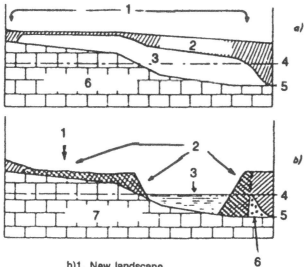

a)1. Exploitable zone
2. Steriles
3. Sands and gravels
4. Water table level
5. Bottom of deposit
6. Bedrock

b)1. New landscape
2. Permanently dumped steriles
3. Water surface
4. Water table level
5. Bottom of deposit
6. Sands and gravels
7. Bedrock

Fig. VIII.4. Complex deposits.
a) before working; b) after working

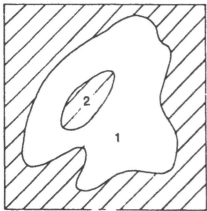

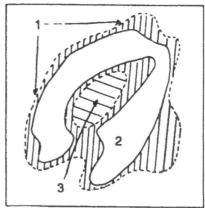

a) 1. Exploitable zone
   2. Unexploitable islet

b) 1. Zone with permanently dumped steriles
   2. Water surface
   3. Raised landscape on islet of steriles

Fig. VIII.5. Partially submerged deposit: Case of facile execution of an 'eminently landscaped' reclamation.
a) before working; b) after working

ment of 'eminently landscaped' reclamations (Fig. VIII.5) in the case of partially submerged deposits.

2.1.2 ECONOMIC COMPATIBILITY OF SITE/RECLAMATION

The search for good economic compatibility (or coherence) amounts to arriving at a scheme or reclamation which, among other things, enables minimising once and for all the distance of transport of steriles from the zones where stripping operations take place to the place or zones of their permanent dumping.

From this point of view, transport distances less than or equal to 500 m are ideal, though distances up to 1000 m are tolerable.

Occasionally transport distances touch much higher figures. These generally result from external constraints arising from a stringent scheme of reclamation imposed a priori by public pressure groups and, of course, entrain extra high costs of exploitation.

2.1.3 SIMULTANEOUS EXECUTION OF STRIPPING AND RECLAMATION

Devising such a method of working a deposit is the last problem to be solved. For this it is necessary to work out a very detailed phased general mining plan with due regard to all the constraints attendant upon extraction of materials, which concomitantly enables simultaneous (or parallel) advancement of the zones (or faces) under stripping, under extraction and under reclamation (Fig. VIII.6).

This phased plan then takes effect in practice at the level of each 'face' through a logical sequence of contiguous operations, say in the order:
— stripping,
— extraction,
— reclamation.

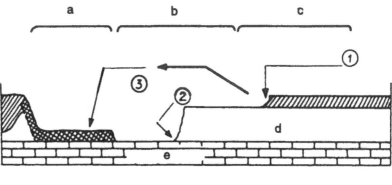

1. Stripping steriles
2. Extraction of sands and gravels
3. Steriles permanently dumped for site rehabilitation

Fig. VIII.6. Simultaneous execution of stripping and reclamation operations
a) Face under reclamation; b) Face under extraction; c) Face under stripping; d) Sands and gravels;
e) Bedrock, f) General direction of exploitation progress.

2.2 Equipment Deployed

The steriles overlying alluvial deposits are loose materials, such as topsoil, silts and clays, which are more or less intermixed with sand, alluvia which are by and large coarse and highly clayey, and more rarely peat.

When the 'overburden' is wholly 'outside water' (deposit's top level wholly above the water surface), the machinery deployed for carrying out the operations of scraping the steriles can vary considerably and cover the entire gamut of existing earthmoving equipment. They are a great deal more limited, complex in operation and costly when the 'overburden' is 'partially submerged' (water level midway between the natural ground level and the top of the deposit).

Thus, a thorough knowledge of the respective levels of the top of the deposit and the water table at all points of the deposits is imperative. In fact, such knowledge dictates the choice of machinery.

2.2.1 'ABOVE-WATER' OVERBURDEN REMOVAL (DRY) (Water table level at least 0.5 m below the top of the deposit)

Most of the classical earthmoving machinery can be used in this case. So the choice of the machinery to be used will mainly depend on:

- volume of steriles to be scraped,
- distance between stripping zone and permanent dumping zone.

We shall not take up here the case of small-scale operations, which for the most part are less than 50,000 m^3/yr and carried out in an artisanal fashion, as the scraping progresses, on small-size workings for which a surplus capacity of extraction and transport machinery is available. Unfortunately, there are as yet far too few examples of machinery of this nature simultaneously executing acceptable reclamation operations on exploited zones. Hence we shall confine ourselves only to operations involving an *annual volume of > 50,000 m³ overburden*, which justify an annual campaign (of limited duration) on a working site, subcontracted if need be to a specialised firm, and executed by means of appropriate machinery.

Under these conditions, and for transport distances not exceeding 1000 m between the stripping and the permanent dumping zones for steriles, a motorscraper is the machine technically and economically most suitable[5]. Among other things, it enables directly achieving relatively complex relief landscapes on the dumping zones of steriles, without necessitating major additional machinery.

When the transport distances exceed 1000 to 1200 m, recourse should be had to a *dumper* for transporting steriles. Then the extraction and loading are done either by 'scraping', most often by *loaders* and, as need be, by means of a specially

---

[5]Unless the geotechnical characteristics of the overburden indicate particularly unfavourable loading conditions, owing to which adoption of a discrete type of solution (extraction-loading of a discrete type of solution (extraction-loading machinery/transport machinery) becomes preferable.

equipped hydraulic shovel, or by 'digging', most often by means of a *back-acting hydraulic shovel.*

The choice of extraction mode, namely 'scraping' or 'digging', depends directly on the loading conditions of the dumpers or, more rarely, the particular demands of the site.

The conditions of hauling and permanent dumping are more difficult and less flexible with this type of equipment than with motorscrapers.

Under the present economic conditions, the cost of removal per m^3 of overburden by a motorscraper and its permanent dumping at a transport distance of 750 m, is of the order of 2 to 3 US\$/m^3.

. For distances of the order of 1200 to 1500 m, implying the use of dumpers plus shovel or loader, the coast of permanent dumping per m^3 of overburden, under the present economic conditions, is of the order of 4 to 5 US\$/m^3.

2.2.2 'PARTIALLY SUBMERGED' OVERBURDEN (Water level midway between the top of the deposit and the natural ground)

With some deposits it is possible to partially lower the water level, thereby enabling the steriles under water to totally emerge in the zone to be scraped, and to deal with them by conventional means (= 'above-water' overburden) thereafter.

This presupposes a relatively low degree of perviousness of the underlying exploitable alluvia, however (pumping out rate < 1000 m^3/h per panel of 2 to 5 ha), and also major preparatory jobs (deep drainage channels around each working panel).

When this procedure is found to be impracticable, the choice of available means narrows markedly. There can be two cases in practice.

• *Total thickness of alluvial deposit does not exceed some 10 m (steriles + sands and gravels)*

In most of these cases scraping is accomplished by a dragline shovel. This is the most suitable machine for this kind of work (back-acting hydraulic shovel is only employed if the submerged bench does not exceed 10 to 20% of the total thickness of the overburden, and does not exceed one metre), especially when the *submerged bench is between one and a quarter metre.*

On such a site, for transporting the steriles to permanent dumping zones, recourse is most frequently had to 'all-terrain' and 'all-wheel drive' types of vehicles. Nevertheless, haulage often poses serious problems due to waterlogging at the loading points of part of the haulage tracks (and of the dumping zones) and the proximity of the water table. Disruptions on the site are frequent and prolonged.

In the worst cases some deposits display such characteristics (water table at ground level, essentially peaty steriles) as render hauling almost impossible for the greater part of the year. A new technique of transporting steriles, especially well suited to this kind of site, is currently under development in France. Widely employed in Holland and northern Germany, it consists of adding water to the steriles extracted by the drag shovel, in a specially designed receiving hopper, and then forcing them out by pumping in slurry form through a pipework connected to the permanent dumping zone. This zone is thereafter specially laid out as a settling pond with damming embankments.

So long as one has dumping zones, howsoever small, which are easy to dyke and pose no problems of water backflow, this technique, relatively simple to implement, is remarkably efficient when the steriles contain no coarse elements (> 5 mm) in an appreciable quantity (accelerated wear of pumps, substantial loss in pumping capacity). Further, it enables resolution of problems which are nearly unsolvable by conventional methods (haulage).

Under the present economic conditions, and for transport distances of the order of 500 to 700 m, the cost of permanent dumping of the cubic metre of partially submerged overburden is of the order of:

— 4 to 5 US$/m^3 in the case of conventional haulage solution;

— 2 to 3 US$/m^3 in the case of hydraulic transport and filling.

(In both cases overburden extraction is accomplished by drag shovel.)

• *Thickness of alluvial cover far above 10 m (steriles + sands and gravels)*

If submergence of the overburden bench does not exceed 3, even 4 m, the situation is as in the foregoing (see Section 2) and the same means may be adopted.

If submergence of the overburden bench is higher than 3 m, and hence permits the use of floating vessels, the overburden can be extracted by means of a multi-bucket dredger or a grab dredger. The dredger then loads boats with shutters, which take care of the transport of steriles over to portions totally stripped of the deposit, which serve as permanent dumping zones. This procedure cannot be used on a deposit *ab initio*. On the other hand, it can serve to extend the life of a deposit worked by dredger (bucket, grab or suction type) whereby zones with a thick overburden initially considered as steriles become exploitable.

Cases of this kind are relatively rare.

3. EXTRACTION AND LOADING

The operations of extraction and loading are generally always indistinguishable in working of alluvial deposits (general case with sands and gravels). Partially submerged deposits constitute the sole exception (see Section 1.1.2) wherein these two operations are most often separated to enable beneficial dewatering of the pit run. This separation becomes practicably obligatory when the pit run has a high sand content (> 50%) and/or when the sand is relatively fine and water-retentive by nature.

3.1 General Approach

Optimisation of the extraction-loading operations is often more complicated than one would generally admit.

The machinery used are costly and do not yield good efficiency except when employed under very specific conditions.

The difficulty is thus traceable to the fact that these conditions do not depend solely on the equipment per se, but also and largely on the characteristics peculiar to each deposit.

The main factors which prima facie help determine the type(s) of machines suitable for a given exploitation are as follows:

• *Deposit-related factors*

— *Water table* level (wetter thickness) vis-à-vis the deposit's own top and bottom levels. *This is one of the predominant factors governing the choice;*
 — exploitable reserve of the deposit;
 — mean thickness and limits (minimal and maximal) of the pit-run layer to be extracted;
 — mean grading and limits of the pit run to be extracted (D_{max});
 — presence or absence of grain size heterogeneities in the vertical plane over the thickness of the exploitable layer;
 — compactness of the pit run to be extracted;
 — presence or absence of big block or hardened crust in the exploitable layer;
 — presence or absence of intercalated clay in the exploitable layer and morphology of the clay layers;
 — abrasiveness of the pit run to be extracted.

A precise assessment of these various qualitative and quantitative parameters can emerge only from a prior and detailed *geotechnical investigation* of the site.

• *Rhythm of extraction*

— necessary extraction-loading capacity and possible fluctuations depending on local constraints.

Since the pit runs extracted are most often meant to be processed by wet methods in the plant (crushed, washed, screened aggregates), the extraction-loading capacity should always be higher (by 10 to 20%) than the processing plant's production capacity of marketable products.

Losses of fines (< 100–150 μm) consequent upon treatment (washing) are as a matter of fact generally of the order of 10 to 15%, but can touch 20% in the case of highly polluted deposits.

Recourse of highly special-purpose, *heavy-duty extraction machinery*, implying huge investments, is most often *not attractive*, when the extraction rhythm adopted or imposed is such as *to exhaust the exploitable reserves in less than 6 to 8 years.*

This then quite obviously limits our scope (see Section 1.2) in this section to examining the means to be adopted for extraction capacities < 800 t/h, say about 400 m^3h of material in situ.

• *Mode of transport adopted between working face and first place of receipt of materials at the processing plant:*

— either *continuous handling* system (conveyor belts—hydraulic transport through pipes),
 — or *discrete handling* systems.

The choice of this option is determined vis-à-vis the extracting-loading machinery. Excellent co-ordination is perforce essential between the machinery and the transport vehicles (trucks and dumpers) in order to ensure perfectly balanced extraction-transport cycles.

• *Machinery types*

— a thorough knowledge of the specific situations of use of each machine;'

— a reasonable and realistic appreciation, largely based on experience, of the results from extended timing on the site, of the efficiencies and outputs (or capacity) of each machine;

— a thorough knowledge, or in its absence, as in the case of new machinery, a reasonable and realistic appreciation of the operating costs of each machine. At present, depending on individual cases and the particular machinery deployed, these costs worked out on a per-tonne basis can vary notably by a factor of 1 to 2.

In practice, before making a final choice in the matter, it is eminently sensible to make several visits to active sites of proximate characteristics at which similar machinery are installed.

3.2 Equipment to be Employed

The level of the water table with respect to the top of the deposit constitutes one of the predominant factors in the choice of machinery to be employed.

3.2.1 SURFACE DEPOSITS

In this type of deposit the level of the water table is below the bottom of the sands and gravels.

Two extraction-loading options are possible:

— extracting by 'digging' from above,

— extracting by 'scraping' from below.

In either case the extraction face can hardly exceed 8 to 10 m. If the exploitable thickness far exceeds these values, extraction is done at two or more levels, each level having one extraction face. Such cases are relatively rare in France, barring deposits situated in Tertiary formations (industrial sands) or of glacial or fluvioglacial origin, wherein the thickness can reach, indeed exceed 50 m.

Drag shovels are the most commonly utilized 'digging' extraction machines, with back-acting hydraulic shovels are accessories, when the pit run to be extracted is compact or present in hardened zones, and when the height of the faces does not exceed 3 to 4 m (Tables VIII.2 (A) and VIII.2 (B)).

The digging mode of extraction, however, remains rare for *surface deposits* though it enables, especially in conjunction with a drag shovel, the best homogenisation of the pit run extracted when the deposit has pronounced granulometric heterogeneities in the vertical plane.

Pneumatic-tyred loaders are the most widely used 'scraping' type extraction machines. 'Scraping' type shovels (hydraulic) are rarely used (hardened zones, presence of large blocks) (Table VIII.3).

Table VIII.2(A): Drag shovels. Extraction of sands and gravels. Partially submerged deposits. Estimated mean hourly outputs (orientation angles 90° to 120°) (piled in rows)

Capacity of bucket (m³)	100% efficiency		85% efficiency	
	m³ in situ/h	t of crude material/h	m³ in situ/h	t of crude material/h
0.75	65	130	55	110
1.0	90	180	77	155
1.15	100	200	85	170
1.35	120	240	102	205
1.5	130	260	110	220
1.9	155	310	132	265
2.3	175	350	150	300
2.7	200	400	170	340
3.1	220	440	187	375
3.8	260	520	220	440
4.6	305	610	260	520
5.4	345	690	293	585

1 m³ in situ $\simeq$ 2.0 t crude materials $\simeq$ 1.7 t marketable products (washed, crushed, screened)

Since the majority of surface deposits are worked by the 'scraping' mode, *pneumatic-tyred loaders* constitute the *majority in the fleet of machines used for this type of deposit*.

To be usable at their optimal efficiency, these machines, which offer quality and concomitantly the drawback of being too 'flexible' in use, should not be employed in loading operations more than 30 to 40 m away from the working face.

Continuous 'digging' and 'scraping' extraction machines, such as the *bucket wheel type excavator* and the *multibucket excavator*, are not used in France on this type of deposit (abrasiveness of materials often incompatible with these machines—capacities required less than the optimal outputs of a bucket wheel type excavator).

3.2.2 SUBMERGED DEPOSITS

In this type of deposit the water table level is higher than the bottom of the sands and gravels.

We may recall at the outset that when the characteristics of the deposit permit (impermeable bottom, clay, marls, materials of low perviousness), it may be possible to economically effect *total evacuation of the water body from the site*. The *deposit*, totally exposed, is then *exploitable like a surface deposit* with all the advantages this implies.

Several examples of this technique can be seen in the Paris region, where exploitations spread over tens of hectares.

When the technique cannot be adopted, recourse should be had to *floating extraction machines, dredgers*.

Table VIII.2(B): Back-acting hydraulic shovels. Extraction of sands and gravels. Estimated mean hourly outputs (orientation angles 90 to 120°). Excavation at 70 to 90% of the machine's maximum dipping-depth capacity (loading trucks)

Capacity of bucket (m³)	100% efficiency		85% efficiency	
	m³ in situ/h	t of crude material/h	m³ in situ/h	t of crude material/h
0.7	80	160	67.5	135
0.9	100	200	85.0	170
1.1	110	220	92.5	185
1.5	135	270	115.0	230
1.9	150	300	127.5	255
2.3	180	360	152.5	300

1 m³ in situ ≃ 2.0 t crude materials ≃ 1.7 t marketable products (washed, crushed, screened).
Note: Outputs capable of increasing very substantially with excavation reduced to less than 50% of the machine's dipping-depth capacity and orientation ≤ 90°.

Table VIII.3: Pneumatic-tyred loaders. Sands and gravels (extraction, transport, loading, distance: 20 to 30 m). Estimated mean hourly outputs

Bucket capacity with (dome) (m³)	100% efficiency		85% efficiency	
	m³ in situ/h	t of crude materials/h	m³ in situ/h	t of crude materials/h
0.7–(1.0)	65	130	55	110
0.9–(1.15)	75	150	65	130
1.15–(1.55)	100	200	85	170
1.9–(2.2)	140	280	120	240
2.05–(2.4)	160	320	135	270
2.6–(3.1)	190	380	160	320
3.5–(4.0)	240	480	205	410
4.5–(5.4)	310	620	265	530
8.6–(10.3)	500	1000	425	850

1 m³ in situ = 2.01 t crude materials = 1.71 t marketable products (washed, crushed, screened)

• Draught less than or equal to 8–10 m

Barring river dredging leases, deposits of this type are rare. In this case the submerged thickness does not suffice to permit use of floating extraction machines of the pump dredger or grab dredger type under optimal efficiency conditions.

The floating extraction machine still commonly used is the *bucket dredger* when the draught is sufficient (≥ 3 m) (Table VIII.4 (A)).

Table VIII.4(A): Bucket dredgers with onboard facilities for materials processing (screening, washing, crushing)

Nominal capacity of buckets (litres)	Mean hourly output (sands and gravels) (t/h)
60	60 to 90
100	100 to 140
200	200 to 280
300	300 to 420

These machines have certainly been in use longer than any other for extraction from floatable deposits when conditions are favourable. They are simple equipmen whose technology has undergone no change for quite some time except for re placement of steam powering with diesel-hydraulic, disesel-electrical or electrica systems.

The 'extraction' part of a bucket dredger comprises:

— A metallic pontoon, with a hollowed-out portion called the 'sump' in front through which a chain with buckets passes. Installed centrally on the pontoon i a metallic frame or 'gantry' supporting the bucket chain's drive system, as als the joint of its beam support or 'dredging ladder'. A frame, generally overhanging at the front of the pontoon, supports the lower part of the dredging ladder throug a swing bar.

— A chain of steel buckets on the back of which two links are affixed. The buckets are connected to each other by means of these links to form a chain. Move ment of the chain is impelled by the 'upper tumbler' (= double wheel with 4, 5 o 6 sides; length of one link = length of one side) affixed to a shaft directly connected to the chain-driving motor. The lower portion of the dredging ladder has a 'lowe tumbler' rotating freely on its axis.

The materials extracted and raised on the chain of buckets discharge by gravity when the upper tumbler is tilted into a stationary chute or 'receptacle' situated immediately below (dispatch to the materials-processing line).

The most commonly used dredging rates are of the order of 20 buckets/min.

As a general rule, these dredgers are not equipped with autonomous propul sion. At the time of extraction, all the movements needed by the machine are ef fected through hoists and cables, one of whose ends is firmly attached to a fixed point, or 'fixed mooring'.

The extracted materials are loaded onto boats (by means of conveyor belts and mobile chutes), either for dispatch to the consumption zones in the form of marketable products (screened, crushed, washed) if the dredger has onboard processing facilities, or in the form of crude materials to a regular plant located on land.

Bucket dredgers are also used in closed excavations as simple extraction machines, and are also connected to land by a system of floating belt conveyors which serve to transport materials up to the processing plant (such applications are rare: Eure, Gironde).

• Draught greater than 10 m

Depending on the situation, three different types of machines are used preferen- tially:

— grab dredgers,
— suction dredgers,
— pneumatic dredgers.

• Grab dredgers

These are the most commonly used machines when the pit run to be extracted is relatively coarse (D_{max} 200 to 500 mm) with a law mean sand content (< 55%) and the draught is more than 12 to 15 m on average.

They enable dredging at great depths (60 to 80 m). They are equipped with hoisting buckets with hydraulic closing action when the materials are slightly compacted or cave in with difficulty (Table VIII.4 (B)).

Table VIII.4 (B): Grab dredgers. Extraction of sands and gravels. Estimated mean hourly outputs (dredging depth 10–15 m, evacuation of extracted products by continuous handling)

Capacity of grab (with dome), m^3	Mean hourly output	
	m^3 in situ/h	t of crude materials/h
1.5–(1.75)	50 to 60	100 to 120
2.1–(2.6)	100 to 110	200 to 220
2.7–(3.4)	125 to 135	250 to 270
3.4–(4.2)	160 to 180	320 to 360
4.8–(6.0)	230 to 245	460 to 490

1 m^3 in situ = 2.0 t of crude materials ≈ 1.7 t of marketable products (washed, crushed, screened).

These simple, rustic devices of recent design are energy effective (0.5 to 0.7 hp/t/h). They came into use in the early 1950s in the Rhine valley. The three main models are described below:

— Fixed gantry model:

Generally has two parallel pontoons, or one pontoon with a central shaft enabling descent of the hoisting bucket. The runner (travelling crane type) swings the hoisting bucket suspended from it into a discharge position over the hopper receiving extracted materials (situated on one of the two pontoons).

— Inclinable jib model:

Has a single pontoon with a central shaft enabling descent of the hoisting bucket. In the filling position the jib enables positioning the hoisting bucket above the central shaft, and in the discharging position, over the hopper receiving materials.

Slewing-jib model (jib's inclination fixed):

This model enables extraction from three sides of the pontoon and thus covers a larger area without the dredger having to shift.

In all these types of dredgers the hopper for extracted materials, equipped with an extractor of variable feed rate, feeds a dewatering-scalping screen. The larger-diameter elements are either straightway discharged with the water or crushed (jaw crushers), as the case may be. The slurry drained from the materials is either discharged with the water or, in the case of deposits rich in sand and poor in fines, treated by hydrocycloning for maximal preclusion of loss of useful fines. On exiting from the dewaterer-scalper screen and after crushing and recovery of fines as necessary, in most cases the materials are straightaway loaded on the floating conveyor belts and sent to the processing plant located on land.

Transports are sometimes handled by self-discharging hopper boats, which discharge directly at the head of the plant at the quay.

Except where imperatively needed (case of river dredging) recourse to this machine has lessened, with direct loading of coffered boats or with shutter boats destined for a land-based processing plant (collections costly, loss of materials etc.).

• *Suction dredgers*

These machines are employed when the pit run to be extracted is relatively fin
(D_{max} < 50–100), rich in sand (> 55%), non-abrasive and when the draught doe
not exceed 15 to 20 m. For the rest, they require materials which slump well, fc
good yields.

Sites possessing such characteristics are relatively rare in France.

Materials are extracted by pumping, using a suction dredger, and forced ou
in the form of a slurry (mixture of water + materials). Depending on local cond
tions, i.e., dredging depth, grading, ease of pit run slump and pumping distanc
the percentage of solids in the slurry may vary from 6 to 12%. The pumps ar
generally characterised by the diameter of their pumping pipework, which mo:
often varies from 200 to 400 mm, for slurry pumping discharges of 400 to 230
m^3/h (Table VIII.5.).

The metallic pontoon of these dredgers has a small 'sump' in front, throug
which a dredging ladder passes which supports the pipework for suction. Th
dredging ladder, articulating with the pontoon, is suspended from a metallic ji
through a tackle block. The dredging depth is regulated through a hoisting winc

When necessary (hardened or compact deposits—clay layers), a *rotary or chai
clay cutter* is installed at the end of the dredging ladder, ahead of the orifice of th
suction pipework.

The suction pipework is connected through a flexible element and a fixe
pipework to the dredging pump, driven by a diesel or electric GMP.

The dredging pumps are specially designed, liberally dimensioned an
adapted to the pumping of slurries that are abrasive and contain large-size soli
elements.

On leaving the pump, in most cases the slurry is pumped out through a floa
ing pipework and sent to the processing plant on land (the maximum pumpin
distance for a pump is 500 to 1000 m).

The pontoon, in addition, has two metallic piles behind it, which are manipu
lated by hoists or hydraulic jacks. The suction device advances by means of thes
piles, turning around the one driven into the bottom. Thus it dredges by successiv
arcs of a circle around one of the two piles.

Movements of the dredger around the pile at the bottom are effected throug
'hoists' and 'cross-cables' according to the 'flutter' principle.

The presence of *hardened or compact* zones and discontinuous *intercalated cla
beds* in the pit run to be extracted can *considerably reduce the extraction capacity* an
the efficiency of these types of machinery.

But the working of a suction system, by ensuring a vigorous stirring of th
materials, facilitates suspension of the clays, which results in *clean marketabl
products* from *totally polluted deposits* that are difficult to exploit by other means.

As for the other types of machines, even though they have the advantage c
being able to straightaway transport the extracted materials to the processing plar
(within a radius of < 500 m), they are relatively subject to greater wear and cor
sume considerable energy (1.5 to 2.5 hp/t/h).

Considering the situation of French deposits, use of these extracting machine
generally remains limited to specific cases.

Table VIII.5

Ø mm pipeworks sucking and pumping	Slurry output (m³/h)	Pumping rate (m/s)	Maximum power (hp)	% of solids in slurry*						
				6	7	8	9	10	11	12
200	400	3.5	150	24	28	32	36	40	44	48
250	700	4.0	250	42	49	56	63	70	77	84
300	1100	4.3	400	66	77	88	99	110	121	132
350	1600	4.6	600	96	112	128	144	160	176	192
400	2300	5.1	800	138	161	184	207	230	253	276

*m³ of solid elements of density = 2.5:
— maximum pumping distances for a pump = 500 to 1000 m,
— percentage of solids in the slurry varies with D of the materials, sand content and pumping distance. Under these average conditons (D and distances) the admissible solids quantum in the slurry is of the order of 8 to 10%.

Airlift dredgers

These machines are used for extracting pit runs of continuous grading and average particle size (D_{max} = 100–200 mm), the sand content of which is relatively constant (too sandy or too coarse materials causing a fall in efficiency), and with draughts which can touch high values (< 60 m). Most of the machines of this type used in France are equipped for dredging maximum depths of the order of 20 m (compressed air at 2.5 bars). An airlift dredger is a relatively simple machine and quite resistant to abrasion by materials but relatively energy intensive (1.5 to 2 hp/t/h on average).

With an airlift dredger, extraction of materials is effected by *suction*, by virtue of the *'airlift' phenomenon*.

The general design of the machine closely resembles that of a suction dredger but has a compressor instead of a suction pump, which blows compressed air at the base of the suction pipe through a separate pipe.

Installed at the upper end of the suction pipework, on the pontoon, is a decompression system (pressure breaker); on leaving it the slurry (water + extracted materials) is sent into a liberally dimensioned dewatering wheel.

On leaving the dewatering device, the materials are directly fed, in most cases, onto floating conveyor belts to be transported to the aggregated processing plant on land, or more rarely into boats. All these dredges are equipped with a *clay-cutter* at the *dredging-ladder end*, at the head of the suction pipework. But like pump dredgers, they too are subject to *considerable loss in extraction capacity* when the pit run contains *hardened or compact zones* and discontinuous *intercalated clay beds*. However, stirring the materials is less intense in this case.

3.2.3 PARTIALLY SUBMERGED DEPOSITS

In this type of deposit the water level is midway between that of the top and bottom of the sands and gravels.

This exploitation situation is encountered most often in France.

If in view of the characteristics of the deposit, an economical evacuation of all the water from the site is found impossible extraction can still be effected by machines used on land provided the thickness of the submerged portion does no exceed a maximum of 8 to 10 m.

For extraction of deposits of greater thicknesses, recourse should be had t floating machines of the types described earlier for submerged deposits (see Sectio 3.2.2).

But when the thickness of the 'exposed' portion does not exceed on averag 2 to 3 m, a single extraction machine (land or floating) can achieve extraction ove the deposit's entire thickness. In the contrary case the upper 'exposed' portion o the deposit has to be extracted separately, as in a surface deposit (see Section 3.2.1)

Hence we shall limit ourselves in this section only to land machines usabl for extracting by digging under water, that is, the back-acting hydraulic shovels drag shovels and bucket excavators.

The back-acting hydraulic shovel can be usefully deployed only on deposit of small thickness (3 to 4 m at the very most), compact, poor in sands and o which the submerged portion does not exceed 1.5 to 2.5 m (see Section 2.2.1 an Table VIII.2(B)).

Its use in a different context (deposit fine, quicksand type or rich in sand, o greater thickness) is invariably attended by significant losses of materials.

Consequently, this type of shovel is hardly ever used aside from small work ings in which its versatility is attractive (stripping, diverse earthworking jobs).

The drag shovel is an extraction machine of great flexibility in use and most com monly employed for partially submerged deposits (see Section 3.2.1, Table VIII.2 (A))

This shovel is eminently suited to this type of deposit (except for compact an hardened materials) and enables extraction of sizeable submerged portions (up t 6 to 8 m in thickness) without notable losses of materials, subject to an adequat boom length (3.5 to 4 times the maximum extraction depth) and a skilled operator

Since pit runs extracted under these conditions generally require dewaterin before transport to the processing plant, they are most often piled in rows to b subsequently reclaimed (loader) or, more rarely, poured into *movable hopper* equipped with adequate devices (dewaterer-scalper) which directly feed portabl belt conveyors.

Bucket excavators constitute a class of machines whose use is expanding fo working partially submerged deposits.

Movable, caterpillar-mounted, this machine has a raisable ladder equippe with a chain of buckets working in a plane perpendicular to the machine's directio of movement along the excavation.

The materials are extracted by slow scraping (15 to 23 buckets/min) fron bottom to top throughout the entire layer under exploitation, discharged into a buffer hopper, then loaded by a mobile conveyor onto a portable belt conveyo (or dumpers). Where necessary, the machine can be fitted with devices for scalpin and dewatering of the extracted materials.

The maximum depth of extraction generally reached is of the order to 10 n (submerged face).

This type of machinery, eminently suited for homogeneous deposits, with a continuous medium grading (D_{max} 20 to 100 mm), enables an optimal recovery o the materials if the morphology of the bottom of the deposit is uniform. Contrarily it is ill-suited for heterogeneous deposits (intercalated clay, hardened zones) o those relatively rich in coarse (elements $D_{max} < 200$ to 500 mm).

Lastly, it should be noted that optimisation of use of this type of machine, especially energy effective (0.15 to 0.25 hp/t/h) and least sensitive to abrasion, calls for simultaneous continuous materials handling (belt conveyors = transport to the processing plant), which in turn implies availability of relatively extensive exploitation sites, of simple geometry and unbroken by private estates (see Section 4.2).

3.3 Adaptations for Special Situations

3.3.1 'LARGE' OR 'MEANDER' TYPE HETEROGENEOUS DEPOSITS

Deposits answering to this definition are exceedingly vast and display horizontal variations in grading, which are generally low and progressive (Fig. VIII. 7).

They are the type of alluvia encountered in meandering valleys, such as those of the Seine, downstream of Paris. Such deposits may occur, however, in the three situational types described earlier.

The need to feed processing plants with materials as homogeneous as possible calls for *carrying out extraction* simultaneously *on two faces*, each extraction face being developed in zones where the pit runs possess complementary gradings.

Such a procedure, not infrequently employed, is often difficult to perfect. It calls for as thorough a knowledge as possible of the distribution of sand contents across the entire deposit; a detailed preliminary geological and geotechnical investigation should furnish the needed information.

1. Inner zone of meander
2. Outer zone of meander
3. Upstream
4. Downstream
5. Median terrace
6. Lower terrace
7. River
8. Isosand content curve, 40% sand content

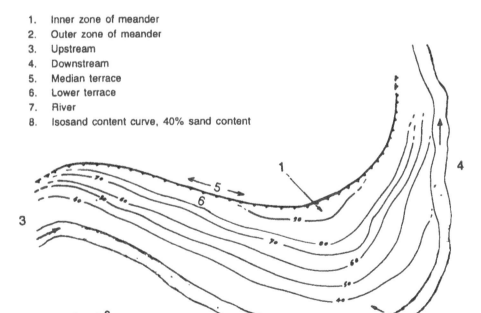

Fig. VIII.7. Schematic representation of sand contents in a meander deposit (Lower Terrace, the Seine)

The procedure is applicable only for surface or partially submerged deposits.

3.3.2 DEPOSITS POLLUTED BY INTERCALATED CLAY LAYERS

The magnitude and frequency of intercalated clay layers in a deposit can often become an insurmountable obstacle in exploitation unless one has at his disposal (for technical or economic considerations) special and adequate means of removing mud and washing at the processing plant.

In the case of submerged and partially submerged deposits, this problem is practically unsolvable when the clay beds are situated at a level below that of the water table.

On the contrary, when it comes to *surface deposits* a solution exists, namely carrying out selective extraction of the clay and the pit run.

Considering the discontinuity of these beds, their lenticular form, and their position, which is apt to vary in the mass of the pit run, it is almost impossible to employ conventional extraction machinery. The *motorscraper* is the only kind of machine which is sufficiently flexible and capable of achieving such selective extraction. Given operators specially trained in the handling of this machinery, this method yields meritorious results.

4. TRANSPORT

Transport operations in the exploitation of alluvial deposits (and of loose rocks) consist, after loading the crude materials at the working face, of reaching them to the receiving area of the processing plant.

4.1 General Approach

The economic optimisation of these operations is a priori simple: it consist of maximally reducing for the entire duration of exploitation, the distances (and level differences) between extraction faces and the processing plant.

Whatever the transportation means selected, such an objective can only be achieved by judicious choice of location of the processing plant and by laying down an appropriate overall phased plan of working the deposit, with due regard for the constraints associated with stripping and reclamation operations.

Thus optimisation is complex in practice. It emerges from the best possible compromise between all the prevailing constraints, that of the transport distance more often playing an overriding role beyond 500 to 700 m.

Here again the availability beforehand of a detailed geotechnical investigation of the deposit is indispensable.

In the case wherein external constraints (frequent as they are) dictate location of the processing plant far away from the deposit's centre of gravity, one will of course favour an overall phased minning increase in the transport distances, and not the other way around, so as to stagger the investments in machinery over a period of time.

Most cases are by and large open to two types of technical solutions, some based on continuous materials handling methods, others on discrete methods.

The former type of solution (continuous handling), except where impossible for particular reasons, is increasingly more widely employed for alluvial deposits (and loose rocks), both surface and partially submerged types, upwards of a certain threshold of production ($\geq$ 200 to 300 t/h).

4.2 Continuous Handling

This comprises:

• **Transport of extracted materials through a hydraulic medium** in the form of slurry (water + solid mixture) by pumping and forcing through pipes.

This solution is employed exclusively for submerged or partially submerged deposits, when extraction is effected by a suction dredger directly pumping off the materials to the processing plant (through floating pipework).

In this case the extraction and transport operations constitute an inseparable whole; accordingly this should be resolved by the choice of a machine compatible with all the constraints incidental to exploiting the deposit.

• **Transport of extracted materials by belt conveyors** (belt conveyors, see Chapter XIV).

This latter solution is fast expanding in major workings of alluvial materials; we shall examine it in greater detail.

This solution is preferred in the case of deposits which have a sufficient exploitation life (> 8 years) and in which the average thickness of the exploitable bed is at least 6 or 7 m (> 10 t of marketable materials/ m^2 = 100,000 t/ha), which implies a none-too-fast advance of the extraction faces.

In other respects, to be in a position to implement this solution well, it is necessary for the deposit not to have too complex a shape (in plan), nor pockets of steriles, nor include vary big private estates.

In all cases the exploitation phasing study should be particularly thorough and appropriate. The rhythm of progress of the working faces should be worked out with great precision (planning of extension of the belts).

Case No. 1

With submerged or partially submerged deposits extracted by grab dredger or airlift dredger, this technique is almost invariable adopted. The dredger is then linked to the processing plant through a chain of conveyor belts held upon floaters and mobile, whereby the dredger is in a position to 'flutter' freely while carrying out extraction.

Case no. 2

With partially submerged deposits extracted by a 'digging' dragline, the pneumatic-tyred loader, reclaiming the materials drained of their water in the pile, loads a mobile feed hopper for the belts. These *belts*, installed on top of the deposit, have *stationary elements* and *shiftable elements* (which can be moved by 'shifting' at right angles to the stationary elements). The shiftable elements are installed parallel to the extraction face (30 to 60 m behind) and should move along with its progress, albeit periodically falling behind because of shifting operations (case of deposits extracted by digging) (Fig. VIII.8).

The *movable feed hopper* is regularly moved along the shiftable elements so as to always maintain the hauling distance for the loader below 40 m between itself and the point of reclamation.

The movable feed hopper should be equipped, almost as a matter of course, with 'scalper grids' in order to preclude transport of elements larger than 200 or 300 mm on the belts.

If such elements are present in too high a proportion (> 2 to 5%) in the deposit, the solution of transport by belt has to be discarded.

The adoption of such a course of exploitation implies the formulation of an extremely 'geometric' phasing plan based on rectangular panels. The working faces advance consecutively 'in a square pattern' in a well-defined and uniform manner.

The successive shifting operations of the belts are relatively fast and easy (400 m of belts can be shifted by 40 m in an 8-hour shift, with a loader specially adapted for this kind or operation) when the surface of the terrain (top of the deposit) is relatively plane and uniform. If it is not, a gross levelling operation (pneumatic-tyred loader and leveller) is generally done before each shifting.

Case no. 3

With surface deposits two options are available: either extraction 'by digging' or extraction 'by scraping'.

— In the former case the conditions are similar to those in partially submerged deposits except that the extracting machine itself is capable of feeding the movable feed hopper of the belts (which generally entails a drop in extraction efficiency).

— In the latter case extraction by scraping and loading of the movable hopper are most often looked after by a pneumatic-tyred loader.

The belts are installed on the bedrock of the deposit. The shiftable elements, positioned parallel to the progress of the face under extraction, go along with it due to successive shifting operations (Fig. VIII.9).

The movable feed hopper is regularly moved along the shiftable elements so as to always maintain the hauling distance between the loader and the scraping extraction point to below 40 m.

Whatever the type of deposit, use is preferably made of conveyor belts which are 800 mm broad for extraction rates varying from 300 to 600 t/h, and belts 1000 mm broad for extraction rates varying from 600 to 900 t/h.

The most commonly adopted belts speeds are of the order of 2 m/s.

Whenever it can be adopted on a site under exploitation, this mode of transporting the materials is by far the most economical.

4.3 Means of Discrete Materials Handling

These are the most conventional and to this day still the most widely employed on sands and gravels deposits, barring special cases of deposits extracted by suction dredgers, grab dredgers and dredgers with facilities on board for processing.

Transport is accomplishing in such cases by mobile machines shuttling between the extraction face and the receiving point of the processing plant.

In the case of *submerged or partially submerged deposits* extracted by means of bucket dredgers (equipped with a simple extraction machine, without onboard

1. Extracted pit (under water)
2. Drag shovel
3. Extracted pit run piled in rows
4. Loader
5. Mobile hopper
6. Shiftable conveyor belt
7. Stationary Belt
8. To plant

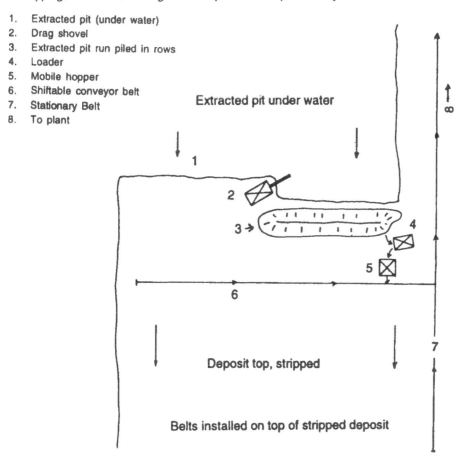

Fig. VIII.8. Transport by conveyor belts. Partially submerged deposit (extraction by digging)

processing plant), transport to the processing plant is done either by floating crafts such as the *shutter boat* or a barge (implying a reclamation machine on the quay, such as a grab-bucket hoist, a crane or a grab pontoon), or by *self-discharging hopper boats*.

By way of recall, mention may also be made of the case wherein *pneumatic-tyred loaders* are utilised as *extraction-cum-transport machines* on *surface deposits* for transporting materials to the receiving point of the plant.

In the latter case a solution of this type is projected only for low volumes of production and for distances within 150 m between the extraction face and the primary hopper of the plant. However, this solution may also be adopted in the case of temporary exploitations, with portable processing units (portable plants for co-ordinated overlay jobs or highway jobs...).

However, as a general rule, for *surface deposits* or *partially submerged deposits* the discrete materials handling means adopted is the *dumper* when the deposits are extracted by digging. Use of this type of machine having already been detailed

1. Shiftable felt
2. Movable hopper
3. Loader
4. Extraction face
5. Stationary belt
6. To plant

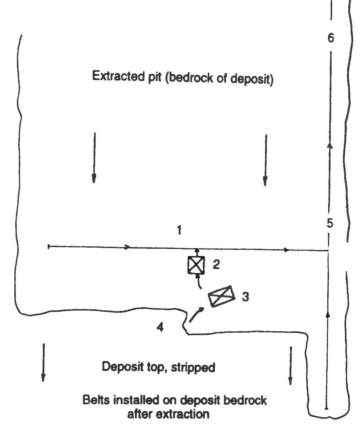

Fig. VIII.9. Transport by conveyor belts: surface deposit (extraction 'by digging')

in Chapter VII, we shall limit ourselves to recalling the main constraints to be taken into account on choosing and deploying it in exploitation (see Tables VIII.6 and VIII.7):

1. good compatibility between the loading machine and the dumpers:

— discharging height of the loading machine at least equal to the loading height of the dumper bucket;

— capacity of dumper buckets matched with capacity of the loading machine's bucket; the dumper bucket should be capable of being fully loaded with 3 loading buckets minimally and 5 maximally.

2. Minimum idle period for loading, between 2 and 4 minutes.

Table VIII.6: Dumpers. Capacity of transport as function of distance to cover (t/h)
(medium haulage conditions)

Distance to cover	500 m	750 m	1000 m	1250 m	1500 m	2000 m
Dumper capacity						
20 t payload ...	140	120	105	95	85	70
27 t payload ...	190	160	140	125	110	95
32 t payload ...	230	195	170	150	135	110
45 t payload ...	320	270	235	210	190	160

Table VIII.7: Optimal capacities (t of payload) of dumpers for different ranges of
output at the working

Hourly capacity of output at the working	200 to 300 t/h	400 to 500 t/h	600 to 800 t/h
Optimal unit capacity of dumpers (t of payload)	20–22 t	30–32 t	45–50 t

3. Minimum idle period for discharging, which implies a design specially tailored to the primary reception point at the processing plant, which maximally minimises the risks of waiting.,

4. Minimum transport time, which implies paying special attention to the alignment and maintenance of the haulage tracks at the working site.

5. Good matching (regularity of cycles) of the number of dumpers deployed with the travel distance and transport capacity, to limit the risks of waiting at the loading point.

This type of transport machinery becomes an *obligatory choice* when the extracted material has a *high proportion of* elements larger than 200 to 300 mm φ (blocks).

IX

Crushing and Grinding

Vincent Duthoit

1. GENERAL

Crushing and grinding figure amongst the oldest activities which man has ever invented. Already during the Paleolithic period our ancestors were manufacturing tools, weapons and projectiles using broken and sharpened stones while their female companions were grinding, for hours together, roots and cereals in stone mortars, which were the forerunners of modern household appliances (Fig. IX.1).

Of course, there was no distinction between crushing and grinding then; even today this distinction is not always properly understood.

Crushing is the division of hard materials into coarse chips; grinding is the reduction of a material into finer chips.

At present, crushing and grinding are two essential functions in the treatment of all rocks and minerals, whatever be their end use, and in particular for the production of aggregates.

The enterprise which undertakes crushing or grinding of rocks which can reach hardnesses and strengths comparable to those of certain metals does not appear mechanically, at first sight, compatible with advanced engineering appliances. However, we have come a long way from the beginning of the century when crushing was still carried out by hundreds of men and women equipped with sledge hammers.

This work was all the more despicable because generally assigned to convicts.

After generations of simple and rustic crushing tools, modern appliances not only make use of advanced mechanical engineering, but also hydraulic and even electronic engineering, which enable automatisation of their function with considerable reliability.

Research is presently underway to obtain fragmentation of rocks by means other than mechanical. Yet it appears that the energy involved is so great that the industrial application of such a process would not, at least for the time being, be profitable.

After an overview of the fundamental laws of fragmentation, we shall successively study the different families of crushing and grinding appliances. We have classified them into three categories depending on whether they operate:

Fig. IX.1. Beginnings of crushing

— By crushing between two parts, one stationary and the other mobile or both mobile;

— By impacts and projections on fixed anvils;

— By gravity.

After a few observations on the technology of crushing and grinding applian-ces and a few thoughts on future prospects in this field, we shall give reasons for selecting a particular appliance or chain of appliances in dealing with a specific problem.

2. OVERVIEW OF FUNDAMENTAL LAWS OF FRAGMENTATION

In scientific terms crushing and grinding are called fragmentation and communica-tion.

Industrial operations of fragmentation consume much energy, representing about 5% of the total consumed at the national level, with an extremely low output. For this reason scientists have looked into the problems of the energy required for fragmentation of solids since the middle of the last century.

Simply speaking, we can say that fragmentation of a solid into two element consists in defeating the forces of cohesion which bind these two elements togethe before rupture.

2.1 Rittinger's Law

Fragmentation produces new surfaces; we can therefore expect that the energy consumed will depend on these new surface areas. It is from this assumption tha the German Von Rittinger proposed a law in 1867 according to which the energy absorbed during fragmentation is directly proportional to the increase in overal surface area of all the grains formed during grinding. Let us assume an initia block of size D and volume D^3 to be reduced into N particles of size d and volum d^3.

$$N = \frac{D^3}{d^3}.$$

If the energy consumed is proportional to the difference in the specific areas we shall have

$$W = K \left[S_D - \frac{D^3}{d^3} S_d \right]$$

$$= K_1 \left[D^2 - \frac{D^3}{d^3} d^2 \right]$$

$$= K_1 D^3 \left[\frac{1}{D} - \frac{1}{d} \right],$$

i.e., by reducing to the unit of volume:

$$\frac{W}{V} = K_1 \left[\frac{1}{D} - \frac{1}{d} \right].$$

According to Rittinger, the coefficient K_1 is constant for a given material tha is supposedly homogeneous.

This law shows that with equal reduction, the smaller the initial particles, the more energy required to fragment them; similarly, the greater the reduction, the higher the energy.

Studies have confirmed Rittinger's law in the mid-area of crushing, assuming that we obtain grains of uniform size, but this law may not be correct in the case of fine grinding due to reagglomeration.

2.2 Kick's Law

The above law does not take into account the compressive energy of the elements before rupture. That is why in 1885, Friedrich Kick from Germany proposed a new law according to which the energy required for crushing is proportional to the reduction of the volumes of all the elements from the precrushing size D to the post-crushing size.

This law is based on the stress-deformation diagram during compression of a cube.

Let an initial cube of volume D_o^3 be compressed on one of its faces; the elementary energy required for a small compression is:

$$dW = K_2 [D_o^3 - D_o^2 (D_o - dD_o)]$$

$$dW = K_2 D_o^2 [D_o - D_o + dD_o]$$

$$dW = K_2 D_o^3 \frac{dD_o}{D_o},$$

i.e., by reducing to the unit of volume

$$dW = K_2 \frac{dD}{D}$$

and for a reduction from size D to size d:

$$W = \int_D^d K_2 \frac{dD}{D}$$

$$W = K_2 (\log D - \log d)$$

$$W = K_2 \log \left(\frac{D}{d}\right).$$

Unfortunately, this law is only valid for homogeneous bodies and is not applicable to fine grinding.

2.3 Bond's Law

The above two theories had their supporters and detractors but it was only in 1951 that the American Fred C. Bond proposed an intermediate law linking the energy consumed by fragmentation not with D^2 as Rittinger did, nor with D^3 as Kick did, but with $D^{2.5}$.

Bond was of the view that the energy consumed is proportional to volume D^3 before fragmentation during elastic deformation since it becomes proportional to the area D^2 when cracks begin to appear, which means that in order to bring to size D a theoretically infinite block the energy required is:

$$W = K_3 D^{5/2},$$

that is, reduced to the unit of volume:

$$W = K_3 \frac{D^{5/2}}{D^3},$$

$$W = \frac{K_3}{\sqrt{D}} \cdot$$

In order to grind a material of which 80% will pass through a mesh of size D into elements of which 80% will pass through a mesh of size d, the energy required is therefore:

$$W = W_i \left(\frac{1}{\sqrt{d}} - \frac{1}{\sqrt{D}} \right),$$

where W_i is Bond's work index, i.e., the energy (in kWH) required to reduce a short tonne of material of theoretically infinite initial size to that size of elements of which 80% will pass through a 100 μm sieve, D and d being expressed in microns.

Table IX.1

Materials	Density	Work index, W_i
Andesite	2.84	18.25
Clay	2.51	6.30
Barite	4.50	4.73
Basalt	2.91	17.10
Bauxite	2.20	8.78
Limestone	2.66	12.74
Silicon carbide	2.75	25.87
Coal	1.40	13.00
Cement	2.67	10.51
Cement clinker	3.15	13.45
Coke	1.31	15.13
Diorite	2.82	20.90
Dolomite	2.74	11.27
Emery	3.48	56.70
Feldspar	2.59	10.80
Ferrochromium	6.66	6.64
Ferromanganese	6.32	8.30
Ferrosilicon	4.41	10.01
Gabbro	2.83	18.45
Pebble	2.66	16.06
Gneiss	2.71	20.13
Granite	2.66	15.13
Graphite	1.75	43.56
Gypsum	2.69	6.73
Slag	2.74	10.24
Magnesia	3.06	11.13
Copper ore	3.02	12.73
Tin ore	3.95	10.90
Iron ore		
Haematite	3.53	12.84
Specular haematite	3.28	13.84
Oolite	3.52	11.33
Magnetite	3.88	9.97
Taconite	3.54	14.61

Table 1 (contd.)

Materials	Density	Work index, W_i
Manganese ore	3.53	12.20
Nickel ore	3.28	13.65
Gold ore	2.81	14.93
Lead ore	3.35	11.90
Titanium ore	4.01	12.33
Zinc ore	3.64	11.56
Zinc and lead ore	3.36	10.93
Molybdenum	2.70	12.80
Phosphate	2.74	9.92
Potassium	2.40	8.05
Pyrite	4.06	8.93
Pyrrhotine	4.04	9.57
Quartz	2.65	13.57
Quartzite	2.68	9.58
Rutile	2.80	12.68
Siliceous sand	2.67	14.10
Schist	2.63	15.87
Bituminous schist	1.84	15.84
Flint	2.65	26.16
Fluorspar	3.01	8.91
Syenite	2.73	13.13
Trap	2.87	19.32
Glass	2.58	12.31

Bond, using a laboratory grinder (Fig. IX.2), determined the grindability indices of a large number of materials (Table IX.1).

Bond's theory is superior to the earlier two, because it is applicable to fine as well as coarse grinding; however, a major drawback is that it does not take into account the grading of the aggregates.

2.4 Gaudin's Law of Granular Distribution

If P is the percentage of undersizes at dimension x, we shall have for most grading curves:

$$P = 100 \left(\frac{x}{\Delta}\right)^{\alpha}$$

where Δ and α are two parameters.

This law has been proved for about 80% of all under-sizes.

2.5 Works of Schumann and Charles

Starting from the fact that all the laws mentioned above regarding the energy absorbed by crushing are of the form:

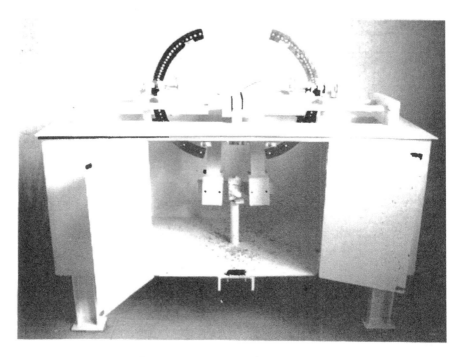

Fig. IX.2. Bond's test with laboratory grinder

$$dW = K \frac{dx}{D^n}$$

$$W = \frac{K}{n} \frac{1}{D^{n-1}},$$

Charles observed that the exponent n–1 was very close to the exponent of Gaudin's law of distribution.

Schumann thus arrived at the following equation:

$$W = \frac{K}{\alpha} \frac{1}{D^\alpha}$$

where K is a coefficient specific to the material and α the manner of granulometric distribution.

Under these conditions, taking into account the fact that α varies between the final state (f) and the initial state (i), we arrive during a grinding from D to d at the following relation:

$$W = K \left(\frac{1}{\alpha_f d^{\alpha f}} - \frac{1}{\alpha_i D^{\alpha i}} \right).$$

More recently, Capdecomme measured the evolution of parameter d_{50} (size of mesh giving 50% undersizes) as a function of the duration of grinding and showed the proportionality between this parameter and the specific areas of the material, thereby confirming Rittinger's law for fine grindings.

Capdecomme's study enables us to realize the complexity of the problems of crushing and grinding while studying them theoretically.

Let us now review the various equipment normally used in practice, it being understood that in each particular case only the experience of the designers and users makes possible the best choice of the most suitable machines.

CRUSHING AND GRINDING EQUIPMENT

3. CRUSHING EQUIPMENT

These are:

— either gyratory or jaw crushers with one grinding part stationary and the other movable;

— or double-roll crushers in which both grinding parts are movable.

3.1 Gyratory Crushers

These are the most modern and the most elaborate equipment. In fact, because of their suitability for solving most problems and the reduction ratio they can affect and that of the outputs they are capable of, gyratory crushers are the most commonly used. That is why they have been subjected to constant improvement since their invention at the end of the last century. Their general principle continues to be the same but often with very different versions depending on the use to which they are put.

3.1.1 PRINCIPLE OF THE GYRATORY CRUSHER

The bottom of the main shaft being engaged in an eccentric, it describes a circle; its head being fixed, the shaft describes a cone around the vertical axis of the apparatus. At mid-height of this shaft a part called the mantle or cone, oscillates according to the movement of the shaft inside a fixed part called the bowl (Fig. IX.3).

During operation of the equipment underload the shaft is driven in a slow rotation and the mantle rolls on the material, which reduces wear of the wearing parts.

When the material is introduced between the two crushing parts, one portion is subjected to compressive loads while the remainder is decompressed, which allows it to move by gravity to the base of the equipment.

Thus we see that the work of gyratory crushers is a continuous and balanced one.

In the geometry of gyratory crushers two techniques are involved; one with suspended shafts with the fixed point on the top, and the other with shafts embedded in the eccentric of the mechanism.

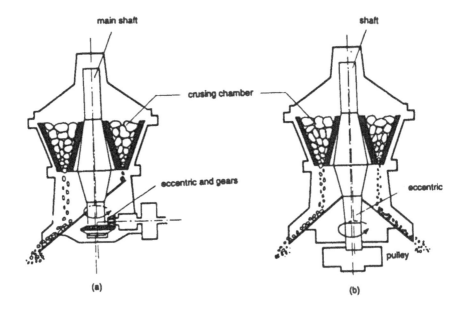

Fig. IX.3. Primary gyratory crusher

At first sight, the principle of suspended shafts appears simpler and more rational from the mechanical point of view, the forces coming into play being lower. But equipment without a fixed head has the great advantage of a completely free intake opening. This device would be of very great importance to primary crushers; however, for obvious reasons of construction, no one has ever ventured to manufacture a headless primary gyratory crusher. On the other hand, this arrangement has definite advantages for secondary equipments.

Originally, all gyratory crushers were mounted either on antifriction metal or on bronze rings. For about thirty years now a few manufacturers have been trying to improve the efficiency and capability of their equipment by mounting the mechanism on roller bearings lubricated with oil or grease. Due to the brittleness of these bearings, however, most users prefer crushers mounted on rings, the reliability and strength of which have been amply tried and tested.

Among the gyratory crushers, we shall distinguish:
— primary equipment,
— secondary equipment,
— tertiary and quaternary equipment.

3.1.2 PRIMARY GYRATORY CRUSHERS

All primary gyratory crushers generally have approximately the following configuration:
— Shaft suspended on top in a spider.

— Crushing chamber whose bisector of the angle of the jaws is practically vertical.

— Shaft driven at the base by an eccentric mounted either on bearings or on bronze bushings.

— Mechanism driven either directly by a vertical axis pulley placed under the eccentric or by means of a horizontal countershaft with gears serving as speed reducers.

— Discharge arrangement either through two lateral spouts (Fig. IX.4) located on either side of the base of the equipment or through annular openings under the equipment which enable a direct discharge of the crushed material (Fig. IX.5).

The design of the primary gyratory crushers highlights the factors which may be important in efficiency and production.

— Speed of rotation of the eccentric may vary. The output of the equipment is directly proportional to the speed of rotation, at least in a certain speed range.

— The Nip angle between the wearing parts should not exceed 28°, otherwise the blocks may jump in the grinding bowl several times before being absorbed.

— Eccentricity at the bottom of the shaft should be sufficient for movement clearance at the level of the wear parts to enable efficient crushing. The output is directly proportional to this eccentricity.

— Position of the fixed point for suspending the shaft with respect to the level of the crusher intake opening is of great importance. In fact, it governs the amplitude of movement of crushing in the entry zone of the largest blocks, i.e., where it is most required. So preference will always be given to equipment with a high spider with well-inclined arms. This arrangement facilitates the design of large intake openings and in addition accepts vertical loads.

— Adjustment of conventional equipment is done by means of a nut screwed onto the upper part of the shaft. Some manufacturers have replaced this system by a hydraulic jack located either on top or at the bottom of the shaft, which permits almost instantaneous setting without having to stop the equipment and dismantle the top parts. Thus the mechanical parts of the latter always remain protected from dust.

The size of the largest primary gyratory crushers is taking on more importance. The largest crushers presently in service have intake openings reaching radially more than 2 metres, which allows introduction of blocks of unit weight 10 to 15 t and thus production that can exceed 3500 t/h (Fig. IX.6).

Gyratory crushers are usually designated by the radial size of the intake opening expressed in inches followed by the diameter of the swinging mantle in its lower part, also expressed in inches. In fact, these are the two values which best characterise the capacities of the equipment since the first governs the size of the blocks absorbed and the second governs the output which the equipment is capable of at a given setting.

3.1.3 SECONDARY GYRATORY CRUSHERS

These are units meant to take up the material precrushed at the primary stage in order to bring it down to a size permissible for fine crushers.

By and large, secondary gyratory crushers are not used for manufacturing finished materials except for ballast. So such units are expected to have an intake

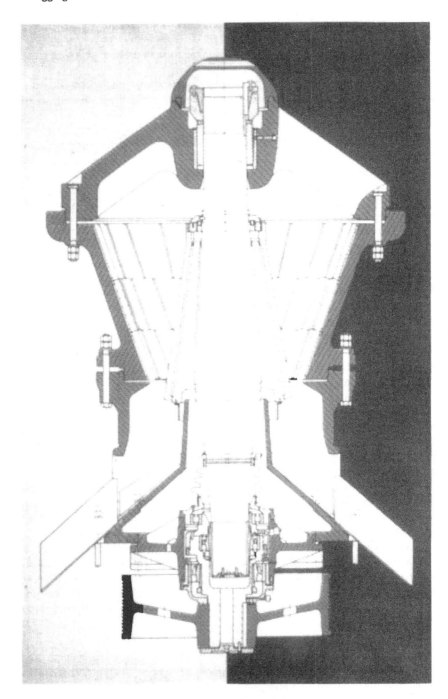

Fig. IX.4. Double-discharge primary gyratory crusher

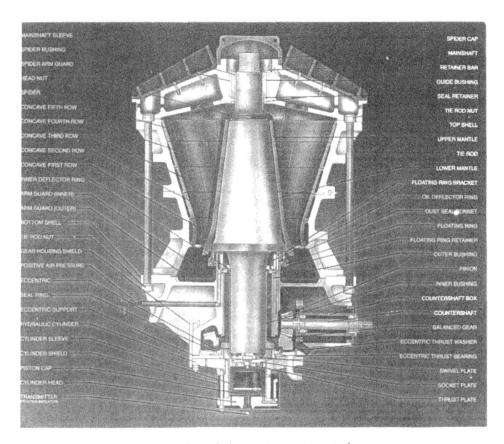

Fig. IX.5. Direct-discharge primary gyratory crusher

opening as large as possible to expedite feeding as well as a large output since the output of the production plant is governed by them.

Furthermore, they should effect a significant reduction in material without producing too high a percentage of fine elements whose shape factor runs the risk of not being acceptable.

Secondary gyratory crushers are based on the same principle as primary ones. They should permit a proper intake of materials and should have a grinding chamber that facilitates introduction and evacuation.

Depending on the model, secondary gyratory crushers can be adjusted in two ways as soon as the cavities show wear, either by lowering the bowl with respect to the main shaft (the level of which remains fixed) or by raising the shaft with respect to the fixed bowl.

In the first case the most commonly used system is to screw the bowl inside the thread provided on the upper part of the framework, either mechanically or hydraulically (Fig. IX.7).

Fig. IX.6. Primary gyratory crushers 61–110
Radial opening 1.50 m
Height 8.2 m
Diameter 5.4 m
Mass 430 t

In the second case the shaft can be adjusted either by a nut located on top, or by a hydraulic jack situated at the top or at the bottom. This jack can also be used as a safety device against accidental overloads (Fig. IX.8).

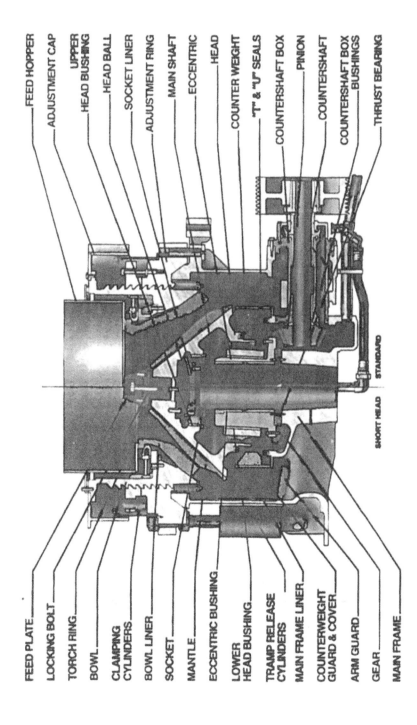

FEED HOPPER

ADJUSTMENT CAP

UPPER
HEAD BUSHING

HEAD BALL

SOCKET LINER

ADJUSTMENT RING

MAIN SHAFT

ECCENTRIC

HEAD

COUNTER WEIGHT

"T" & "U" SEALS

COUNTERSHAFT BOX

PINION

COUNTERSHAFT

COUNTERSHAFT BOX
BUSHINGS

THRUST BEARING

FEED PLATE

LOCKING BOLT

TORCH RING

BOWL

CLAMPING
CYLINDERS

BOWL LINER

SOCKET

MANTLE

ECCENTRIC BUSHING

LOWER
HEAD BUSHING

TRAMP RELEASE
CYLINDERS

MAIN FRAME LINER

COUNTERWEIGHT
GUARD & COVER

ARM GUARD

GEAR

MAIN FRAME

SHORT HEAD   STANDARD

Fig. IX.7: Cone crusher designed for adjustment by 'screwing' the bowl with complete hydraulic control

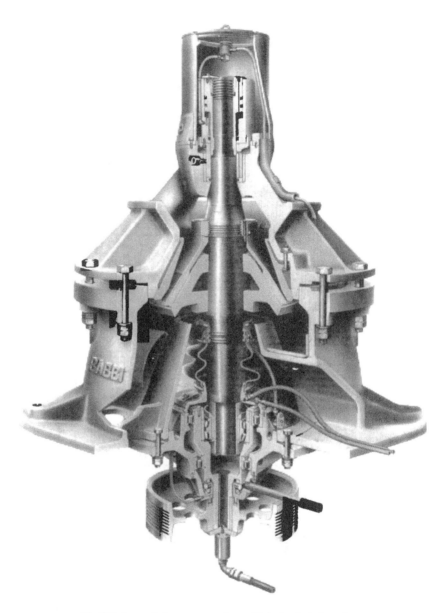

Fig. IX.8: Suspended gyratory crushers with hydraulic jack at the top

The first device is used in headless cone equipment. Safety is then ensured by raising the whole assembly (bowl and fixed jaw) formerly on a battery of springs or on a battery of hydraulic jacks connected to gas accumulators. The advantage of this device is that the fixed jaw can instantly return to its initial position after the passage of an uncrushable body or a momentary overload.

Manufacturers offer therefore hydraulic device which makes it possible to adjust the setting without manual intervention.

Most of the modern equipment, regardless of whether the safety device is secured through a hydraulic jack or a battery of springs, can start up under load after any untimely stoppage (one due, for example, to a power failure). Start-up is effected either by activating the adjustment hydraulic jack or by compressing the battery of springs via the hydraulic device in-built in the equipment.

Let us not forget that in the older models of gyratory crushers, half an hour to one hour of work was often required to manually clean the equipment before it could be restarted.

Like primary equipment, secondary gyratory crushers are direct driven either on a vertical axis pulley (Fig. IX.9a), placed under the mechanism, or by a horizontal intermediate shaft which drives the eccentric by means of a pinion and a gear (Fig. IX.9b). The advantage of the second device is that the bottom part of the equipment is left free for evacuating the material.

3.1.4 TERTIARY OR QUATERNARY GYRATORY CRUSHERS

When dealing with hard and abrasive materials, especially for roadworks, if we want to obtain stone chippings with a good shape, only gyratory crushers can be used. In fact, they are the only equipment capable of dealing with large outputs with a satisfactory reduction ratio (of the order of 5 to 6) and a relatively modest wear even with very abrasive materials containing a high percentage of silica.

The general principle of the equipment is similar to that of primary and secondary gyratory crushers but in this last case various manufacturers have been concentrating over the past fifteen years on technical improvements that would create ideal conditions for production of good shape aggregates.

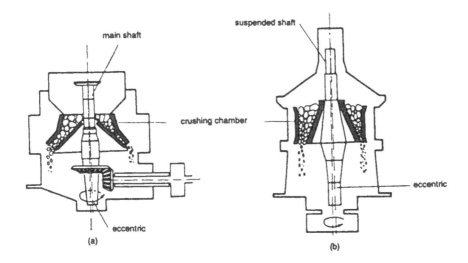

Fig. IX.9: Diagram of secondary gyratory crushers

The following are those conditions:
— proper speed of crushing,
— large number of impacts on all the faces of the products,
— grinding in dense, compact and homogeneous environment,
— obtained by good design of the crushing chamber.

• *Mechanical characteristics*

To obtain the best quality, both from the point of view of shape and granulometric curve of the aggregates produced by the equipment, it is very important that the grinding bowl, especially in the bottom part, be completely filled so as to obtain a compactness favouring autogrinding.

Crushing by attrition or autogrinding is based on fragmentation due to the simultaneous action of bending, shear, punching and crushing stresses in a highly compressed milieu.

So, independent of even the design of the crushing chamber, it is very important that the ratio between the speed of rotation and the eccentricity at the level of the crushing parts be perfectly suitable to the problem to be solved.

An equipment in which the rotational speed is too high may cause an insufficient intake of the material in the feed zone and thereby a reduction in throughput that does not allow correct feeding of the lower portion of the crushing parts. On the other hand, a too high eccentricity at the exit point engenders too much evacuation of the products and thereby reduces compactness and consequently autogrinding.

So, it is very interesting to be able to vary, to some extent, these two parameters—rotational speed and eccentricity—according to the problem to be solved.

• *Characteristics of the crushing cavities*

A very important concept is that of the elementary volume of the crushing chamber per unit length in the direction of movement of the product.

This elementary volume is the product of the elementary area of the vertical section by the length of the circumference.

Primary and secondary crushers have crushing cavities with a highly decreasing volume. Tertiary equipment have crushing cavities with a constant volume or slightly decreasing volume in the particular case of equipment with a short head. This leads to cones that are very wide-mouthed towards the bottom. This arrangement helps autogrinding and fragmentation of the long and flat elements. In fact, the crushing operation already reduces in a significant manner the apparent volume of the material by filling the voids with fine elements. In addition, the products should be slightly slowed down at the exit in order to favour crushing by attrition.

The grinding chamber of a tertiary gyratory crusher can be divided into three separate parts:
— entry zone,
— precrushing zone,
— parallel finishing zone.

Entry zone of materials: In this zone where only the largest elements are broken, a too large intake angle should be avoided in order to facilitate feeding even at a high speed. Further, one should always be careful about equipment with an opening too large compared to the exit opening because most of the time the grinding angle is too big, which is harmful to the proper feeding of the finishing zone.

Precrushing zone: Reduction in apparent volume takes place in this zone by the filling of voids.

Parallel finishing zone: Impacts are the largest in number in this zone and here, too, autogrinding and rupture of the long and flat elements take place.

Thus we see that the mechanical characteristics of the gyratory crusher and the physical characteristics of its crushing chamber have an extremely important influence on the output and quality of the products. Also, while taking into account the broad principles mentioned above, it may be necessary, when experience is lacking, to adapt the characteristics of the equipment to the materials to be processed. Hence it is advantageous to choose equipment for which the manufacturer has originally provided for a range of speeds and eccentricity, and crushing design which are readily adaptable (Fig. IX.10).

Like primary and secondary equipment, most tertiary gyratory are equipped with a hydraulic adjustment device and a protective device against overloading, which for this type of equipment have several advantages:

— Frequent adjustment may be necessary on this crushers both for compensating wear of the grinding pieces and for modifying the grading curve of the finished materials.

— The tiniest metal piece or the smallest uncrushable body can cause considerable damage to equipment functioning with a setting of a few millimetres and only a reliable safety device ensures efficient protection. It is very important that after the passage of an uncrushable body, the equipment automatically return to its original setting without intervention (Fig. IX.11).

— As in the case of secondary equipment, it is very important for this equipment also to have a device that enables cleaning and restarting it after an unexpected stoppage.

— Manufacturers already offer automation of the working and adjustment of crushers, especially tertiary crushers.

This automation is possible only in the case of equipment with a remote-control hydraulic adjustment device.

The principle consists of constant check on the power consumed by the equipment, output grading, setting of the jaws or the output produced. This data is received by a computer which, whatever be the wear of the grinding pieces, enables production of a material for which the granulometric curve is constant.

It has been observed that for most of the equipment it is advisable to carry out frequent but minor adjustments, thereby maximally restricting the wear of the grinding pieces.

3.1.5 GYRATORY CRUSHERS FOR PRODUCTION OF SAND

Some twenty-five years back a new technique came into being. It consists of using gyratory crushers fitted with a grinding chamber of special shape for large-scale

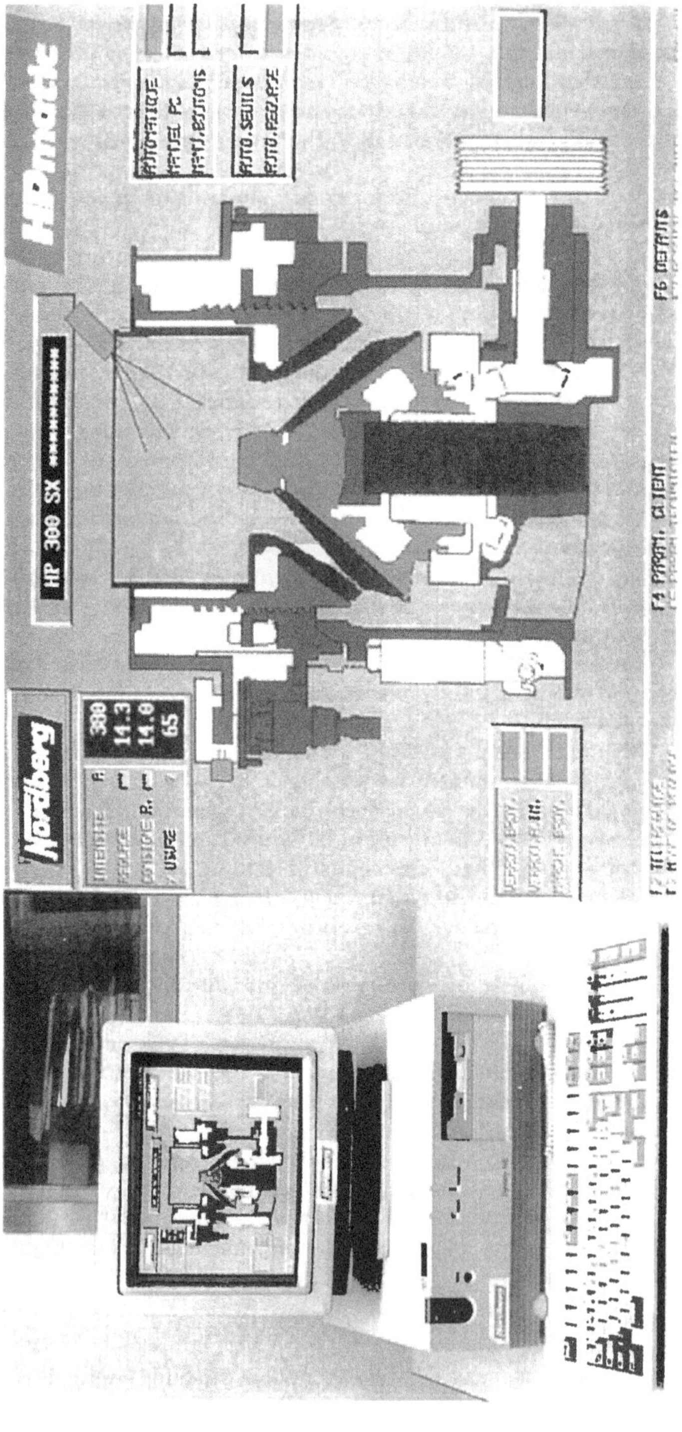

Fig. IX.10. 'Omnigraw' window. Fully hydraulic system.

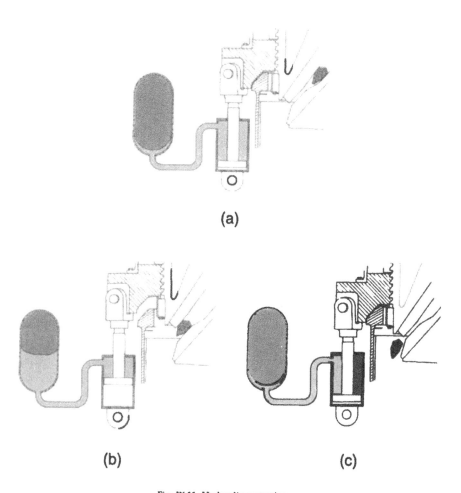

(a)

(b)   (c)

Fig. IX.11. Hydraulic protection
Protection against ungrindable elements is obtained by means of maintenance jacks. The nitrogen
cylinders absorb the overfeeds and reposition the adjustment ring on the frame. The adjustment
remains unchanged.
a) Normal position of grinding. Adjustment ring rests on frame. b) Passage of an ungrindable body.
Elevation of adjustment ring. The jack sends back the oil to the cylinders by compressing the nitrogen
reserve. c) Return to normal position. Adjustment unchanged.

production of sand under interesting economic conditions compared to those of a
rod mill. The sand produced is constantly filtered of 14 or 15% fines for 0/2 mm
fraction.

Such equipment work at once by impact and attrition (Fig. IX.12).

The 'attrition chambers' of this equipment are characterised by the absence of
an entry zone and precrushing zone. There is a single zone corresponding to the

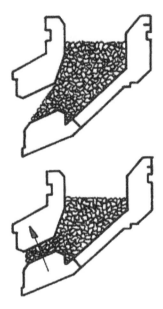

Fig. IX.12. Gyratory sand crushers, the 'Gyradisc'. The closing-in motion of the jaws causes a sudden increase in compression inside the chamber and mainly at the end of the jaws. Literally crushed on top of itself, the material undergoes autogrinding through attrition

finishing one in which the grinding pieces come into a forceful striking action. The slope of the jaws in this zone is only 25°, due to which the material cannot escape by virtue of its own weight. It is the large striking action of the swinging jaw which causes the forward motion and evacuation of the crushed material. At each impact of the swinging jaw, a thick sandwich layer of material is created inside the crushing chamber. The compression of this layer is suddenly increased and the feed material particles are crushed against each other; this results in production of a substantial percentage of sand and a remarkable cubic configuration of the stone chips.

Since crushing takes place in a thick layer, the crusher setting has little influence on the size of the products obtained. A major percentage of the feed material does not come into contact with the jaws, a noteworthy fact in economising on wearing parts.

This equipment, which accounts for a considerable production of sand, should wherever possible, be operated in a closed circuit with substantial recycling to facilitate autogrinding and attrition.

3.2 Jaw Crushers

The general principle of this equipment is as follows: The material is crushed between two jaws, one swinging and the other fixed. Movement of the swinging jaw differs according to whether the equipment is single or double toggle (Fig. IX.13)

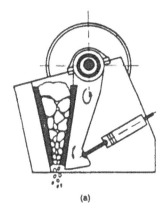 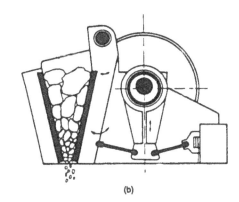

(a)   (b)

Fig. IX.13. Jaw crushers. a) single toggle; b) double toggle

3.2.1 SIMPLE TOGGLE JAW CRUSHERS

This equipment consists of a frame of cast steel or, as in modern equipment, mechanically welded. The fixed jaw is mounted on one inner face of the frame. The pitman carrying the swinging jaw is suspended on an eccentric shaft rotating between two roller bearings placed in the upper section of the frame (Fig. IX.14).

This shaft is provided with two flywheels carrying counterbalancing masses. Owing to the eccentricity, the upper section of the pitman describes a circle while the lower section describes only little arcs of a circle as it butts upon an oscillating toggle plate having two toggle seats. The horizontal amplitude of the swinging jaw in its lower part is essentially governed by the inclination of the plane of the toggle plate to that of the pitman. If this inclination is inadequate, a neutral zone results in the crushing chamber where no flow takes place and the crusher yields no output.

In one complete cycle the swinging jaw starts by moving away from the fixed jaw in an upward direction, then comes close to the latter in a downward motion which carries the products along towards the exit.

We shall see later that the geometry of the jaw crushers usually exerts a certain influence on their capacity (Fig. IX.15).

3.2.2 DOUBLE-TOGGLE JAW CRUSHER

Two pitmans are fitted in a frame supporting the fixed jaw: one, the driving one, is suspended on an eccentric shaft carrying two flywheels. This pitman is inserted at its base between two articulation toggles, one of which butts against the back of the frame, the other upon the bottom of the pitman carrying the swinging jaw. The driving pitman, drawn along by the eccentric shaft, describes circles in its upper section, which engenders an up-down motion in the lower part. This vertical motion is transformed into a reciprocating horizontal motion transmitted to the bottom of the pitman supporting the swinging jaw which is suspended at the top

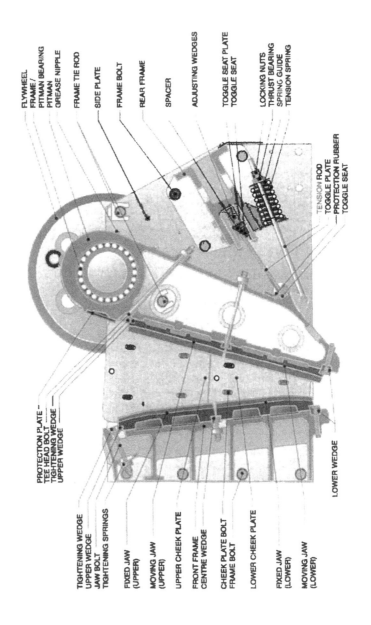

FLYWHEEL
FRAME /
PITMAN BEARING
PITMAN
GREASE NIPPLE
FRAME TIE ROD

SIDE PLATE

FRAME BOLT

REAR FRAME

SPACER

ADJUSTING WEDGES

TOGGLE SEAT PLATE
TOGGLE SEAT

LOCKING NUTS
THRUST BEARING
SPRING GUIDE
TENSION SPRING

TENSION ROD
TOGGLE PLATE
PROTECTION RUBBER
TOGGLE SEAT

PROTECTION PLATE —
TEE HEAD BOLT
TIGHTENING WEDGE
UPPER WEDGE

TIGHTENING WEDGE
UPPER WEDGE
JAW BOLT
TIGHTENING SPRINGS

FIXED JAW
(UPPER)

MOVING JAW
(UPPER)

UPPER CHEEK PLATE

FRONT FRAME
CENTRE WEDGE

CHEEK PLATE BOLT
FRAME BOLT

LOWER CHEEK PLATE

FIXED JAW
(LOWER)

MOVING JAW
(LOWER)

LOWER WEDGE

Fig. IX.14. Primary jaw crushers with hydraulic adjustment

Fig. IX.15. Cutaway view of a single-toggle jaw crusher. It can be seen that this equipment is provided with a convex fixed jaw and a concave swinging jaw.

on a second shaft, whereby it undergoes a swinging motion around this shaft. The motion of the swinging jaw is thus much simpler than for the single-toggle jaw crusher. As there is no vertical reciprocating motion of the swinging jaw, wear is less than in single-toggle equipment. That is why this type of equipment has long been preferred for highly abrasive materials, albeit there are some drawbacks:

— For the same intake openings, its price is nearly double that of single-toggle crushers.

— Owing to its proximity to the shaft's axis, the motion at the top of the swinging crusher is of very small amplitude. Hence it is currently recognised that a block of rock should be 20–30% smaller in size that the nominal dimension of the opening for proper entry into the machine.

— The weight of this double-toggle equipment prohibits installation on a mobile set.

3.2.3 SECONDARY AND TERTIARY JAW CRUSHERS

It may be recalled that this equipment is on the way out in modern crusher in
stallations given its low output and relatively high wear when processing abrasive
materials. The machines are almost identical to primary crushers but generally
have a wider intake opening and a higher rotation speed.

The jaws may have specially shaped teeth.

3.2.4 SHAPE OF CRUSHER JAWS

For primary as well as secondary crushers, flat jaws are often used. At present
the manufacturers supply curved jaws.

— Either both jaws are convex which, by reducing the angle of intake of the
material in the lower part of the crushing chamber, favours autogrinding while
retaining the same feed opening dimensions. But the angle of admission at the top
is often too big for the intake of large-size elements, especially if they are rounded

— Or one jaw is concave and the other convex, especially in the case of single
toggle equipment, which facilitates fragmentation even while reducing wear.

In most cases the jaws consist of one or more reversible parts so that the two
extremities of each can be used in turn.

3.2.5 ESSENTIAL FEATURES FOR CHOOSING A JAW CRUSHER

— Large intake opening (measured in closed position between the peaks and valleys of the teeth in the plane perpendicular to the bisector of the angle of the jaws)

— Jaws that are reversible and of large height.

— Intake angle of 19 to 22° maximum in open position.

— Mechanism mounted on very large-dimensioned roller bearings.

— Relatively square intake opening to accept the biggest blocks possible.

— Position and inclination of articulation toggles giving the maximum movement at the bottom of the swinging jaw.

— Proper protection of the upper part of the pitman which bears the impact
of big blocks.

3.3 Comparison between Jaw and Gyratory Primary Crushers

When a modern crushing installation is to be designed, the first aspect to be
decided is the type of primary crusher to be installed.

The fixed factors influencing the choice *a priori* are as follows:

— Petrographic nature of the rock.

— Its simple compressive strength.

— Its abrasiveness.

— Its degree of fragmentation by blasting.

— Dimension of the loading machinery and capacity of the buckets; these two
factors determine the maximum dimension of the blocks which can be delivered
to the primary crusher station.

— Average flow rate of material to be processed.

— Cleanliness of the deposit.

The first five of these parameters determine the dimension of the blocks to accept, that is, the intake opening of the crusher, so that the final choice of the type of primary crusher depends on the intake opening and the raw material flow rate.

In the great majority of quarries, for the primary crushing station the plant manager gives priority to a rugged machine with an adequate intake opening for accepting the biggest blocks from quarrying. The output of the crusher then becomes a secondary criterion.

Just the opposite scenario unfolds for very high-capacity plants; the plant manager chooses his crusher with due regard to the raw material flow rate and the intake opening size must then be proper.

We shall categorise the primary crushing plants on the basis of their outputs, pointing out nevertheless that the limits indicated obviously have no absolute validity (Fig. IX.16).

For output below 600 t/h, essentially single-toggle jaw crushers are used.

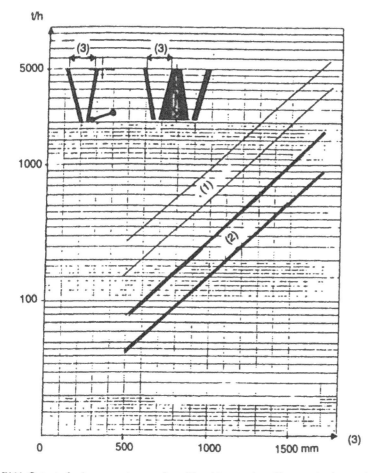

Fig. IX.16. Output of primary gyratory crushers (1) and jaw crushers (2) versus intake opening (3)

For output between 600 and 1000 t/h, the choice is more debatable. Howevei single-toggle jaw crushers are generally preferred if the quarrying tends to produc large boulders.

In respect of materials which are at once very hard and highly abrasive gyratory crushers are preferred.

However, it must be mentioned that the investment cost is much higher fo a gyratory crusher than for a jaw crusher, albeit its operating cost is lower.

For outputs above 1000 t/h, which go with very huge quarries, only gyrator crushers are suitable.

In conclusion, the choice of type of primary crusher calls for a careful comparativ study of the specificities and advantages of the different types of equipment.

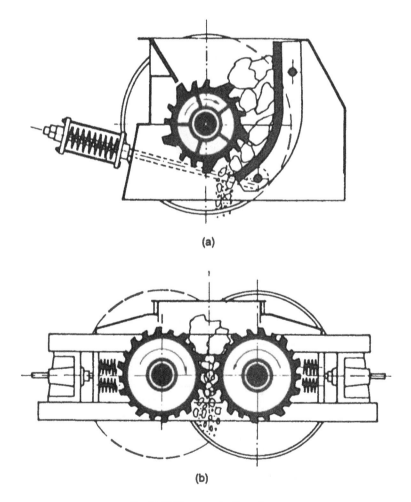

(a)

(b)

Fig. IX.17. Diagram of roll crushers

3.4 Roll Crushers

One should distinguish between double-roll crushers (Fig. IX.17b) and crushers with one roll and one jaw (Fig. IX.17a).

3.4.1 DOUBLE-ROLL CRUSHERS

These essentially consist of two rolls with parallel axes rotating in opposite directions and fixed inside a very rigid frame. Each line of shafting rests on babbitt swivel bearings or roller bearings, depending on the size of the equipment.

To prevent serious accidents in the event of oversized uncrushable tramp bodies finding their way into the equipment, one line of shafting or, better still, both lines ·of shafting are held in place by packs of springs which facilitate some movement in the horizontal plane. These lines of shafting return to their original position after the uncrushable material exits. Displacement of these lines of shafting does not necessarily take place in a parallel manner; therefore it is indispensable for the bearings to be mounted on swivels.

Depending on the end use envisaged, this equipment can be provided with:
— Liners with slugger teeth (sheeps-foot type) if they are primary crushers.
— Corrugated liners if they are secondary crushers.
— Smooth liners if they are finishing mills.

Double-roll crushers have a rather limited scope for two main reasons:
— The reduction ratio is low, 4 at the most for primary and secondary crushers and 3 for finishing mills with smooth rolls. This is because of the fact that the larger the blocks fed, the higher the angle of intake becomes for a given roll diameter. But then one cannot indefinitely increase this diameter, considering that the price of the machinery rises as the third power of this dimension.
— These crushers undergo a considerable wear even when processing moderately abrasive materials. Replacement of the liners being a long and difficult operation, it is preferred to resort to build-up through electrical welding, although the procedure is very costly.

Use of double-roll crushers becomes inescapable if the material to be processed is soft and sticky since, given its tendency to choke, it cannot be processed in any other machinery.

Under such conditions, sometimes one is downright obliged to permit and meet the costs of a certain wear.

If build-up welding of the teeth of the equipment is well programmed and done with an appropriate metal, it becomes possible to permit wear of up to a few tens of grams per tonne of material processed.

We may emphasise the fact that double-roll crushers, with proper feed, are capable of extremely high outputs up to several thousand tonnes per hour, and that too without the need for prescreening since there is no risk of blinding. Further, when highly sticky materials are processed, combs can be provided under each roll and possibly slightly different rotation speeds for each roll.

The rotation speed of the rolls should be perfectly adapted to the size of the blocks, the type of material to be processed and the tooth system chosen. It may vary from 4 to 6 and even 8 m/s at the periphery of the rolls.

3.4.2 CRUSHERS WITH SINGLE ROLL AND JAW

This machinery consists of a corrugated roll rotating about a horizontal axis opposite a fixed jaw carrying vertical grooves on its surface, the corrugations of the roll being capable of penetrating the grooves. Generally the jaw has a concave profile in the vertical plane. The setting of the exit aperture is done by moving the jaw closer by means of compensating rods. Such machines are hardly ever used, however, though often reserved for processing tender and soft material which has no more than a moderate tendency to blind.

4. MACHINERY WORKING ON IMPACT PRINCIPLE

The general principle is always the same. A rotor revolves at high speed inside a closed chamber provided with impact plates or anvils. A certain number of fixed or articulated impact pieces on the rotor throw the material on the anvils (Fig. IX.18).

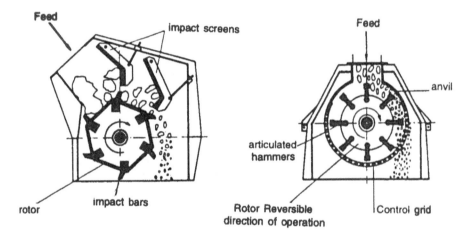

Fig. IX.18. Diagram of crushers working by impact. a) Impact crusher; b) Hammer mill

Crushing is effected by:
— impacts of the impact bars or of hammers of the rotor on the material,
— mutual impacts of the material particles themselves,
— impacts of the material particles on the anvils or on impact screens.

4.1 Impact Crushers

In these machines the impact components of the rotor are made up of fixed bars called 'impact bars' affixed in the rotor. They generally slide into a groove and are reversible so that their different surfaces can be used.

The rotor can carry 4 to 6 rows of impact bar

Impact crushers can be of the primary or secondary type.

Fig. IX.19. Horizontal shaft impactor

4.1.1 IMPACT-TYPE PRIMARY CRUSHERS

These have a hopper with a huge intake opening to permit feeding of big blocks. The feeding front is generally adjustable in inclination so that the angle of incidence of the blocks with respect to the rotor can be changed. The peripheral speed of the impact bars varies from 25 to 30 m/s depending on the size of the blocks to be fed (Fig. IX.19).

These machines have to be extremely robust because they have to endure considerable forces; for this reason often a hydraulic coupler is introduced between the rotor and the transmission with a view to absorbing the forces in the event of accident.

4.1.2 IMPACT-TYPE SECONDARY CRUSHERS

This size-reduction equipment finds are downstream of a jaw-type primary crushe
when materials which are practically or completely non-abrasive are to b
processed. These crushers are almost identical in design to primary crushers. Th
admissible feed size is often limited to 250–300 mm when the grinding pieces ar
made from special cast iron, which while being relatively fragile nonetheless be
haves quite well in abrasion.

In both primary and secondary crushers of the impact type, the only parameter
that permit variation of the granulometric curve of production are rotation spee
and setting of the opening left free between the anvils and the impact bars.

4.2 Hammer Mills

In this machinery the impact components of the rotor are hammers which are no
interconnected, generally distributed over 6 rows and fixed on axes situated at th
periphery of the rotor.

These hammers are either of a single piece or have an interchangeable heac
that permits changing just the head when it wears out.

Hammer mills are used either as primary crushers, in which case they hav
a double rotor, or more often as finishers.

Finishing mills are equipped with control grids enveloping the rotor to at leas
180°, which literally enables sizing of the products obtained. In these machines i
is possible to change:
— the rotation speed,
— the spacing of the grids
— the number of hammers,
— the arrangement of the hammers.

The high efficacy of the hammer mill is dependent on the attacking angle o
the hammers; with wear, this angle rapidly becomes blunted. Consequently, some
manufacturers offer crushers whose rotation direction can be reversed.

Thus when one of the angles of the hammer is worn out, the opposite angle
is sharpened and the rotation direction changed every 3 to 4 days. This procedure
enables a very substantial saving while concomitantly maintaining a relatively constant grading of production (Fig. IX.20).

4.3 ADVANTAGES AND DRAWBACKS OF IMPACT CRUSHERS

By their design, impact-type primary crushers are equipment which enable very
high reduction ratios, especially when brittle materials are processed. This
machinery in particular permits production of a high percentage of fine elements.

On the contrary, their use is absolutely forbidden:
— When material that is ever so little abrasive (more than 5 to 8% quartz or
any other abrasive mineral) is to be processed;
— When the material to be processed has very little sensitivity to impacts (soft
limestone, chalk etc.);
— When production of fine elements at the primary stage is not desirable in
order to be able to eliminate by means of a screening, the contaminating material
before the secondary stage.

Fig. IX.20. Hammer mill

Furthermore, this machinery demands very high installed power to cope with oversized blocks.

Control of the size of the blocks at feed should be rigorous. It is not possible in practice to extricate a block stuck in the feed inlet without opening the equipment after stoppage. One should not forget that given its inertia, such equipment often entails more than half an hour to be shut down without feed material.

Given their high reduction ratio and the direct production of fine elements, numerous impact type crushers have been installed in cement works, though with occasional disappointments due either to excessive wear, or poor output in the case of too hard or too soft materials.

Hammer mills are currently employed for secondary crushing in quarries for producing calcareous aggregates since, in addition to the large production of quality sand, they permit obtaining coarse aggregates of unmatched shape.

The grading of the material produced by this machinery varies widely with rotor speed.

4.4 VERTICAL SHAFT IMPACTORS

A new type of crusher which works by jet action has appeared in recent years.

Its working principle is as follows: The aggregates to be crushed are introduced inside of the vertical axis rotor which is similar to a turbine. This rotor revolves at high speed and throws the material to the equipment's periphery either

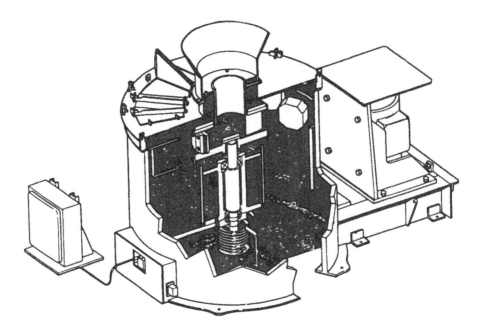

Fig. IX.21. Vertical-axis crushers. Operation by jet action

on a bed of the material which serves as a crushing anvil, or on interchangeable anvils made from abrasion-resistant metal.

The rotor itself is internally protected by a layer of the material retained in each blade by means of an ejector made of tungsten carbide rods (Fig. IX.21).

This machinery is capable of processing moderately abrasive and relatively less hard materials.

The output comprises a substantial percentage of fine elements and the shape of the crushed material is excellent.

Sometimes this machinery is solely used for correcting shape. The other side of the coin is that the edges of the chips or sand are too blunt and have a bad effect on the stability of the roads.

5. EQUIPMENT WORKING BY GRAVITY

This essentially comprises rod mills, ball mills and autogenous grinding mills. The machines are constituted by a cylindrical drum of horizontal axis revolving around this axis. The drum has liner plates affixed inside that serve as anvils. Grinding pieces in the form of steel rods or spherical balls or small cylinders also of steel are placed inside.

Fig. IX.22. Rod mills resting upon trunnions

The material to be crushed is introduced at one end through the centre of the tube. In some cases it can also be introduced simultaneously at both ends. The material is crushed by impacts and friction with the grinding pieces. Once reduced to the desired fineness, it is discharged either through openings situated at the periphery of the tube at its end or in the middle section, or through the central opening facing the feed entry.

In mechanical engineering terms, these machines are of the following types:

— Conventional crushers resting upon two smooth bearings forming the trunnions and seated on concrete blocks. Feeding and ultimate extraction of material are effected through central apertures provided inside these trunnions (Fig. IX.22).

— Tyre-mounted crushers, which are much lighter and easier to install. In this type of machinery the tube rests solely on two series of wheels with a horizontal axis, provided with tyres that at once ensure suspension and carriage of the tube. The smaller wheels with vertical axis serve the purpose of lateral guidance (Fig. IX.23).

In these machines entrainment is effected either by driving a reducer on the transmission shaft of a rear axle of truck supporting the wheel train, or by an electrical motor and a reducer whose two small-speed terminals are equipped with tyred wheels.

Tyre-mounted rod mills are generally mounted upon a chassis with treads that permit their installation directly on a properly compacted soil, i.e., without

Fig. IX.23. Tyre-mounted rod mills

prior foundation. Hence this type of machinery is often employed on job sites requiring mobility or semi-mobility of equipment (see Chapter XVII).

Rod mills can work by the dry or wet method depending on the kind of job in question.

Those working by the dry method are meant for producing very fine elements. They are utilised especially for obtaining sands rich in fines.

Contrarily, rod mills working by the wet method are meant to produce sand for concrete which contains a minimum of fines. The water introduced into the tube along with the material carries the crushed elements faster between the rods, thereby precluding overgrind.

Autogenous grinders, which are used for processing relatively soft and heterogeneous materials, are based on the following principle: the grinding media of steel are replaced by harder and larger pieces of the material to be processed. This type of equipment is basically employed in cement plants or for crushing flint pebbles.

A hot airstream is passed through some autogenous crushers which enables pneumatic classification of the crushed particles and drying of the material. Generally these are drums of very large diameter and short length.

Lastly, a few words may be said about vibrating rod mills. This equipment is generally made up of two parallel tubes integral with a mechanism of eccentrics or a counterweight which sets the whole assembly into a circular vibratory motion perpendicular to the longitudinal axis of the tubes. The tubes are filled with rods. Feeding them is done either in parallel or serially depending on the grading desired.

These machines are generally reserved for the production of very fine elements. They have two drawbacks: only small throughputs can be handled and they are extremely noisy.

The use of a rod mill or ball mill affords a number of parameters whereby operating conditions can be varied:
— Magnitude of grinding load.
— Size of grinding media (diameter of rods, size of balls).
— Rotation speed (the range of variation of this speed is relatively narrow since it should always remain well below the critical speed beyond which the centrifugal force keeps the grinding pieces adhering to the wall of the tube).
— Setting of exit apertures.
— Output and grading of feed material.
— Water feed when the machine is operated by the wet method.

Furthermore, the choice of diameter, length, manner of feeding (single or double entry) and type of discharge (central or peripheral) is very important depending on the problem to be solved, and the right decision as to which rod mill to use comes mainly from years of experience.

However, certain broad guidelines can be enunciated:
— The larger the diameter, the greater the dimensions of the feed particle size admissible, since the grinding media have a greater height of fall.
— The longer the tube, the finer the products obtained for a given grading of feed, since the grinding duration is longer.

If fine products are the objective, it will always be worthwhile to mix rods or balls of different diameters in order to ensure better internal packing of the grinding load.

Double entry with median peripheral discharge will generally be reserved for processing of soft products with a small reduction ratio.

Ball mills are basically employed for obtaining very fine sized particles. The discharge will always be central at the end, where a grid stops the balls from exiting.

6. GENERIC TECHNOLOGY OF CRUSHING AND GRINDING MACHINERY

As was seen at the beginning of this chapter, crushing and grinding of hard rocks are touch tasks requiring ruggedness from both the men and equipment employed.

Add to this already tough job the considerable efforts which have to be expended in attending to the equipment when it is overcome by choking or the entry of an uncrushable or incompressible body, which unfortunately happens quite often in quarries. Just by way of recapitulation we might mention the whole range of tramp bodies which could obstruct smooth flow of the products—teeth of shovels, drill bits, pieces of caterpillar tracks, bits of wood or wet plant root as well as contaminating or clayey elements. It is for this reason that manufacturers

of these machines have had to invent a number of safety devices upstream of th
crushers or in the crushers proper.

We have also seen that some of this equipment operates with moving par
that reach several tens of tonnes in mass and are subject to a relatively fast eccentri
motion. This entrains considerable forces on the structure of the machine assembl
so much so that designing crushers and grinders calls for vast experienc
since the forces engendered in their principal components are often difficult t
calculate.

The first consequence of this great ruggedness is the exceptional service li
of this type of machinery. While machines such as earthmovers have a service li
that seldom exceeds 4 or 5 years, it is not rare to come across crushers which, we
maintained, continue to work perfectly well beyond 30 years of good and reliabl
service.

It should also be noted that it is much more difficult for a manufacturer t
design a tertiary crusher than a primary one.

Indeed, the forces engendered by the passing of big blocks are proportionall
much less in magnitude than those necessary for tertiary or quaternary size redu
tion (Rittinger's law), and without reverting to the theories of grinding, let u
remind the reader that for a given rock the energy needed for crushing increase
in direct proportion to the difference in specific surfaces of materials fed to
crusher and the ground products emerging from it.

7. FUTURE TRENDS

As a final point to the foregoing exposition on the different machines currentl
employed for crushing and grinding in the production of aggregates, we shall no
present a *résumé* of the research presently ongoing in the field of new techniqu
of crushing.

7.1 Automation (see Chapter XVI)

For purposes of reducing manpower requirements and securing consistency i
production, efforts are underway to make the regulation and monitoring of th
operation of secondary and tertiary crushers wholly automatic. Some manufa
turers have already developed gyratory crushers that are completely automate
and run by remote control. A central control panel receives the diverse data comin
from the equipment: hydraulic pressure, setting of the equipment, output achieve
power consumed by the motor as well as information pertaining to safety, suc
as maximum permissible pressure in case of overfeed, temperature pressure c
flow of lubricating oil, level of this oil in the storage tank etc.

Synthesising the different operational data, the control panel dictates to th
crusher the parameters which will cause it to give the grading envisaged.

What is more, if accidents, such as entry of uncrushable tramp bodies or ir
advertent overfeeding occur, the equipment, after release of the opening, will aut
matically return to its original setting. Lastly, if for some reason or the other, th
equipment is partially or fully starved of feed material, it will automatically sto
and restart only when sufficient material to crush has been fed.

7.2 Crushing by Ultrasonics

Experiments were conducted in the USSR for breaking up blocks by applying an ultrasonic wave generator on one of the faces. Apart from the fact that this process calls for relatively complex instrumentation it entails an excessively high energy consumption vis-à-vis conventional methods.

7.3 Crushing by Hyperfrequency Waves

This method consists of passing materials through the field of a hyperfrequency wave generator. The excitation of interstitial water molecules brings about a high rise in temperature, which creates a microfissuration inside the blocks.

When these materials subsequently pass through the crushers, this prior weakening manifoldly boosts the output of the equipment.

Equipment based on this technique is already being employed for demolition of concrete.

8. FACTORS DETERMINING CHOICE BETWEEN A SINGLE MACHINE OR MACHINES

When confronted with a problem of crushing or grinding, the following data is absolutely essential:

— petrographic nature of the rock to be processed, i.e., degree of abrasiveness as reflected by the percentage of the hardest minerals in it (see chapter VI);

— mechanical strength of the rock;

— size of the feed materials;

— desired particle size in the product (particle size distribution);

— throughput to be processed;

— moisture content and freedom from contamination of the raw material, where necessary.

Most of the time the petrographic nature of the rock enables selection of the right type of equipment: gyratory crushers, jaw crushers, impact crushers or autogenous crushers etc.

The feed size range and the output will then enable determination of the size of the equipment in the range selected. Generally one of these two criteria (feed size or output) will dictate the choice of equipment, which may be overdimensioned for the other criterion.

It is then ascertained whether the equipment chosen will, under normal conditions, provide the desired reduction ratio, failing which it becomes necessary to envisage two or more machines in succession.

In any case, to achieve the highest efficiency it is always advisable to operate the equipment in closed circuit with a circulating load of about 20% elements larger than the control screen size of the target product.

One should also take into account the efficiency and intake capacity of the equipment selected with due regard to average wear of the grinding wear parts.

In effect:

— In a gyratory crusher and especially in a tertiary one, the intake openin constricts as wear progresses, and few are the manufacturers who state the rea functioning size of this opening at the end of wear as the nominal one. Further more, the output of this equipment also diminishes due to lengthening of the paral lel crushing zone.

— In a jaw crusher the size of the largest elements produced increases quit rapidly as the corrugations of the jaws disappear consequent to wear.

— In a impact crusher or hammer mill the production rate of fine element falls off markedly with wear of the impact bars or hammers.

— In a corrugated roll crusher the gripping action on the blocks lessens notabl when the corrugations become worn.

9. EXAMPLE OF GRAIN SIZE STUDY

A granulometric study in a crushing plant for a dock project in Africa i reproduced below.

The material comes from a granite quarry, of which the largest blocks reacl $1000 \times 800 \times 800$ mm in size.

Production targets are as follows:
— 91,500 t of 0/2—2/4—4/6 and 6/10 materials for bituminous mixes;
— 61,000 t of 0/4 and 6/25 concrete aggregates for structures such as bridges
— 225,000 t of 25/50 ballast for service railway lines.

The undertaking is expected to last 17 months. Applying a moderation facto of 0.8 for the installation's efficiency, the effective working time comes to 236(hours, which leads to designing the installation for a nominal output of 160 t/h.

The first part of the granulometric study takes into account the total require ments and the means therefore in each aggregate sizes.

Based on the distribution curve of the run of mine material, the second par enables deciding the machines necessary for each of the stages of crushing witl respect to their standard granulometric curves.

REFERENCES*

*Adamski, L. Rational crushing with the IBJ crusher. *Aufbereitungs-Technik* (August 1965).
*Andreas, E. Princep and possibilities of crushing with impactors. *Zement, Kalk und Gip* (Nov. 1965).
Anonymous. Hammer failures in a crushing mill. *Engineering* (15 Sept. 1961).
Bauman, V.A., V.A. Strelstson, A.I. Kosarev and S.S. Gusakov. Six articles (in Russian) or percussion crushers. Central Institute of Research on Construction Machiner) (Vniistroidormach). Moscow (1966).
Beke, B. *Principles of Comminution.* Budapest (1964).
Bergstrom, B.H. The electrohydraulic crusher. *Jour. Eng. Mining* (Feb. 1961).
*Blanc, E.E. Rod mill chambers. *Revue de l'industrie Minérale* (August 1960).
*Blanc, E.C. Comminution of solids. Fasc. A 901–902 of Engineering Technics. In: *Genera Aspects* (Oct. 1964).

*Entries marked with an asterisk are in French—General Editor.

*Blanc, E.C. Calculation of the output of jaw crushers. *Le Génie Civil* (January 1971).

Bond, F.C. Action in red mill. *Eng. Mng. Jour.* (March 1960).

Bond, F.C. *Crushing and Grinding Calculations*. Allis Chalmers Mfg. Co., Milwaukee, Wisconsin (1961).

Bond, F.C. More accurate grinding calculations. *Cement, Lime and Gravel, London* (May 1963).

*Capdecomme, L. Reflections and experience of mineralogists on crushing and abrasiveness of minerals. *Revue Ind. Min.* (March 1976).

*Chenais, J. *Mechanical Equipment. Quarries and Materials. Crushers and Grinders. Aggregates, Theoretical Account and Practice*. Cie Française d'Edition, Fasc. X, Paris (1970).

*Colenot. Machinery for producing aggregates in quarries: Crushing. *Eg. Méc. Car. Mat.*, no. 159 (August 1977).

Comminution. Proc. 1st Symp., 1 vol. VDI-Verlag, Dusseldorf.

Comminution. Proc. 2nd Symp., 2 vols. Verlag-Chemie, Weinheim, Bergstrasse, Amsterdam (1966).

Comminution. Proc. 3rd Symp., 2 vols. Verlag-Chemie, Cannes (1971).

*Cook, N.C.W. and N.C. Joughin. Comminution of rocks by mechanical, chemical and thermal methods. *Proc. 6th Internat. Cong. Mineral Industry*, Madrid (1970).

*Coste, J. *Machinery for Extraction and Processing of Materials*. Eyrolles, Paris (1968).

Erhard, H. Crushing by percussion. *Equipment Mécanique* (Oct.–Nov. 1971).

*Franking, H. Research on stresses in crushing machines. *Ecole Technique Supérieure d'Aix-la-Chapelle* (1960).

*Guillot, R. *Problem of Grinding and Its Development. Grindability of Materials*. Eyrolles, Paris (1960).

Hagtorn, B. Selection and application of crushers. *Quarry Managers Journal* (July 1966).

Johnston, H.A. The mechanism of electrohydraulic crushing. *Proc. 2nd European Symp. Comminution*, Amsterdam (1966).

Kluchantsov, B.V. Jaw crushers. Central Institute for Research on Construction Machinery (Vniistroimach). Moscow, ser. VI (1962) (in Russian).

Krahenko, V.S., A.P. Obrastsov and V.V. Ustinov. Dustless breaking of rocks electrically. *Mining Cong. Jour.* (May 1961) (French trans. from Russian).

Krucher, E. Differential and synchronous roll crushers. *Die Ziegel-industrie*, nos. 1–2 (1970).

*Leveque, B. Study of a percussion crusher for processing hard abrasive rocks. *Eg. Méc. Car. et Matér.*, no. 139 (May 1975).

*Maldonado, A. New methods of processing materials. *Equip. Méc. Car. Matér.*, no. 155 (March 1977).

*Masson, A. Current theories on the energy needed for grinding. *Silicates Indus.*, no. 10 (1964).

*Millier, J.P. Contribution to the theory of cone crushers. *Proc. 3rd Symp. on Comminution*, Cannes (1971).

*Miranda, F. *Mechanical Preparation of Minerals and Coals*, Madrid (1965).

Maeder, J. Considerations about crushing with impactors. *Aufbereitungs-Technik* (Feb. 1970).

Modern Crushing Technics. ANRT Committee on Crushing, vol. 1. Eyrolles, Paris (1965).

Monument, W.E. Production of sand and sand classification. *Quarry Managers Jour.* (Jan. 1960).

*Patat, F. Kinetics of crushing. *Génie Chimique* (March 1960).

*Planiol, R. Centrifugal crushers and vacuum. *Revue des Materiaux de Construction* (Feb. 1962).

Pryor, E.J. *Mineral Processing*. London (1960).

Puffe, E. Selective crushing with impactors considered as a way to decrease the running costs and to increase the production in crushing plants. *Glückauf* (January 1960).

Raab, A. Cubical and dust-free aggregates obtained by electrohydraulic comminution. *Highways and Public Works* (April 1968).

Remenyi, K. *The Theory of Grindability and the Comminution of Binary Mixtures*. Akadémiai Kiado, Budapest (1974).

Rose, H.E. and J.E. English. *Theoretical Analysis of the Performance of Jaw Crushers*. British Inst. Mining and Metallurgy (1967).

Ruhl, H. Die Gestaltung von Zerkleinerungsmaschinen durch rechnerische Voraus-Bestimmung der Zerkleinerungs-Kräfts unter Zerkleinerungsarbeit. *Aufbereitungs-Technik* (May, July, October 1964).

Rumpf, H. Struktur der Zerkleinerungs-wissenschaft. *Aufbereitungs-Technik* (August 1966).

Sarapun, E. Electrical fracturing and crushing of taconite. *Proc. VII Symp. Rock Mechanics.* Univ., Pennsylvania (1965).

Schonert, K. Modellrechnungen zur Druck-zerkleinerung. *Proc. 2nd European Symp. on Comminution,* Amsterdam (1966).

Scrom, E.C. Dustless breaking of rocks electrically. Special report, General Electric Co. America (Sept. 1961).

Sommer, H. Toothed roll crushers. *Aufbereitungs-Technik* (May 1964).

Steffon, T. Crushing using a double cylindrical roll. *Aufbereitungs-Technik* (June 1970).

Stein, E. Analyse des Zerkleinerungvorganges in der Hammermühle. *Aufbereitungs-Technik* (August 1966).

Supp, A. Über die Möglichkelten due electrothermische Beans pruchung für Zerkleinerungsaufarben zu nutzen. *Aufbereitungs-Technik* (April 1973).

Taggart, A.F. *Handbook of Mineral Dressing.* New York-London (1945).

Taggart, A.F. *Elements of Ore Dressing.* New York-London (1951).

Wilmanns, F. and H. Wolf. Grosszerkleinerung mit Backenbrecher, Kreiselbrecher und Backenkreiselbrecher. *Aufbereitungs-Technik* (May 1964, April 1965).

Winter, H. Die Lagerung der Antriebs-wellen in Backenbrecher Berechnung und Gestaltung. *Aufbereitungs-Technik* (April 1973).

Wohlbier, H. Grundlagen und Anwendungs-möglichkeiten der elektromagnetisch-thermischen Zerkleinerung von Gestein mithilfe von Mikrowellen. *Proc. 6th Internat. Cong. Mineral Industry,* Madrid (1970).

Zemlika, J. Zerkleinerung von klebriger Braunkohle in Hammerbrechern mit Gegenwalzen. *Bergakademie* (Dec. 1968).

X

Screenings

Jean-Pierre Delille and Jean-Paul Moutot

1. INTRODUCTION

Screening is the operation which enables separation of a body of aggregates into two subgroups, one containing elements of dimension larger than a given value and the other those that are smaller.

The two main objectives of this operation are:

— Technical screening meant to render the aggregates acceptable for a processing machinery which may either be a grinder or final screening equipment.

— Screening for grading the products according to given size specifications into marketable categories.

What with user requirements on the grading and quality of products becoming ever more stringent on the one hand, and the distinction in denomination of the products, assessment of their quality and grading, besides determination of the screening machinery performance being unclear on the other, it becomes imperative to lay down some definitions.

2. DEFINITIONS

The set of definitions given below was extracted from the European standards for size classification of equipment published by the European Committee for Civil Engineering equipment.

2.1 Distribution Curve $q(x)$

The curve which for any size fraction Δx gives the percentage of this fraction present in the total quantity.

2.2 Standard Retained Fraction

Proportion of the retained fraction which is correctly classified.

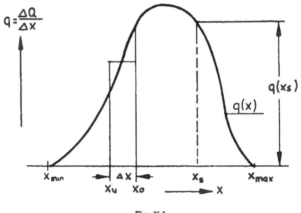

Fig. X.1

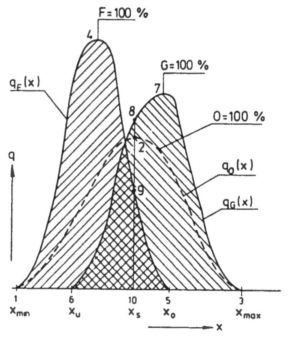

Fig. X.2

2.2.1 STANDARD COARSE FRACTION g_g

Proportion of the retained fraction which is correctly classified as coarse (see Fig. X.2, area 10.8.7.3).

2.2.2 STANDARD FINES f_f

Proportion of the passing fraction which is correctly classified as fines (see Fig. X.2, area 1.4.9.10).

2.3 Misclassed Fraction

Screened product which should not be present in the size fraction which has been analysed.

2.3.1 MISCLASSED COARSE g_f

Proportion of the passing fraction which remains in the coarse fraction (see Fig. X.2, area 10.9.5).

2.3.2 MISCLASSED FINES f_g

Proportion of the coarse fraction which has found its way into the fines (see Fig. X.2, area 6.8.10).

2.4 Organic Yields

2.4.1 Organic yield of fines V_F (Figs. X.2 and X.3)

$$V_F = 100 \; \frac{f_o - f_g}{f_f - f_g}$$

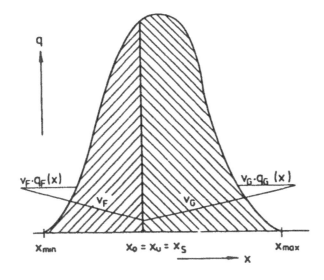

Fig. X.3

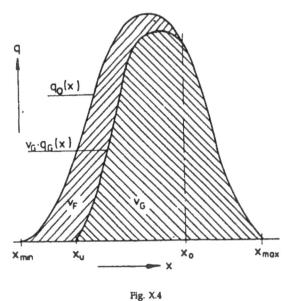

Fig. X.4

2.4.2 ORGANIC YIELD OF COARSE V_G (Figs. X.2, X.3, and X.4)

$$V_G = 100 \, \frac{g_o - g_f}{g_g - g_f},$$

where g_f is the proportion of fines contained in the all-in and g_o is the proportion of coarse contained in the all-in.

Instead of these mathematical formulae, we prefer to define:
— V_F as the % of the product (in the total quantity of all-in) passing through the screen;
— V_G as the % of the product retained on the screen.
Which automatically means $V_F + V_G = 100\%$.

If the cut-off is ideal, the diagrammatic depiction will be as in Fig. X.3; if not, as in Fig. X.4 or X.5. It can be seen that these two yields are by no means a qualitative reflection on the screening operation, but only a quantitative proof of the outcome of this operation.

2.5 All-in

2.5.1 ALL-IN CORRECTLY SCREENED INTO FINES V_F, f_f

The proportion of the initial all-in which is found in the fines (Fig. X.5, area 1.4.9.10).

2.5.2 ALL-IN CORRECTLY SCREENED INTO COARSE V_G, g_g

The proportion of the initial all-in which is found in the coarse (Fig. X.5, area 10.8.7.3).

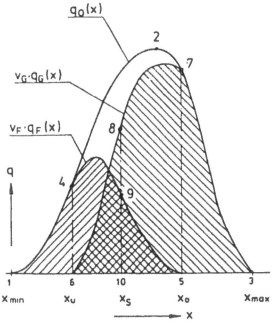

Fig. X.5

2.5.3 MISCLASSED ALL-IN IN FINES V_F, g_F

The proportion of the initial all-in found in the fines, whereas it should have been in the coarse (Fig. X.5, area 10.9.5).

2.5.4 MISCLASSED ALL-IN IN COARSE v_g, f_g

The proportion of the initial all-in found in the coarse, whereas it should have been in the fines (Fig. X.5, areas 6.8.10).

2.6 Separation

2.6.1 DEGREE OF SEPARATION $T(\Delta x)$

For a given size fraction Δx, this is the relative quantity of product present in the retained compared to the original quantity in the all-in.

2.6.2 SEPARATION CURVE

The curve which depicts the variation in degree of separation with respect to the grain size X (Fig. X.6). It will be ideal when vertical (Fig. X.7), otherwise a curve whose ordinate will be 0 for the point X_U (smallest grain in the coarse) and 100 for the point X_0 (largest grain in the fines, grains of larger size being found to 100% in the coarse).

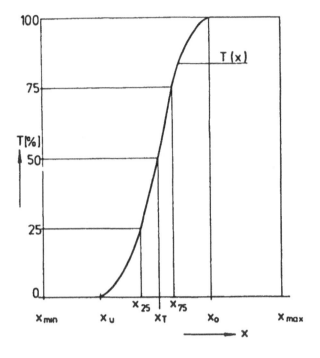

Fig. X.6

2.6.3 THEORETICAL CUT-OFF MESH X_T

The grain size X_T for which $T(X_T) = 50\%$ (Fig. X.6). The imperfection is then defined by the ratio:

$$\frac{X_{75} - X_{25}}{2X_T}$$

The smaller this ratio, the closer the separation will be to the ideal state.

2.6.4 NOMINAL CUT-OFF MESH x_s

The square mesh of the control sieve used for defining the separation.

2.6.5 AVERAGE CUT-OFF MESH X_M

The mesh at which the amounts of graded-out (misclassed) coarse and fines will be equal (in Fig. X.9, a will be equal to b).

2.6.6 EFFECTIVE CUT-OFF MESH X_A

The mesh at which the misclassed all-in found in the fines is equal to that found in the coarse (in Fig. X.8, a will be equal to b) $V_F, g_g = V_G \cdot f_g$.

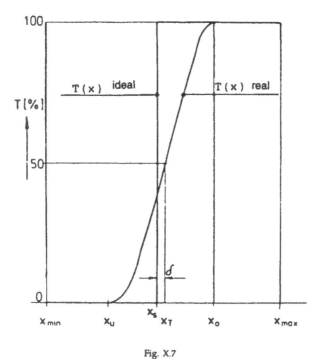

Fig. X.7

2.6.7 PRECISION OF CUT-OFF

Δ will be defined as the difference between the desired cut-off mesh x_s and any of the other three meshes X_T, X_M or X_A which will be chosen as the criterion of precision. It is obvious that the choice of the precision criterion will depend on the problem posed.

The theoretical cut-off mesh X_T will essentially serve to determine the coefficient of imperfection, which in turn will reflect the quality of sizing of the screening apparatus and not the size of the sieve used. Indeed, one might think that in most cases slightly modifying the size of the mesh used on the machine will hardly, if at all, change this coefficient of imperfection.

The average cut-off mesh X_M will be more meaningful at the dimensional level when a comparable quality is desired in the products both for the passing and the retained size fractions (percentage of fines in the coarse equal to the percentage of coarse in the fines $f_g = g_t$).

In effect, the effective cut-off mesh X_A cannot be significant (Fig. X.8). For example, if the retained V_G are much larger in proportion than the passing V_F, the ratio b/V_G could be very small, hence giving a good quality in the retained, while the ratio a/V_F could be very high and give a passing (fraction) in which proportion of the misclassed coarse is large.

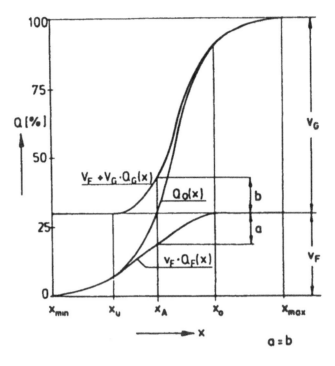

Fig. X.8

2.7 Yields

The yield of fines will be defined as the ratio of the quantity of fines actuall present in the fines to the quantity of fines present in the all-in, and the yield coarse as the ratio of the quantity of coarse actually present in the coarse to th quantity of coarse present in the all-in.

Yield of fines $A_f = \dfrac{V_F\ f_f}{f_o}$

Yield of coarse $A_g = \dfrac{V_G\ g_g}{g_o}$

These two concepts are preferred to the concept of efficiency which, measure at the effective cut-off mesh, is defined as the following ratio:

$$e = \frac{100 - V_F\ g_f}{100}.$$

(The higher this ratio, the better the quality of screening will be.)

Efficiency thus defined favours the product which is present in a larger quar tity in the all-in (retained or passing).

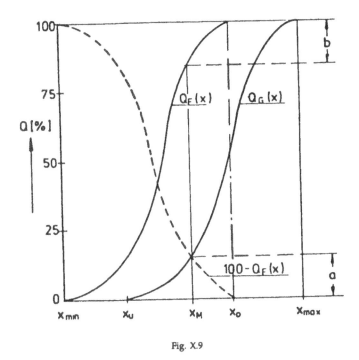

Fig. X.9

It is obvious that if the retained accounts for 80% of the all-in, a 90% efficiency will correspond to a yield of coarse of:

$$100 \cdot \frac{80 - 10}{80} = 87.5\%$$

and a yield of fines of:

$$100 \cdot \frac{20 - 10}{20} = 50\%.$$

Therefore it will be well to choose the criteria of measurement with reference to the actual problem faced.

3. TYPES OF SCREENS

We shall call as screens all equipment capable of helping to grade (or classify) into two or more size categories particles of a material placed on perforated surfaces which let pass everything which is smaller than the aperture of the perforations and retain that which is bigger. Therefore a screen is a device which serves to grade a mixture of particles according to their dimensions.

3.1 Stationary Screens

Leaving aside manual sorting, these are the simplest devices still used by mason who throw gravel pellets onto a wire mesh for a coarse separation.

In gravel quarries stationary grids made up to diverging rods inclined 20 (40° are often connected to receiving hoppers and help scalp the all-in. They prote the installation from the entry of blocks larger in size than the intake opening (the comminution equipment they form part of.

These blocks constitute the stocks which one day justify the investment in larger crusher.

Stationary grids can also be used to stop big blocks meant for stone-beddin;

These grids are generally fed directly by lorry or dumper but their use : justified only when the retained are limited in quantity, which permits hig throughput of passing.

3.2 Trommels (Rotary Drum Screens)

Mentioned solely for the record, since they are rarely made any more, trommel were universally employed before the advent of ball-bearings which paved th way for vibrating screen development.

A trommel consists of a perforated drum. The drum is mounted on rollers a an inclination and revolves at a linear velocity of 35 to 50 m/min. Holes of mor than one size may be present over the drum's length to obtain several size cut-off: These machines were fairly efficient for products liable to clog but their capacit was limited in relation to their size.

3.3 Screens with Discs or Flyweights

Almost no longer fabricated, they accomplish a coarse prescreening but are ill suited for abrasive materials. They find use rather for coal or limestone.

They comprise one or more shafts with discs or fly-weights at intervals fo the passage of small products. The discs rotate at slow speed and carry the retained fractions (Fig. X.10).

3.4 Inclined Vibrating Screens

In all these machines the inclination of the screening planes favours the forward movement of the material by virtue of the horizontal component of the vibrating action. Depending on the differences in mechanisms or vibration trajectory, screens can be of several types.

3.4.1 ECCENTRIC TYPE VIBRATING SCREENS

These made their appearance with the advent of bearings. They are made by of an eccentric shaft and 4 bearings, two side-plates on the vibrating caisson holding the screening fabrics and two more resting on the frame. Clamped between the bearings are 2 unbalance-wheels meant to balance the eccentric movement of the caisson. A set of springs or rubber cushions between the caisson and the frame ensures that the caisson is secured.

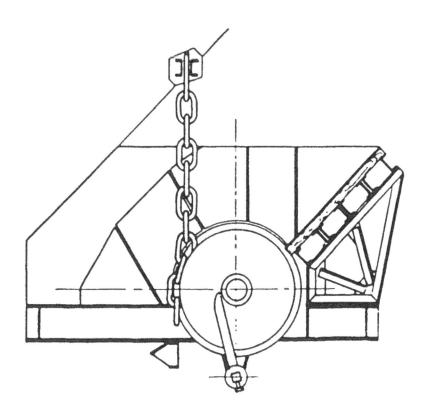

Fig. X.10. Disc screen

The circular throw resulting from the eccentric is therefore fixed and the dynamic reactions on the frame depend on the precision of the balancing.

These screens have been eased out for some 20 years by unbalanced weight vibrating screens. They are rarely made now but a few are still in service in processing sites.

3.4.2 UNBALANCED WEIGHT SCREENS (Fig. X.11 and X.12)

Currently and for long the most utilised screens, they became indispensable vis-à-vis eccentric screens because of their simplicity and lower cost.

They consist of a vibrating box with a unbalanced shaft. The whole assembly is supported by a flexible suspension with springs, rubber pads or air cushions intended to annul the effects of weight.

While rotating, a dynamic equilibrium is established, adjustment of the throw being made possible by increasing the moment of the unbalanced weights.

The mass of the materials influences the amplitude of the throw, which in turn leads to a certain dependence of the vibrating mass vis-à-vis the load.

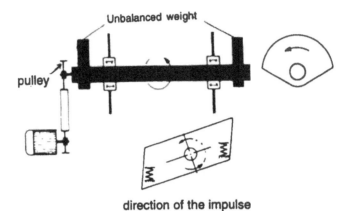

direction of the impulse

Fig. X.11. Principle of the unbalanced weight screen

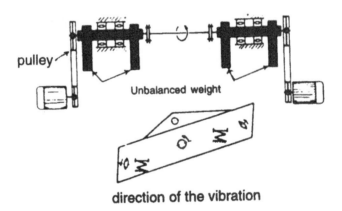

direction of the vibration

Fig. X.12. Unbalanced weight screen with elliptical vibration

To take care of different service categories, these machines have a more o: less heavy structure; a distinction is made between the family of scalpers whicl receive blocks prior to crushing, primary screens which receive materials of larg size coming from crushers, and the secondary and tertiary series used for classifying

Up to 10, 12 m² surface area, these machines are of the same configuratior with a mechanism placed near the centre of gravity and with 2, 3, or 4 screenin decks depending on need.

The speed/throw combinations vary from 750 rpm/12 mm to 1300 rpm/3.! mm depending on the tasks of screening to be tackled. For surface areas large than 12 m², the centrifugal force needed for ensuring movement of such masse necessitates the addition of more bearings. This leads to screens with double o triple shaft lines synchronised by interlinked belts carrying large masses and entail very exacting maintenance sometimes.

Some manufactures have devised original solutions to ensure control of the conveying speed of the materials over the length of the screen by distorting the circular vibration into elliptical vibration. This distortion can be achieved by placing two-bearing vibrator sets above the screening planes.

This concept, which assures very high-quality screening, makes possible the fabrication of large units of up to 20 m^2 surface areas.

3.5 Vibrating Screen with Twin Unbalanced Weights

3.5.1 SCREENS WITH LINEAR VIBRATION (Fig. X.13)

The screening basket is suspended as in single unbalanced weights screens, the screening planes being horizontal.

A semi-linear vibration at 45° inclination is obtained by the action of 2 unbalanced weights shaft lines rotating in opposite directions. Either these 2 shafts naturally synchronise and the twin line of action of the unbalanced weights is situated at right angle to the line joining the centres of the shaft lines in the direction of the common centre of gravity, or a set of gears imparts to them a synchronisation from without which aids the mechanism towards a horizontal disposition.

The cost is higher (gears) and the only advantage of this mounting is the possibility of a single motorisation.

In both cases vibration is obtained by the twin action of the centrifugal forces which cancel each other in one direction and add to each other in the opposite.

The vibratory action being at 45°, the vertical component assured the screening action and the horizontal component causes the materials to move forward. For an equal surface area, this type of screen requires more power but, on the other hand, is of low height and achieves a cut-off closer to the mesh opening than unbalanced weights screens. For these reasons, this type of screen has an assured future though its use should be limited to cut-offs below 50 mm given its sensitivity to bolt fastening.

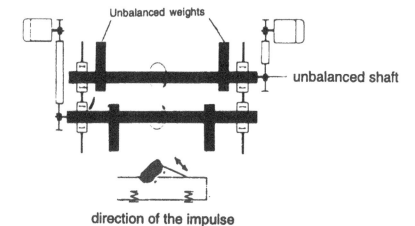

direction of the impulse

Fig. X.13. Principle of double unbalanced weights screen

3.5.2 SCREEN WITH ELLIPTICAL VIBRATION

These vibrating screens are worked by a double unbalanced weights mechanisr or by two sets of vibrators arranged horizontally above the screening planes.

Their relative disposition with respect to the centre of gravity and the cor comitant dynamic equilibrium transmits to the whole suspended unit an elliptica vibration inclined frontward. This inclination is not sufficient for successfully cor veying the materials. Therefore the screening planes are inclined but less so tha in single unbalanced weights screens and the vertical component just barely ir clined towards the screening plane, which generates an effective motion for screen ing and less sensitivity to bolt fastening.

In effect, this type of screen is a good compromise between conventional screen and double unbalanced weights ones, which imparts to long screens of very larg surface area (up to 35 m^2) an optimal efficiency through total control of the vibratior

3.6 Vibrating Screens Driven by Motors with Unbalanced Weights

Vibrators are replaced by motors with unbalanced weights in many assemblies i as much as this offers an economic advantage. As of now, these motors are limite to powers under 8 kW, with the drawback of a fixed speed.

Unbalanced weight vibrating screens with elliptical throw: These are screens witl a single unbalanced weight whose throw is normally circular and counteracte either by an additional mass or by a suspension offering a greater degree c freedom in one direction. The principle has become obsolete.

3.7 Miscellaneous Screens which are Either Less Common or Meant for Specific Application

3.7.1 OSCILLATORY VIBRATING SCREENS

The vibrating box is activated by a crank-rod system. The vibratory motion i directed by leaf springs on which the box rests. The throw can be very large, 3 or 40 mm, with speed of the order of 450 rpm.

These machines have practically disappeared from quarry installations.

3.7.2 RESONANT VIBRATING OR DOUBLE-WEIGHTED SCREENS

These are comparable to the oscillatory type. The eccentric shaft is mounted on counterbalancing rod, also suspended on leaf springs. At equal weights the tw assemblies have identical vibrations opposed to each other. The throws are 12 t 10 mm for 6 to 800 rpm frequencies.

This type of screen, which is costly and delicate to operate, has progressively disappeared from quarry operations.

3.7.3 CIRCULAR SCREENS OR SIEVES

The screening plane, which is slightly inclined towards the direction of flow of th products, is activated by a circular movement which is practically parallel to this plane

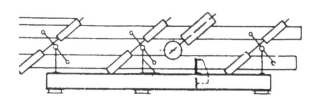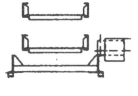

Fig. X.14. Resonant vibrating screen

These devices are rarely used, being limited to fine screenings with low outputs.

3.7.4 SCREENS WITH VIBRATED SCREENING MEDIUM

The bin is stationary; the screening fabrics rest on suspended beams which carry vibrators that work at a steep inclination and a high vibration frequency (Fig. X.14).
This kind of machine is very rarely found nowadays.

3.7.5 UNBALANCED WEIGHT SCREENS WITH MOVABLE FITTINGS

This screen reinforces the effects of a circular vibration of the bin, with support beams for fabrics which move relative to the bin and the deformation of a flexible fitting. This equipment finds justification when fine products with a high clogging tendency need to be screened.

3.7.6 MULTIPLE-DECK SCREENS

The equipment consists of a vibrating bin generally worked by motors with unbalanced weights and achieves a cut-off through multiple screening decks of progressive mesh openings, thereby decreasing the depths of material layer.

3.7.7 GRIZZLY SCALPERS

The vibrating bin is worked by a counterweighted mechanism with unbalanced weights situated behind the machine, the front portion resting on rubber supporting blocks. Thus there is a circular vibration behind the machine which is thrown frontward. The scalping bars are tapered to accelerate the discharge. These two effects in combination impart to this machine a high efficiency at large outputs.

3.8 Vertical Drum Screens

This machine enables size cut-offs in fine products down to 400 micrometres with remarkable output and efficiency. It also takes care of the function of dedusting or defiltration through air circulation inside.

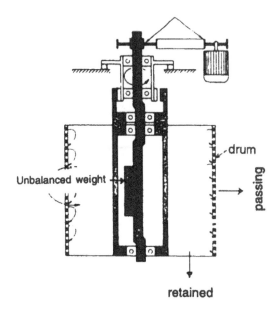

Fig. X.15. Vertical drum screen

The screening surface is made up of a cylindrical drum with a vertical axis. The materials are deposited at the top on a distributor plate that revolves along with the screen and become dispersed over the inner surface of the drum. They travel downward under the effect of their own weight (Fig. X.15).

The drum driven by the shaft is subjected to a gyratory motion around the axis of this shaft. This motion of the gear of the rim on the pinion causes it to rotate on itself at low speed. The combination of these two motions engenders a pulsation whose frequency equals the shaft's rotation speed. Consequently the materials are subjected to about 1000 pulsations per minute as they travel downwards, which facilitates their progress and separation of fines from coarse matter.

Gyration at high speed and rotation of the drum subject the materials to a centrifugal action that is many times greater than their weight and invest this type of screen with a greater capacity and a better efficiency than conventional screens which rely solely on weight for forcing the fines through the openings on the screening surface.

The screening capacity is generally 2 to 2.5 times that of a conventional screen.

3.9 Equipments of Similar Function

3.9.1 FEEDERS WITH PRESCREENING

As a matter of fact, these are scalping screens in which the feeding function predominates and a coarse prescreening achieved.

3.9.2 PNEUMATIC SEPARATION EQUIPMENT

The materials are dispersed into a suspension by a mechanical process and an induced air draught carries off the lightest particles (see Chapter XIII).

3.9.3 HYDRAULIC SEPARATION EQUIPMENT

— curved grid: a highly inclined and incurved grid, generally made of polyurethane and slotted, receives a pulp and the finest materials are entrained by the water.

— Paddle classifier-dewaterer: the ascensional speed of the water entrains the finest particles.

— Water treatment equipment, such as vibrating dewaterers, filter-cyclones etc. (see Chapter XII) which carry out separation of fines.

4. ACCESSORIES OF SCREENING MACHINERY

4.1 Screening Accoutrements

To achieve the desired cut-offs in the particle size distribution of feed, screening machines are fitted with perforated surfaces made from different materials depending on service requirements, or wire screens fashioned from woven or crosslaid wires and presenting specific mesh opening sizes.

These accoutrements are either stretched or fastened by different devices over the screening frames.

4.1.1 METALLIC CLOTHS OR FABRICS

The most commonly used, these consist of interwoven high-strength steel wires, their diameters and mesh openings being of standardised values. The mesh openings are generally square but may be rectangular as well.

The mesh opening sizes range from 250 micrometres to 125 mm. Welded wire screens for larger openings are also available.

4.1.2 WAVY METALLIC FABRICS

The wires are not laid criss-cross but zigzag so as to enclose square or triangular holes. The wires are held together by interwoven networks of wires or by neoprene bands.

4.1.3 ABRASION-RESISTANT RUBBER OF SYNTHETIC TEXTILES

These are rubber or polyurethane panels in which the perforations are diverse in shape: square, circular, rectangular or oblong.

The perforations are made by punching or moulding. The materials can be reinforced by textile or steel wire threads. These panels are either stretched, bolted or clamped. They offer the advantage of a good abrasion resistance; on the other hand, they are costlier and their permeability less compared to metallic fabrics. This makes it imperative to provide better screening surfaces.

The mesh openings best suited for the desired cut-off have to be decided b the shape of the materials to be graded and the grading of the product. The choic should be rational, which means that sometimes experiments have to be undertaker

4.1.4 PERFORATED PLATES

These are probably the foremost screening accoutrements used, first one trommel and then on vibrating screens. The perforations are generally round or square bu any shape is suitable. They are arranged in staggered rows and the overall per meability of a metallic fabric is of the order of 2/3.

Depending on the application in view, perforated plates may be made c abrasion-resistant steel or reinforced with skid bars.

They are most often employed for feeds of large particle size in the case c prescreening, on primary screens for example.

4.1.5 BAR GRIZZLIES

These consist of diverging bars in order to obtain a cut-off of a specific size.

The bars are not interconnected and may be clamped individually onto th cross-members of the screen. Cutting a plate onto which skid bars and a bearing block are welded is equally effective. Bar grizzlies are thus integral in constructio and offer high resistance to impacts.

Bar grizzlies are suitable for scalping as well as prescreening tasks.

4.1.6 SLOT GRIZZLIES

Slot grizzlies are panels made up of either parallel metal rods attached to assem bling rods with slots left between the rods, or spaced panels (strips) of rubber o synthetic material.

The slots/spaces range between 0.5 and 4 mm and are especially used fo straining grits or sands.

4.2 Rinsing devices

To obtain washed products or to facilitate screening under water, as a better way to preclude clogging of the screening media in the case of fine cut-offs, the screen should have provision for fitting watering devices consisting of tube manifolds with holes or diffusers and paddles for regulating the flow.

The volume of water is of the order of 10 $m^3/h/m^2$ surface under a pressure o 1.5 to 2.5 bars.

4.3 Hooding

These systems may be either stationary or vibratory; they reduce dust or muffle noise.

4.4 Dedusting

An induced draught and filtering unit attached to the hoods to maintain the as sembly under vacuum.

4.5 Ball Beater

A declogging system consisting of balls generally of rubber located below the screening fabric. When vibrated, the balls strike against the fabric. The shock waves transmitted dislodge the fines from the woven wires (Fig. X.16).

4.6 Screen Heating

A radiation or heating device that eliminates clogging of the screens by wet fine products. This method is only rarely employed in quarries.

5. SIZING AND SELECTION OF MACHINERY

5.1 Parameters that Affect Screening

5.1.1 SCREEN LENGTH

It is obvious that the longer a screen, the more chances a particle will have of finding an opening in the grid during its travel over the machine. Accordingly, the length parameter will directly affect the quality of screening.

5.1.2 SCREEN WIDTH

On the other hand, width of the screen directly influences the possible output of the machine. It is obvious that if a 2 m² (L = 2 m, W = 1 m) screen correctly screens 50 t/h, a 10-fold wider screen ought to correctly screen 500 t/h. But one would thus have a machine (L = 2 m, W = 10 m) whose width would make proper feeding quite impossible. Hence the operation has to be accomplished with a less wide but longer screen with which it can be confirmed that the depth of layer on the screen and the machine's discharge ends remain within such a limit as will maintain the quality of screening (see Sec. 5.2.4). Generally the length to width ratio of the equipment is 2 to 3.5.

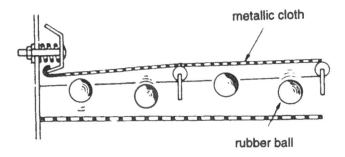

Fig. X.16. Anticlog system

5.1.3 VIBRATION

The amplitude, frequency, pattern and direction of vibration also exert a significar influence on the screening operation.

For example, for fine cut-offs recourse is had to higher frequencies and lowe amplitudes than for coarse cut-offs. The pattern and direction of vibration affec the machine's output, as does the quality of screening. For example, a single ur balanced weight screen with elliptical vibration (see Sec. 3.4.2) will have a positiv effect because of one vibration spreading the product on the feed side and anothe retaining this product on the discharge side of the machine. Thus the depth of th layer will be more uniform over the whole screen length.

5.1.4 MATERIAL TO BE SCREENED

The grading of the material also influences the outcome of the operation. For th same output and quantity of retained, it is clear that a material which is rich ii fines will be easier to screen than a material having a number of large grain approaching the size of the sieve used. This phenomenon will be taken into accoun in the method of calculation by means of the 'passing at half-mesh' coefficient (se Sec. 5.2.1).

The shape of the aggregates again plays a role; rounded material will screei more readily than irregularly shaped crushed material.

5.1.5 SCREENING SURFACE

The type of surface employed and the manner in which it is fastened also pla) specific roles.

The type of weave of the metallic screens, the shape of the holes in the per forated plates, or the rubber or polyurethane fabrics will expedite or curtail th(possibility of the grains passing through the screening medium.

The permeability (or ratio of area of holes to total area) has an obvious in fluence on the capacity of screening in terms of output as much as efficiency.

A very big bulge in a stretched screening fabric tends to distribute the produc unsatisfactorily over the breadth of the screen, producing deeper layers along th(edges of the equipment.

5.2 Sizing Screening Surfaces

Amongst a host of methods available, we shall first describe one which, though by no means the best yet, has the merit of having served to size hundreds of machines in an acceptable manner.

5.2.1 CALCULATION OF SURFACE AREA

The surface area is calculated by the following formula.

$$S \text{ in } m^2 = \frac{P}{A \times B \times C \times D \times E \times F}$$

where

P: output of passing in t/h for a material of density 1.5 (the result is readily corrected for other densities by multiplying it by 1.5/d);

A: base output in t/h/sq m (see Fig. X.17), function of the cut-off mesh size M;

B: coefficient pertaining to percentage of retained in the feed of the deck (Fig. X.18);

C: coefficient pertaining to the quality of screening:
standard screening, 1
summary screening, 1.2
thorough screening, 0.8;

D: coefficient pertaining to percentage of particles smaller than the half-mesh in the feed of the deck (Fig. X.19);

E: coefficient for screening under water (Fig. X.17);

F: coefficient of deck to top deck: 1
2nd deck: 0.9
3rd deck: 0.8
4th deck: 0.7

M	(t/h)/m²		E	opening
mm	Δ	O		%
1	1.5	1.8	1.3	39
1.25	2	2.5	1.4	41
1.6	2.8	3.4	1.5	48
2	3.6	4.4	1.7	48
2.5	4.6	5.6	1.8	48
3.15	5.8	7	2	48
4	7.2	8.8	1.9	44
5	8.8	11	1.8	48
6.3	11	13	1.7	55
7.1	12	14	1.7	56
8	13	15	1.6	58
9	14	17	1.6	56
10	14	18	1.5	59
11.2	15	19	1.4	60
12.5	16	20	1.4	63
14	17	21	1.3	63
16	18	22	1.3	64
18	19	24	1.2	67
20	20	25	1.2	64
22.4	21	26	1.1	67
25	22	27	1.1	69
28	24	29	1	68
31.5	25	31	1	71
35.5	27	33	1	73
40	28	34	1	70
45	29	36	1	72
50	31	38	1	74
63	35	43	1	79
80	40	49	1	79
100	47	56	1	79
	d = 1.5			

Δ : Crushed
O : Rounded

Fig. X.17

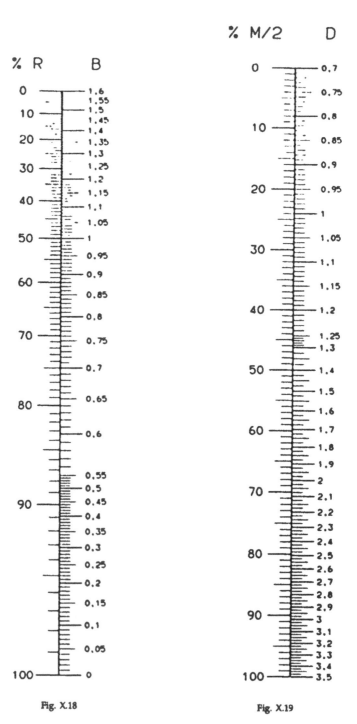

Fig. X.18

Fig. X.19

5.2.2 EXAMPLE OF APPLICATION

Let the screening be for 10 mm mesh size, at 100 t/h, for a product containing 40% larger-than-10 mm and 25% smaller than-5 mm. The product is a dry crushed material of bulk density 1.5.

For 10 mm it will be found that A = 14 t/h/m=2 (Fig. X.17);
For 40 mm of the retained, B = 1.1
We shall choose a thorough screening, C = 1.2
For 25% of the passing at 1/2 mesh opening D = 1
Dry screening, whereby E = 1
1st deck of the screen whereby F = 1
In which case:

$$S = \frac{60}{14 \times 1.1 \times 1.2 \times 1 \times 1 \times 1} = 3.25 \text{ m}^2$$

Should there be more than one deck, the subsequent decks will then be calculated and the largest surface area obtained by calculation adopted. Then the machine whose size is equal to or just higher than the value adopted will be selected but after having confirmed that the depths of layers remain within reasonable limits (see Sec. 5.2.4).

5.2.3 COMPARISON BETWEEN DIFFERENT METHODS

The methods based on 'tonnage fed' and those based on 'tonnage passed' exemplified above, have often been considered contrary. This in fact is not a problem: it is easy to convert one to the other as follows:

Let F be the feed rate, which means

$$F = \frac{P}{1 - r}$$

where r is the proportion of retained.

Then, based on the foregoing method, one can calculate the coefficient of retained B_1 for a tonnage-fed method by the formula

$$B_1 = \frac{B}{1 - r}.$$

The formula as per tonnage-fed will then be (all other coefficients remaining the same):

$$S = \frac{F}{A \times B_1 \times C \times D \times E \times F}.$$

In the preceding example (see Sec. 5.2.2), it would be:

$$B_1 = \frac{B}{1 - r} = \frac{1.1}{1 - 0.4} = 1.83$$

and for F = 100 t/h

$$S = \frac{100}{14 \times 1.83 \times 1.2 \times 1 \times 1 \times 1} = 3.25 \text{ m}^2.$$

Amongst the many methods available, it appears that they generally given com parable results in the zone of retained between about 20% and 70%, while in othe cases the results may diverge by sizeable margins.

Hence it is generally wise to fall back on earlier experience whenever possible trying to look for proximate cases of application for which qualitative and quar titative results are known.

5.2.4 DEPTH OF LAYER

Having calculated the surface area in m^2, it is imperative to make sure that th depth of the product layer upon the screen is compatible with good screening. W shall define this as follows:

$$\text{Mean depth} = \frac{(\text{Retained} + \text{Passing}) - \text{Passing } 1/2 \text{ mesh opening} + \text{retained t/}}{2v \times 1 \times d}$$

$$\text{Depth at end of screen} = \frac{\text{Retained (in t/h)}}{v \times l \times d}$$

where v = speed of advance of products (1100 to 1500 m/h); l = useful width screen in metres; d = density.

The mean depth should not exceed 1.5 times the size of the largest lump or times the mesh opening. For depth at end of screen:

— Horizontal or inclined screen with unbalanced weights and circular vibratior
3 times the installed mesh for dry screening
4 times the installed mesh for wet screening
— Inclined screen with elliptical vibration:
5 times the installed mesh for dry screening
7 times the installed mesh for wet screening

For all other cases the depth at the end of the screen should not exceed th size of the largest lump.

5.2.5 QUALITY

All the formulas can give us no more than a broad indication of the size of th machine, so what exactly is desired in terms of machine output (yield of fines coarse, efficiency etc., see Sec. 2) should be properly specified in the first instanc to actually fulfil the requirement. It would be useless to oversize a dispatching fun tion screen but often worthwhile to provide a safety margin in the case of screer for grading wherein correct sizing is the prerequisite for function quality of th finished product.

REFERENCES

*Blanc, E.E. 1974. *Technology of Comminution and Size Classification of Equipment*. Eyrolles, Pari
*Giroud, A. 1980. Screening. In: *Aggregates*. ENPC, Old Editions, Paris.
European standards for size classification of equipment. CECE (1986).

*Entries marked with an arterist are in French—General Editor

XI

Feeders and Batchers for Crushers

Alain Maldonado

Crushing, which enables size reduction of quarry-run materials, is inconceivable without distributors and feeders which constitute the interface with transport or stockpile.

Feeders and distributors are differentiated on the basis of their kinematics, design (if additional functions are expected of them), size and mass in relation to their power demand, nature of the contact with materials (wear-resistant lining etc.) and so forth.

Judicious selection of this equipment indeed governs the smooth functioning of individual machines as well as co-ordination of the various production operations.

1. FEEDING OF PRIMARY CRUSHERS AND PRIMARY PROCESSING STATION

1.1 Direct Feeding

This method of feeding does not involve a mechanical device but takes into account, from the engineering viewpoint, the special infrastructures at the points of unloading by quarry tracks and the choice of type of crusher. The materials fall into a chute by gravity with no control exercised over the flow. Therefore the crusher's working and the geometry of its feed opening should be well adapted to the peculiarities of the geotechnical properties of the materials coming from the quarry.

The drawbacks of direct feeding without regulation could well be:

— Congestion of the crusher's inlet under an afflux of materials; these effects are exacerbated when the materials are very flaky.

— Risk of vaulting caused by a self-compaction during feeding.

— Risk of blockage resulting from the awkward positioning of an oversized block.

— Untold difficulties or even the impossibility of manual intervention, without prior emptying of the feed hopper.

Consequently, such an arrangement often entrains erratic working, excessive downtime and need for frequent, costly, painful and hazardous manual intervention. In other words, the possible saving in investment is vitiated by the high maintenance or operational cost. To mitigate these risks of erratic working, measures should be devised for bringing about some uniformity in the size of the blocks fed by recourse to:

— Sorting at the working faces of the quarry by means of shovel buckets judiciously sized and with due regard to the size of the primary crusher's opening.

— Control of discharge from the quarry truck to regulate the feeding rate.

— Breaking or elimination of the biggest blocks before loading on the truck.

Direct feeding of crushers was originally introduced for feeding jaw crushers but at present is widely employed for feeding large gyratory crushers because this operation can then be effected from several points of discharge around the machine. This modus operandi is still more an exception than the rule, however, because applicable only to very large-scale exploitation.

1.2 Mechanical Feeders

While primary jaw crushers, as noted above, can sometimes be fed by direct discharge from loader buckets or dumpers, it is generally advantageous to use a mechanical feeder. This feeder becomes indispensable in the case of impact crushers.

Selection of the feeder is dictated by the heterogeneity of the feed since the machine is expected to perform some additional functions as shall presently be seen.

The all-in materials coming from the working face can well be in the shape of very large blocks, which may be mixed to a greater or lesser extent with finer crushed ones and occasionally muds.

1.2.1 RECIPROCATING GATE-CONTROLLED OR MECHANICAL FEEDERS

This equipment, which is very simple in design, comprises a rugged rectangular plate positioned horizontally or with a very slight slope (3 to 5 degrees) below the opening of the hopper receiving the all-in. The plate rests on idler wheels and a regular reciprocating movement is activated by means of a system of eccentrics or disc cranks whose throw can be adjusted. The forward movement of the plate picks up the materials held inside the feed hopper, which on its return are dumped into the opening of the primary crusher.

Adjustment of the crank L and the eccentric's rotation speed make it possible to calculate the feeding rate of the feeder by taking into account the petrographic nature and properties of the all-in blocks (see Fig. XI.1).

To obtain greater operational flexibility through near instantaneous adjustment, some manufacturers have built hydraulically driven trolley feeders. This could become an excellent solution towards developing automation in adjusting the feed rates to the machines.

Other manufacturers have prepared reciprocating feeders which have a pre-elimination bar screen made up, in the simplest cases, of just an extension over-

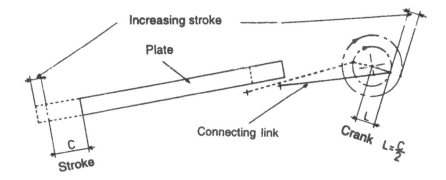

Fig. XI.1. Relation between throw and adjustment of crank

hanging the plate's front. In other cases the portion forming the screen receives a movement independent of that of the plate (dual control, Fig. XI.2), the objective here being an attempt to disconnect the extraction function from the sorting one.

1.2.2 VIBRATING PLATE TYPE FEEDERS

These machines have been the object of numerous applications in recent years. They comprise essentially heavy rectangular plates supported by a spring mounting, subjected usually to vibration through double or quadruple unbalanced mechanisms.

These vibrating tables are installed with an inclination that may vary from 0 to 15°. The flow rate is adjusted by means of a speed adjuster in the case of mechanical vibrators.

These feeders have geometrical characteristics and mass which vary depending on the mass of the blocks and the grading of the feed, which could be more or less in the range of sand.

Accordingly, these machines are called:

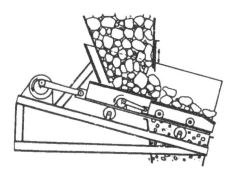

Fig. XI.2. Trolley feeder with reciprocating screen (dual control)

— Scalpers when they accomplish elimination of dry muds and clays on the one hand, and relieve the primary crusher by removing elements of a size lower than the set value.

— Giant grizzlies when employed for screening rocks for riprap.

— Bar grizzlies when the feeder has an attachment and special kinematic that ensure prevention of wedging of blocks (the distance between the bars increasing in the direction of block advance).

The assembly of these machines is illustrated in Fig. XI.3.

1.2.3 METALLIC TABLE-TYPE FEEDERS

These tables are made of articulated metal plates mounted on pins and rollers disposed on rails that form part of the supporting frame.

These plates are driven at low speed through a head boss with control by motor and reducer.

The flow rate is adjusted by varying the speed of the table (0.05 to 0.30 m/s), which in turn is obtained after change of pulleys or by means of a speed adjuster.

This type of feeder can be installed horizontally or inclined at 20° at the most. In the latter case an appreciable decrease in height of the feeding platform is achieved.

Inadequate leakproofing of the plates necessitates installation of a sheet-metal hood and a transporter of fugitive particles recovered under the deck.

These heavy-duty machines are still very costly and given their rather high maintenance cost besides, are being used less and less by aggregate producers.

2. FEEDERS IN PROCESSING PLANT

These machines are capable of feeding secondary and tertiary crushers, grinders and screens at a uniform rate. Firstly, they ensure extraction from under the hopper

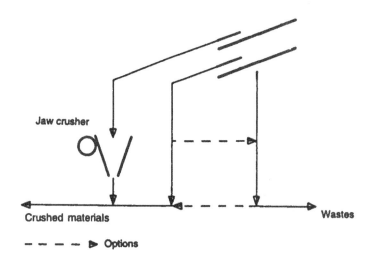

Fig. XI.3. Assembly of a vibrating feeder

or under big piles and secondly, batching of the products meant to go into a mixture. They all work on products of a limited or predetermined particle size (coarse and fine aggregates, fines).

Their utilisation in processing plants is becoming increasingly conspicuous since the quality of the supplies and smooth management of material conveyance are better ensured.

Indeed, optimal working of the machinery (crushers, grinders, screens) at this point is accomplished only if the feeding is uniform in rate of flow and grading.

Therefore it has become more and more imperative that every feeder-batcher be equipped upstream with regulatory hoppers incorporating a flow-regulating mechanism. This would compensate change in performance of crushing machinery with the passage of time, since homogeneity of the resultant gradings depends on the uniformity of the material fed to the crushing equipment.

The feeders in a processing plant are often and mainly to two types: those subjected to electromagnetic vibrations or those of the belt type.

2.1 Electromagnetic Feeders (Fig. XI.4)

These are frequently employed and comprise an electromagnet (3) energised from alternating current. Pressing on a counterweight (2), the electromagnet induces vibrations in the trough (1) into which the material flows.

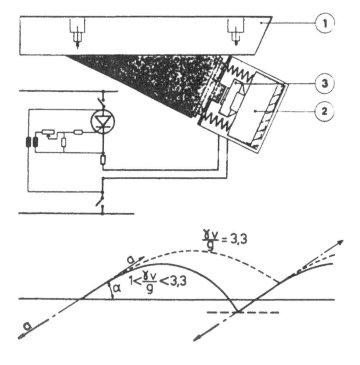

Fig. XI.4. Microprojection

The movement of a layer of material inside a vibrating chute is complex and hence we can at best study the behaviour of a single discrete particle by setting forth the hypothesis that the impacts with the trough are inelastic.

Let us consider a transport trough on which a sinusoidal vibratory movement is impressed by electromagnetic means along a direction forming an angle α with the trough, the said angle α being termed the angle of work (Fig. XI.4). Under this effect a point of the trough moves between two extreme positions separated by a distance 2a (where a is the amplitude).

A physical point (particle) placed on the trough accompanies it during the rising phase, then in full deceleration, up to the moment when the vertical component of this deceleration assumes a value first equal to g, then higher than the acceleration of the weight. The physical point then leaves the trough with the latter's speed at the moment of separation and continues its motion along a parabola. It again makes contact with this trough, which meanwhile has moved away, and accompanies it once more up to a new microprojection, when the cycle starts all over again.

From the foregoing it follows that in the case of a horizontal chute the microprojection takes place only for such vibratory movement as impresses upon the trough a vertical acceleration exceeding that of weight. Further, it can be shown that the time of contact between the trough and the point is zero for a vertical acceleration value equal to 3.3 g. This is the value which leads to the optimal speed for a given work angle and we may also note that in this case if no sliding takes place between the matter and the chute, the wear of the latter is reduced to a minimum if the duration of projection is less than or equal to the period of vibratory motion. For the general case the theoretical speed of the matter is given by Böttcher's formula:

$$V = \frac{g}{2} \times \frac{n^2}{f} \cot g\alpha$$

where α is the angle of work; f the frequency of vibratory motion in Hz; n an auxiliary coefficient which depends on vertical acceleration; V the speed of advance in mm/s.

The speed of the product's advance, governed by the dimensions of the section of the vibrating chute, enables calculation of the volumetric flow of the matter transported.

Control of this speed can be effected by changing:

— The amplitude of vibration which leads to a vertical augmentation of the vertical acceleration $y = a\omega^2 \sin \alpha$, where ω is the pulsation of the vibration.

— The angle of work, keeping all other parameters constant.

— The frequency of vibration: when the frequency decreases, it is necessary to increase the amplitudes of vibration in order to have sufficient acceleration. But too high amplitudes can give rise to mechanical disorders. The frequencies most often adopted are 50 Hz or 25 Hz.

— The trough's inclination which can accelerate the speed of displacement of the particles. However, sliding on the chute has always to be avoided as it might increase deterioration due to wear.

As a matter of fact, the speed of the layer also depends upon the geotechnical properties of the materials (grading of the aggregates, moisture content etc.), lateral friction of these materials on the trough's sides and behaviour of the dilated matter under vibrations.

2.2 Belt-type Volumetric Batchers

In a withdrawing belt extractor, the belt is flat and moves between a pair of guide-plates, carrying upon itself a uniform layer of material whose depth is predetermined by means of a gate. Should large variations be necessary, the belt can be driven by means of a variable speed drive or a variable speed motor.

Often, several belt conveyors arranged in parallel serve for batching mixes used in making cement or bituminous concretes or treated aggregates. In installations of this kind it is advisable to safeguard against possible disruption in the arrival of material into one or the other batcher (empty hopper or vaulting in a hopper) by fitting the latter with sensors or contact gauges which will interrupt the recomposition in case of a problem.

This machinery can also be equipped with a weigh bridge (see Chapter XIV).

3. AUTOMATIC CONTROL OF FEEDERS

Automatic control of the working of feeders by downstream machinery has been introduced in the last few years.

Now the primary crushing station in a quarry, which comprises a feed hopper, a feeder and a jaw crusher, can work without manual intervention.

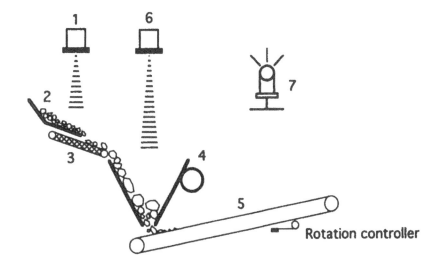

Fig. XI.5. Automatic control of a feeder by the working of a crusher (after D. Crouzier)

A typical primary crushing station equipped with many sensors is shown in Fig. XI.5. Two ultrasonic sensors help to gauge levels. The first (1) measures the filling in the receiving hopper, while the second (6) measures the level of filling in the primary crusher; 1 rotation controller and 1 control help to vary the feeder's rate of feeding.

Let us consider an installation in operation in order to understand how it works.

When the sensor (1) monitoring the hopper (2) detects a low level (with a cushion of materials), it stops the feeder (3).

When a truck discharges its load, the sensor above the hopper registers a variation in the level and causes the feeder to resume.

The materials then fall into the jaw crusher (4), are crushed and collect upon the conveyor (5).

The ultrasonic sensor (6) of the crusher records the variations in level in the crusher's mouth. When the level is high, it causes the feeder to stop. If the level persists high for 15 seconds, the feeder is stopped and an audio-visual signal activated to signal the occurrence of blockage (7).

The stepped intensity relay connected to the conveyor circuit confirms that the jaw crusher is tamped if the intensity is minimal or stops the feeder when the intensity of overloading is reached.

After cleaning, the crusher's sensor records a low level which activates the feeder to resume.

4. CONCLUSION

With the recent advances in sensors, which can be inducted in all the unit operations of a processing chain, feeders are going to be used more and more widely and in greater number. They will play a more conspicuous role in controlling material flows in processing plants and more directly influence quality assurance of the products.

XII

Washing of Aggregates and Treatment of Fine Aggregates

Jean-Claude Exbrayat and Alain Mishellany

1. GENERAL

Wet treatment of aggregates seeks to improve their cleanliness by a vigorous washing that carries away polluting substances, in particular clayey fines. A major portion of these substances is removed immediately after extraction but further treatment is often needed for upgrading the cleanliness of the aggregates.

Another aspect of wet treatment of aggregates is classification of the sands they contain into finer size grades for which screening is not feasible on an industrial scale. Present-day techniques make it possible to obtain sands of the desired grading. These processes are under constant improvement since aggregate users are more and more aware of the merits of controlling the internal curve of sands.

Wet treatment uses large quantities of water which obviously becomes polluted by fines; hence recycling and clarification of the water are very often indispensable complementary operations.

Specification of the equipment needed for the above purposes should take into account the nature and the pollution of the deposits upstream, as well as the qualities expected for the aggregates produced and the levels of pollution for rejection downstream.

1.1 Pollutions in Deposits

Pollutions vary according to the nature of the deposits, amongst which the principal kinds are:

— Presence of substances coating clean materials

These are clayey compounds adhering to the surface of aggregates or constituting a binder of conglomerates which are dislodged by mechanical action.

— Presence of fine elements

These products are troublesome, either because of their nature (clays) or solely because of their fineness (alluvial deposits) when present in excess.

— Presence of impurities:

vegetable wastes: grass, wood, roots, lignites

animal wastes: shells

fragile, porous and friable materials (soft limestones, slags, schists, slates).

— Surface moisture

This can be harmful either directly through its entering into concrete or its dripping from materials during their transport, or indirectly because of the dissolved salts it contains (particularly in the case of aggregates of marine origin).

Amongst the polluting substances, clays are the most common and the most difficult to remove. Those encountered are of three types:

— Mass clay—products of weathering which break up entirely in water;

— Gangue—difficult to dilute in washing water since it adheres tenaciously to coarse elements;

— Consolidated lumps which tend to form pellets during gradation or screening, and which are found in fine gravels.

Detailed planning of any treatment process should, at the outset, take into account:

— Firstly, the kind of pollution and

— Secondly, the magnitude of pollution vis-à-vis the quantity of materials to be processed.

Thus it is imperative to properly estimate the variations in the deposit which must be taken into account before deciding the kind of treatment needed and its scale.

The types of pollution found in deposits of massive rocks differ from those found in alluvial rocks.

• At the quarry, pollution may occur at three levels:

— The overburden, which when poorly ripped, can cause pervasive pollution of clean blasted rock.

— Faults or clayey pockets which have to be purged to preclude the aforementioned problem.

— An interbedded, widespread and diffuse pollution, for example in sedimentary rocks.

In the first two cases, it is often possible to eliminate the main element of the pollution through extraction or by elimination at the primary section. In the third case, on the contrary, removal of the impurities is much more difficult and the wet method of treatment appears to be the only solution.

• At the borrow pit, pollution can be either localised in benches, lenses and pockets or be dispersed throughout the deposit.

— The quality of stripping influences the cleanliness of the underlying aggregates except when the deposit itself is totally polluted.

— In some cases pollution from scraping by the extracting machinery at the base of the deposit should not be ignored.

— The strongest technical hitches could originate from argillo-marly strata, pockets, benches and lenses, which in every case warrants wet screening.

Except with a sound appreciation of a deposit, it is not possible to design an appropriate installation for the treatment of aggregates, which would yield the quality of products demanded by the market.

To assess the extent of pollution and to decide the manner of exploitation, it is imperative to detect any significant localised pollution and to eliminate it straightaway on the spot, thereby precluding contamination of the entire mined material, which would in turn necessitate large-scale treatment for the average pollution of the deposit.

Treatment of the materials in general, and washing in particular, becomes all the more effective since judicious sorting at the extraction and loading stages at the mine face will have already eliminated the major pollutions.

In the matter of treating pollution, there is a rather major difference between quarries of massive rocks and other deposits.

• Quarrying of massive rocks often involves only dry treatment. Reduction in pollution is carried out in successive eliminations. However, one should not ignore the possibility of processing the wastes, when present in large quantities, by the wet method so that they can be reclaimed. This implies certain installations, as it is estimated that a deposit which contains 1.5% clays for 10% scalpings, for instance, should be equipped to treat the latter by a scrubber washing system similar to that for treating a material polluted to 15%.

• In quarries of alluvial deposits pollution is intimately admixed with the materials and the installations are generally designed for a more or less large-scale washing. But the fact should not be lost sight of—and this is equally valid for the removal of finer elements from sands—that the total pollution is found concentrated with sands in the wash water. The percentages are then very high and it is considered, for example, that in a deposit polluted by 2%·of clay and containing 10% of 0/5 sand, the pollution at the level of the sands will be 20%.

1.2 Expected Qualities of Materials

Users of sand and gravel, for concretes as much as road construction, know that cleanliness and appropriate grading are essential criteria for these materials if constructions of quality are to be achieved. It should also be noted that the requirements in respect of materials are now well defined by standard specifications appropriate to the different techniques as revealed by experience. These standards differ for the same aggregate depending on the end use, which enables an optimum utilization of the deposits.

The means for obtaining the desired qualities will vary in thoroughness according to particular needs and above all, the nature of the materials and their pollution.

Studies carried out by the services of the Administration have specified the cleanliness requirements of the materials (Chapter IV).

It is seldom that one comes across sands in nature that directly fit the desired grading envelope. This is why it is necessary to correct the grading. Several methods are possible depending on the grading of the sand to be treated.

• *Gap graded sands*

Such sands are common when they originate from crushing or from fluviatile deposits extracted in water streams. They contain few fine particles, their curve is situated below the lower limit at 0.08 to 0.4 or 0.5 mm (Fig. XII.1).

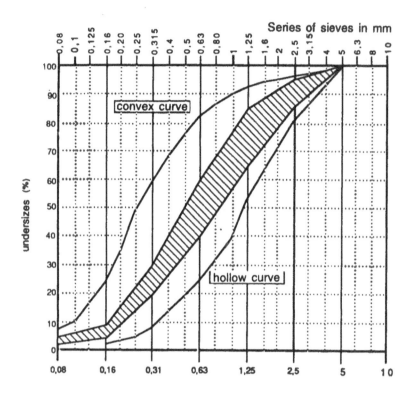

Fig. XII.1. Grading curves of sands

Correction can be made:
— by the addition, of suitable percentages, of fine natural or crushed sands having a convex curve.
— by elimination of a portion of the granular fraction in excess,
— by screening around 0.5 mm, then elimination of a portion of the excess fraction (coarse) and mixing in suitable percentages in order to push up the curve into the grading envelope.

• *Convex curve sands*

This case is frequent in alluvial deposits and corresponds to the characteristics of crushed sands.

Its correction can be effected:
— by addition, in suitable percentages, of natural or crushed sands having a hollow curve,
— by elimination of a portion of the excess fraction,
— by screening situated judiciously between 0.4 mm and 1 mm depending of every individual case, in such a manner as to recompose the suitable percentages

of coarse sands and medium-to-fine sands for obtaining the curve fitting in the grading envelope.

The granulometric curve of sands is usually bounded by:

— The greatest mesh size of the classifying screens between 2 and 6 mm.

— A separation around 0.100 mm with more or less well-selected equipment, not always of high efficiency, but offering the combined possibilities of decantation, assortment and dewatering.

Maximum recovery of the fine but non-polluting elements is aimed at in the plants. Fine elements which should be separated from the wash waters also contain unwanted fine elements in suspension; hence it is advisable to work out schemes of treatment involving equipment which can fulfil multiple functions:

— Simple rinsing for aggregates from least polluted sources.

— Dilution of plastic clays, disintegration of consolidated grains armoured mud ball.

— Scrubbing of aggregates coated with clays.

— Attrition of highly polluted sands.

— Preparation of aggregates and sands for recovery and removal of slurry.

— Classification of the aggregates into different aggregate sizes as needed by the market, by vibration and the hydraulic method.

— Separation of impurities (wood, shells etc.) present in the aggregates.

For considerations of economy or simplification of the treatment schemes, it is common in small-scale exploitations to use a single machine that can accomplish several functions. Frequently, there is one screen for sorting out the aggregates and washing them, or one wheel for washing the sands and separating them from the used water. For treating less polluted deposits, the washing-classification or washing-separation functions can be accomplished in good working conditions. This is not the case with highly polluted materials, which necessitate equipment with well-defined specific functions, such as scrubbing, attrition, removal of impurities and so forth.

2. WASHING OF AGGREGATES

2.1 Definition

The term washing denotes those actions which effectively dislodge by mechanical, hydraulic or vibratory means, the fine elements agglomerated amongst themselves or adhering to the surface of the aggregates, so that these can be removed along with the wash water and the aggregates which have been dewatered and freed of harmful elements then recovered.

Depending on the degree and nature of pollution of the materials, the actions of washing can be incorporated at different stages of treatment of the materials. They may be simple, such as brief washing on a screen, or complex, with multiple operations of treatment accomplished by equipment, such as wash trommels, attrition mills, wash wheels, cyclones and so forth.

Washing is an operation which can be carried out at different stages of the treatment scheme, starting from the initial up to the final level of classification of the sands.

The circuits of washing or classification by the wet route can be represented in three major stages in a schematic manner:

— Bringing the raw materials into contact with clean water put the polluting or surplus substances into suspension.

— Separation of the coarse and fine commercial products from the polluted water.

— Decantation of the slurry-pulpe for clarification with the inevitable conse quent production of muds (Fig. XII.2).

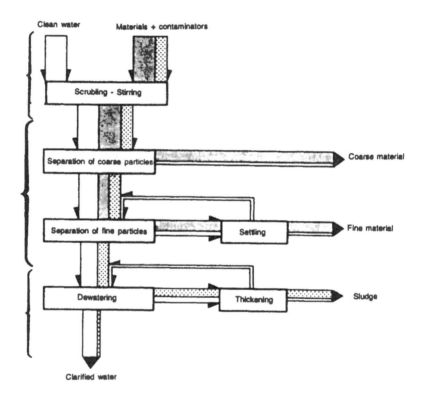

Fig. XII.2. Treatment of polluted materials. Operations carried out in three stages

2.2 Washing After Extraction

Washing of the all-in materials following extraction enables removal of the main polluting elements. Placed at the beginning of the treatment chain, washing ensures a feeding of clean coarse materials to crushers and grinders, an indispensable condition for obtaining crushed and ground aggregates of satisfactory cleanliness. Washing can also facilitate reclamation of materials eliminated at the primary section, thus effecting an increase in production, which can be substantial.

Machinery for washing comprise:

— Equipment enabling rinsing of the materials.

— Attrition equipment and scrubbers which ensure the disintegration and dilution of clays.

— Equipment for separation and recovery of sands and removal of clayey water.

— Classifiers for removing undesirable materials, such as friable or schistose particles, organic wastes, wood, shells etc.

2.3 Washing Coarse All-in Having Low Percentage of Clay

Washing by rinsing on a screen is commonly employed for materials which are easy to wash. The operation of wet screening is resorted to, wherein materials are sprayed with water under pressure or otherwise to obtain the desired level of cleanliness. Washing by rinsing is especially suitable for dedusting and "desanding" aggregates.

Proper selection of the spray bars over the screen is very important. The objective is to obtain the highest washing efficiency for the lowest water consumption (Fig. XII.3).

Fig. XII.3. Spray bars with holes

The following equipment is based on rinsing:

• *Screen equipped with spray bars with holes or slots*

Spray bars with slots (Fig. XII.4) generally give better results than the earlier ones with point holes. Slot-type spray bars wash the aggregates by creating continuous water sheets in a transverse direction. Like the hole-type spray bars, they are often liable to clogging, which is a drawback inherent in the operation and the environ-

Fig. XII.4. Slotted spray bars

ment of the screens, as well as the particles present in the source of rinsing wate
a river or wells.

• *Screen fitted with spray bars with cyclone nozzles*

These bars offer the same advantages as those with water supplied through slot
but if the type of nozzle is appropriately selected for the size of the grain
suspended in the wash water, clogging can be practically eliminated. Uniform dis
tribution of the flows on the surface of the screens is facilitated by the pressur
required for the functioning of the cyclone nozzles.

• *Screen fitted with spray bars with spray nozzles*

Spray bars are often fitted with pulverization devices, available commerciall
working under high pressure (Fig. XII.5). These spray nozzles are designed to cr
ate flat jets in the shape of a water curtain striking the screen meshes cross-wis
Some cyclone-type spray nozzles form hollow cylindrical jets. These device
calibrated for flow and pressure, permit adjustment of water flow for washing th
required tonnage of materials with maximum efficiency.

When clays adhered tenaciously to the all-in material, a chute with baffles i
incorporated upstream of the screen for softening the materials before they reac
the screen. This simple operation can greatly facilitate removal of clays.

Washing materials by wet screening is effective only if the equipment, by vir
tue of good regulation of the vibratory system, ensures dispersion of the aggregate
on the grids and uniform advance of the grains by short throws, thereby exposin
the entire surface of the grains to the action of the water jets.

The consumption of water generally reaches 0.5 to 0.8 m3 per tonne of ag
gregates washed by wet screening. The wash water, only slightly dirty in this cas

Fig. XII.5. Spray bars with spray nozzles

nevertheless contains all the sands and should be treated in the separation equipment forming part of the treatment chain for the materials.

• *Washing gravels in trommels*

This old method is still employed in some installations treating aggregates with very little clay. The spray bars for rinsing are disposed in a judicious manner inside the trommel.

The rolling and tumbling of the material on itself and on the trommel bottom causes a. slight attrition action by the larger elements. The fine gravels and sands, which quickly pass through the trommel holes along with the rinse water, should be treated in a subsequent phase if they are not totally cleaned from clay.

2.4 Washing All-in Having a High Percentage of Clay

The main function of washing by scrubbing and attrition is to dislodge the admixed clay. To do this, the forces of cohesion which hold the agglomerated clay particles in the form of balls or cause its adherence to the aggregates should be broken so that the particles are diffused in the wash water. The mechanical action of attribution which operates by friction of the grains amongst themselves and against the walls of the equipment or internal fixtures provided for the purpose, frees the aggregates from the clayey gangue which envelops them. To complete the washing, it is imperative thereafter to drain the scrubbed aggregates on the screen and then rinse them to remove clay traces.

The washing equipment involved in scrubbing and attrition is as follows:

• *Wash trommel*

This is a slowly rotating drum made from perforated sheets with paddles affixe
to its inner periphery, which triturates the pulp under spray bars at high pressui
(Fig. XII.6).

Fig. XII.6. Sectional view of wash trommel

The clay-laden water flows through the perforations of the drum, carryin
along the fine elements, while the washed aggregates are impelled by the paddle
to the end of the machine.

• *Vibratory scrubber*

This comprises two identical horizontal wash tubes arranged on either side of
central tube to which they are clamped by means of a spectacle-shaped stirruĮ
The whole assembly is suspended by a device provided with shock-absorbin;
springs. A shaft with an unbalanced weight forming part of the central tube i
driven through a flexible coupling by an electric motor, imparting to the assembl·
a circular movement of about 5 mm radial amplitude (Fig. XII.7).

The pulp introduced in the two wash tubes on the coupling side is subjecteᵣ
to vibratory movement which at once produces an attrition of the particles amongs
themselves and their advance by dispersion. The polluted water is evacuated b·

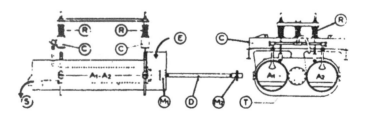

Fig. XII.7. Principle of vibratory scrubber.
A_1-A_2 — Wash tubes; C — Spectable-shaped stirrup; B — Shock-absorbing springs; T—Central tube;
D—flexible coupling; E—Entry of materials; S—Exit of materials.

the perforations provided on the underside of the tubes and the rinsed product flows out by the downstream weirs while passing over the wire draining trays.

This equipment is employed in particular for treating 20/70 mm and 20/200 mm aggregates, with a flow rate of 1 to 5 m3 wash water per tonne of materials treated.

• *Revolving scrubber mill*

This consists of an entire cylinder (i.e., not perforated) provided with internal paddles for advancing, lifting and churning the materials. Some models also have baffles, buckets or partitions for enhancing residence time and promoting material attrition. The cylinder rests on tracks of engaged wheels or on groups of tyres; in most cases it is driven by motor-reducer assemblies, some with truck decks. There are two distinct operational modes for this equipment:

• *Countercurrent scrubber mill*

The materials are introduced through a chute in the axial opening of the equipment's upstream end. The wall carries a circular sheet-metal peripheral weir, then a ring made of a grid or perforated sheet between the weir and entry ferrule.

The aggregates advance inside the equipment towards the downstream end where a lifting device ensures their extraction through the exit ferrule; they thus leave with a low percentage of water. The water is introduced inside the equipment through the opening of the exit ferrule and travels countercurrent to the aggregates to progressively take up clay, fine elements and sands. The overflow takes place over the weir situated at the upstream bottom, through the grid or the perforated sheet, carrying along the clays and a major portion of the sands which will be treated in the separation or classification equipment.

• *Scrubber mill with unidirectional flow*

In equipment of this type the feed and discharge end walls are made of an unperforated sheet, the overflow weir being situated downstream towards the exit ferrule, which has a larger diameter opening.

The water and the materials flow in the same direction, stoppage of the materials being ensured by the lifting paddles and diaphragms.

All the materials to be treated undergo an attrition during the time they spend inside the equipment, under water immersion, which is essential for ensuring the scouring of the aggregates and dilution of the clays.

Outside the exit ferrule, an adjoining conical trommel, with single or double mesh, permits return of the dewatered aggregates. The water carries off the clay and the sands for the appropriate treatment in the separation or classification equipment, as the case may be.

Tube scrubbers can handle substantial tonnages of materials of a wide particle size range 0/500 mm. The flow rates of wash water should be higher when the pollution is higher or the clay difficult to treat; they could vary from 1 to 3 m3 per tonne.

To obtain a consistent quality of clean aggregates, the throughput of an equipment should be adjusted in accordance with the variations in quality and the nature of the pollutions in the deposit exploited.

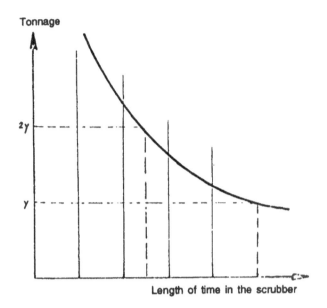

Fig. XII.8. Output curve of a scrubber

The output may vary from one- to twofold depending on the time needed fo the attrition of the aggregates and dilution of the clays (Fig. XII.8).

Every compartment of the equipment is watered. The bottom of first trough is in the shape of slotted grids to allow the loosened clays and sands to pass, whil the middle troughs are of an entire sheet to enable stirring and attrition of th materials under water. The last trough is fitted with a grid to dewater the fin gravels before they are conveyed for classification where they can be rinsed o vibrating classifying screens.

These machines can treat up to 200 t/h of all-in whose particle size may rang from 0 to 300 mm. They are employed in some quarries, especially for treatin all-ins rich in gravels, since the sands get evacuated right from the first grid: which are quite often inadequate for attrition action.

• Scrubber screw

Stirring is accomplished by a helical blade which may be single or double pitcl affixed upon a longitudinal shaft whose head is driven by an electric motor wit a reducer. The helical blade triturates the material in the water trough to remov the unwanted matter, lifts it out of the water, dewaters it by the action of the scre and removes it through the upper exit. The unwanted materials are evacuate through a overflow.

Some machines have an arrangement for raising the screw, which enables ad justing the height of the helical blade above the trough's floor depending on th size of the material.

Though used sometimes for scrubbing fine gravel despite the not inconsiderable risks of wear, washing screws that are specially designed for churning sands should be reserved for the attrition of least polluted sands. The particle size of the material treated is limited to 6 mm at the most. The permissible tonnages are 20 to 150 t/h per machine with a wash water rate of 1 to 5 m3 per tonne treated.

• *Log-washer*

In the log-washer two parallel longitudinal shafts carrying intermeshing paddles revolve in opposite directions. These paddles vigorously triturate the material under water in the lower section of the trough to dislodge the impurities, which are then evacuated through the overflow. The aggregate, carried upwards by the paddles and washed, if need be, by a spray pipe with nozzles, then dewatered, passes through the upper discharge chute. The log-washer is used particularly for scrubbing 4 to 100 mm size aggregates. The paddles and the internals are made of replaceable wearing parts. The rate of wash water is between 0.8 and 1.2 m3 per tonne treated. The rate of flow capacities of the equipment vary from 50 to 150 t/h (Fig. XII.9).

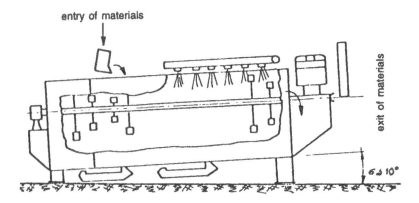

Fig. XII.9. Paddle washer (log-washer)

• *Attrition mill*

The attrition mill is a machine for producing a vigorous rubbing between the particles in a pulp containing 65 to 80% solids, to strip the grains free of the organic gangue, and crush and remove the inclusions of clay and soil (Fig. XII.10).

The path of the pulp is intercepted by a series of vertical and horizontal baffles which set up vertical movements, making the solid particles rub against each other.

The attrition is repeated in a series of octagonal cells, each carrying a rotor and independent motor. Each rotor has horizontal blades of opposing slope at three levels which reverse the pulp flow twice in each cell by producing a controlled turbulence.

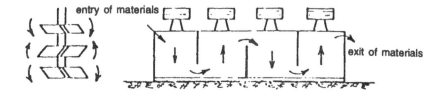

Fig. XII.10. Attrition cells

The cells, which may be even or odd in number, are arranged in zigzag fashion to prevent direct passage from one cell into the other and to facilitate uniform treatment of the pulp in each cell for a predetermined period.

Some attrition mills, though working on the same principle, are equipped with a turbine in place of the rotor at the bottom of the cell to enable stirring of the pulp. Turbo-attrition mills are employed solely for treating 0–5 mm sands that are barely to highly polluted. The output of such a machine varies from 20 to 30 t/h and the water needed for washing is 0.8 to 1 m3 per tonne treated. (Table XII.1)

Table XII.1: Washing of all-in with a high percentage of clay

Equipment	Main function: Attrition		Additional function: Dewatering	Permissible tonnage per machine, T/h	Wash water consumption , m^3/t	Power, kW/t	Wear
	Nature of material	grading, mm					
Revolving tube scrubbers	Barely to highly polluted	0–100 to 0–500	+ 5 by counter-current or by trommel	5 to 700	1 to 3	0.1 to 0.30	Low to medium
			Depending on manufacturer of machine and type of fabrication				
Vibratory tube scrubbers	Highly polluted gravels	0–70 to 0–200	(Naturally) extracted gravels	5 to 150	1 to 5	0.6 to 1	High
Paddle washers (log-washers)	Highly polluted gravels	4 to 100	All (naturally) extracted aggregates	50 to 150	0.8 to 1.2	0.6 to 1	High
Attrition mills	Polluted sands	0–5	Nil	20 to 30	0.3	0.8 to 1	Low
Cyclone-pump assembly	Polluted sands	0–6	Nil	5 to 200	2 to 3	0.3 to 0.4	Medium

2.5 Washing by Rinsing Graded Aggregates and Crushed Aggregates

The installation for treating alluvial materials often has two processing lines:

— One line for the natural materials graded by wet screening starting from the primary section.

— Another line for the crushed and ground materials graded by dry screening.

Natural aggregates should be divided into aggregate sizes demanded by the market, such as the aggregates used in hydraulic concrete.

Washing of natural aggregates originating from a lesser polluted deposit can be done on classifying screens. At the final screening stage the water entrains through the mesh opening the sands, fines and clays which should be subsequently separated by appropriate equipment.

In some installations, which primarily supply materials for building purposes, the crushed and ground aggregates are directed to the processing line for natural aggregates, at least in respect of the fraction containing sands, where they undergo the washing necessary to remove the dusts covering them and the fines produced by the size reduction and fragmentation machinery.

Generally, crushed and ground aggregates are graded by dry screening, especially for the preparation of aggregates for road layers and for ballast. In certain cases (surface coatings) the use of these materials requires a thorough elimination of the dust that adheres to the aggregates, which necessitates special provision of a horizontal screen working under a spray of water, for rinsing the aggregates at the time of their delivery.

Sands emanating from this process line contain a high percentage of fines; they are generally used in road construction works. However, some markets requires removal of a part of these fines, either by air-blowing equipment or by hydraulic separation equipment.

2.6 Washing Sands

Washing sands in a materials processing line assumes a special character.

In any installation, be it of the primary section or the stage of sorting of the aggregates on screens, sands are found entrained by the rinsing, scrubbing and attrition waters, which also carry all the dislodged clays.

Hence it is imperative that the sands be separated from the polluted waters.

This is the function of the separating equipment.

2.6.1 STATIC DECANTATION EQUIPMENT

Simple conical or pyramidal tanks, this equipment can only be used for almost clean sands. The sands extracted call for an additional machine for draining or dewatering.

This equipment does not classify; it simply removes the unwanted materials by overflow.

The cone hydroseparator comprises a conical decanter tank made of sheet metal which is fed with pulp from a box centred at the top. The surface flow

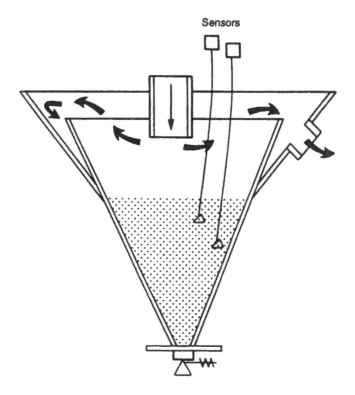

Fig. XII.11. Cone hydroseparator

causes deposition of the product along the walls of the tank and entrains fines and clays towards the peripheral overflow (Fig. XII.11).

The automatic discharge through a lower trap is controlled either by a hydro static system or by sensors recording the level of the materials with reference to the tip of the decanter.

Static thickeners and hydroseparators can achieve a production of sands reach ing 200 t/h for a water consumption rate of 1 to 1.5 m3 per tonne depending on the model chosen. The particle size separation is between 80 and 150 µm provided a clay content of about 40 g/l is not exceeded in the wash water. The sands are extracted with a minimum moisture content of 35%.

2.6.2 SEPARATION EQUIPMENT WORKING BY STATIC DECANTATION WITH MECHANICAL EXTRACTION

Decantation is also accomplished in tanks equipped with systems for extraction of the decanted sands. There are three main systems.

• *Dewatering flight conveyor (Mines)*

This comprises a sloping trough with a classifying tank provided on the lower side. Transverse scraper plates attached to two moving lateral chains scrape the trough as they move upwards. The chains are driven by head tumblers that effect return via the foot tumblers whose axis, situated above the overflow weir to check turbulence, is borne by strain screws (Fig. XII.12).

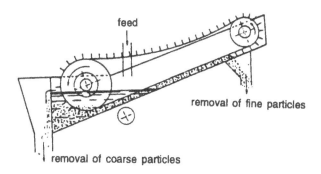

Fig. XII.12. Principle of dewatering flight conveyor

The pulp is introduced through a side pipe upstream of the foot tumblers. The sand drops to the bottom of the tub where it is caught by the moving plates, pushed up the sloping end, dewatered and then discharged in the upper section of the trough. The unwanted materials are removed by overflow.

• *Rake classifier (Mines)*

Here rakes are rigidly attached to two longitudinal plates which have a reciprocating motion. On ascension the rakes scrape the tank's bottom carrying the sand along with them; in descent the rakes are lifted above the sand layer by the plates (Fig. XII.13).

These two machines are eminently suited for clean or barely polluted sands with waters which are not unduly laden with clays and fines. At higher concentra-

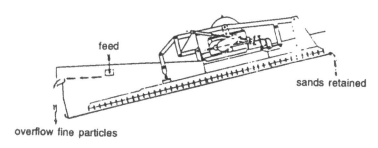

Fig. XII.13. Rake classifier

tions the liquid's viscosity changes, the density gradient due to decantation of the sands increases, with a concomitant discharge of clay into the sands and a loss in sand through the surface currents of water under evacuation. The corrective action often resorted to—the increase in amount of circulating water—may upgrade the cleanliness of the sands but by and large leads to a considerable loss in fine sand elements, between 80 and 400 μm.

• *Screw decanter*

Consists of an inclined canal whose lower section widens to form a decanting tub with an overflow weir.

An Archimedes' screw turns slowly inside the canal which raises the decanted sands at the bottom of the tub to the top section, where it emerges to push the sands out onto a conveyor or hopper, where the water content of the material is 18 to 25% depending on grading of the material. The water flows in the opposite direction, towards the discharge weir, entraining the fine particles and clays (Fig. XII.14)

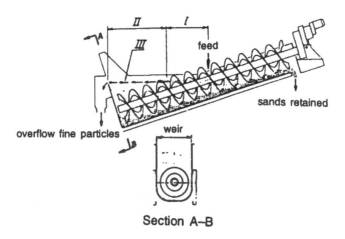

Section A–B

Fig. XIII.14. Principle of a screw decanter

Stirring, done by the screw inside the canal for raising the sands, adds an attritional effect which improves the cleanliness of the materials in some cases.

Various internal accessories, such as pulp feeding devices, adjustable overflow weirs, double pitch and variable pitch screws have brought about definite improvements in working and efficiency. A screw washer separates particles in the size range 80–160 μm and the consumption of water for washing varies from 3 to 6 m3 per tonne of sand. The output range of the machinery is 0 to 200 t/h.

• *Wheel dewatered*

Widely used. This is a wheel of large diameter carrying a series of perforated buckets on its periphery. The lower section of the wheel sinks into a prismatic decantation tank. The buckets scoop the sand lying at the tank's bottom and dewater it during their travel above the water surface. The fines and clays are evacuated by the overflow from the tank (Fig. XII.15).

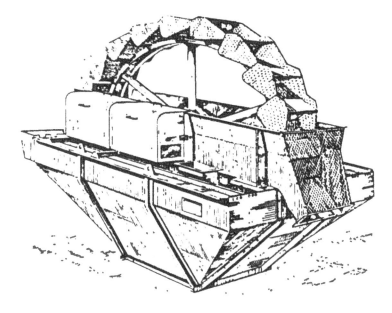

Fig. XII.15. Wheel dewaterer

The speed of the wheel, so chosen as to ensure evacuation of the volume of sand inside the buckets, should not give rise to substantial parasite movement; it should be low enough so as not to hinder dewatering of the sand. There are internal fittings which enable isolation of the decantation zone and extension of the lengths of the overflow weirs, so as to improve the efficacy of washing by increasing the water flow. Normally, particles in the size range 100–150 μm can be separated with little pollution of the water. The water content of the sand at discharge is 20 to 25% and the consumption of wash water varies from 1.5 to 2.5 m3 per tonne treated. The output range of the machinery is 0 to 200 t/h.

There are also machines which combine the effects of the screw and the wheel. The sand is collected by a wheel carrying buckets. The dirty water is evacuated from the side opposite to that of removal of sand. During natural flow of the water, the fine sand settles at the bottom and is pushed again at low speed towards the wheel by a screw with helical blades.

2.6.3 SEPARATION EQUIPMENT BASED ON CENTRIFUGAL FORCES

Hydrocyclones utilise the forces of centrifugal flow to accelerate decantation of particles suspended in water by weakening the effects of viscosity and density. The centrifugal action is caused by tangential entry of the flow inside a conical-cylindrical vessel. The particles are flung against the walls, which facilitates decantation towards the lower tapering end, and the water drained through a

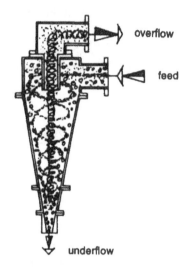

Fig. XII.16: Centrifugal flow separator

central pipe towards the overflow entraining the fine particles, such as clays (Fig XII.16).

Injection of water for washing demands a pressure which is provided by a group of pumps. The greater the speed of the fluid, the finer the particles flung against the walls and hence the lower particle separation size. The separation point can be conveniently placed between 20 and 100 μm by fixing the flow of pulp treated in the equipment for given size characteristics, which have been expressed in formulae by Dahlström and Fahlström.

A number of workers have studied and proposed formulae which enable specifying and calculating the characteristic of cyclones. The most common are those of Dahlström.

$$Q = k \ (D_1 \cdot D_2)^{0.9} \ \sqrt{H}$$

$$d_{50} = 0.6 \ \frac{(D_1 \cdot D_2)^{0.68}}{Q^{0.53}} \ \sqrt{\frac{1}{\rho_s - 1}}$$

where

Q = inflow rate in m3/h;

k = shape-related coefficient of the cyclone;

$D1$ = diameter of the feed pipe in mm;

$D2$ = diameter of the overflow diaphragm in mm;

H = pressure at entry in m;

$d50$ = dimension of separation in μm halfway (i.e., at 50%) of the separation curve (Chapter X Section 2.6.2);

ρ_s = density of the grains.

As a matter of fact, these formulae can be applied only to a certain range of cyclones, namely those which conform to some specified shapes. Moreover, they do not include certain parameters, such as rate of underflow and the solids content at the entry and at the underflow.

Fahlström studied the influence of these parameters and presented new formulae duly incorporating them.

$$d_{50} = d_M (1 - g_u) \frac{1}{n} \text{ if } 0.4 < g_u < 1;$$

$$d_{50} = dM \left(\frac{\log k_1 - \log g_u}{k_2} \right) \frac{1}{n} \text{ if } g_u < 0.4;$$

$$p = g_u \left(\frac{1 - C_u}{1 - C_i} - \frac{C_u}{C_i} \right);$$

$$Q = k_3 S_i \frac{r_2}{r_3} \sqrt{2g \ H},$$

where

$$r_2 = \frac{D_2}{2}$$

d50, H, Q, D2 are the same as in Dahlström's formulae;

dM = maximal dimension of the particles on the grading curve;

gu = yield by weight at the underflow;

n = slope of the grading curve;

k1 = 1.07 and k2 = 0.66;

p = precision or efficacy of separation;

Cu = percentage of grains smaller than d50 in the underflow;

Ci = percentage of grains smaller than d50 in the inlet;

k3 = constant;

Si = section of injection;

$r3 = \dfrac{D_0 - D_1}{2}$ (radius of the injection zone);

D0 = diameter of the cylindrical part of the cyclone.

The operational range is fairly wide for each machine but the effective working point gets restricted by the parameters peculiar to each feeding circuit. The parameters can be maintained constant for a given equipment even when the feed rate varies, by changing the inlet section of the cyclones. In this way the speed of entry into the machine remains constant. Despite the relative variations in the solids content of the pulp, the pattern of the separation curve can be maintained by reducing the inlet section and hence the speed of entry for a constant feed rate. For particle size fractions below 100 μm, quality in separation is achieved by restricting the solids content of the pulp at the entry to the cyclones:

— 250 g/l to 300 g/l between 80 and 100 μm,
— below 100 g/l, around 40 μm,
— below 50 g/l, under 20 μm.

The separation point of a cyclone corresponds to a zone of unstable working of the underflow nozzle for the sand. In practice, due to internal centrifugal flows a vortex forms in the conical section which gives rise to a depression at the bottom underflow nozzle and causes an 'umbrella'-shaped flow outside the nozzle. When the equipment extracts the sand, the pulp concentration rises, forming a 'cylindrical' flow which occurs at the underflow outlets. At this point of separation the flow alternates between the 'umbrella' and the cylindrical' mode. The efficiency of the vortex can be maintained within the permissible limits only by equalising the inlet, overflow and underflow sections.

Equipment which has no provision for modifying the inlet section can be fitted with devices placed outside the underflow of the equipment for regulating the flow and residual water content of the sand without interfering with the internal hydraulic equilibrium of the cyclones (Figs. XII.17 and XII.18):

— Flexible membrane regulator which closes as soon as a drop in pressure occurs.

— Concentric double-chamber regulator which enables direct evacuation when working in the cylindrical mode and recycling of the least polluted water during the 'umbrella' mode of working.

— Tilting chamber regulator with a spoon-shaped chamber attached to an axis and assembled to balance a counterweight. This regulator closes the nozzle when

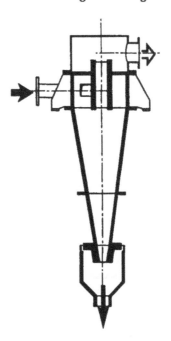

Fig. XII.17: Hydrocyclone with regulator for underflow

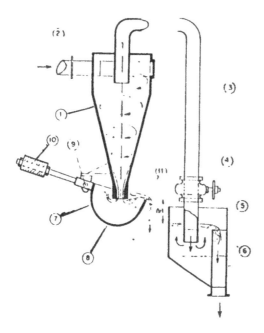

Fig. XII.18. Principle of working of a cyclone with automatic extractor

working in the umbrella mode and opens under the effect of the pulp pressure when the pulp is extracted in the cylindrical mode.

The centrifugal forces utilised in the turbiflux are smaller than in a cyclone. The centrifugal flow is caused by central directional blades.

For an 80-μm particle size separation, the equipment can handle 40 m3/h per m2 surface.

2.6.4 LIQUID-SOLID SEPARATION EQUIPMENT BASED ON VIBRATION AND FILTRATION

The residual water content of the materials should be reduced to the minimum:

— Firstly, to facilitate transport of the aggregates and sand upon conveyors or in trucks for storage.

— Then for subsequent utilisation of the graded materials which should be delivered with a minimum and consistant moisture contant for different manufacturing operations, such as hydraulic concrete and glass-making.

The aggregates generally fulfil the requirements after classification on screens or trommels which restore them to less than 10% moisture content after draining.

Sands treated by screws, wheels, classifiers, stationary or vibratory screens, are evacuated by means of these equipment, dewatered beforehand, with the residual moisture content varying from 18 to 25% depending on the grading of the materials. Sands extracted by cyclones are dewatered beforehand to a moisture content of 30%. Draining follows during storage:

— In hoppers, where it is of brief duration and too brief to be effective.

— In open-air ground areas which enable well-drained sands provided a sizeable stock is built up, fed in turns at several discharge points, and with several points for reclamation to ensure the standing period needed for thorough draining.

Therefore, separation equipment based on filtration and vibration are utilised specially for completing the washing and classification treatment of sands so as to reduce the residual moisture content for:

— shortening the dewatering time of the stocks,

— improving the final cleanliness of the sands by removing the maximum water remaining after washing.

• *Vibratory dewaterers*

These machines consist of an upward sloping vibratory tank provided with slotted dewatering grids. The unit, mounted on a suspension, receives a linear vibration under the action of 2 unbalanced weights revolving in opposite directions. Under this action the sands compress themselves, rising forward in the equipment, and get dewatered. The water thus expelled flows through the grids.

Depending on the size of the equipment, vibrations vary from 750 to 3000 rpm. An electromagnetic mechanism is often resorted to for higher vibrations.

The dewatering surfaces could be made of stainless steel, polyurethane or rubber.

The machines are used for completing the draining of sands and reducing their residual water content after leaving the separation equipment and especially cyclones. Depending on the grading of the sands and the types of equipment, the water content may vary from 12 to 18%; some dewaterers equipped with devices for reducing pressure under the dewatering grid enable reduction of the water content of dewatered sands by 2 to 4% (Fig. XII.19).

• *Filtering dewaterer*

This constitutes a horizontal rotating disc on which the pulp is poured in uniform layers of adjustable thickness. The disc rotates above a conical tub which is subjected to suction. The aspirated water is filtered by the product itself which is extracted during rotation by a fixed plowshare or discharge screw (Fig. XII.20).

This equipment is generally fed by cyclones, that is, a water content of about 30%. At discharge the residual water content is between 4 and 10% depending on

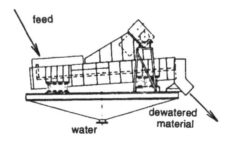

feed

dewatered
material

water

Fig. XII.19. Dewatering grid.

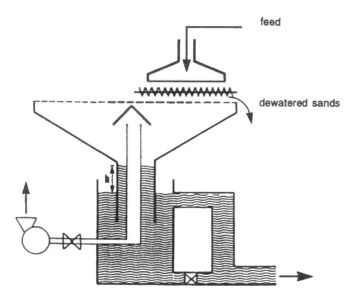

Fig. XII.20. Rotating filter dewaterer with suction

the grading of the sands. The machines are more eminently suited for fine and very fine sands—industrial sand—for glassworks and foundries. Drying can be accelerated by passing steam.

2.6.5 WASHING-CUM-DEWATERING COMPACT EQUIPMENT FOR SANDS

This range of equipment results from a coherent assembly of several machines described above (Fig. XII.21).

Tank, pump and cyclone accomplish the attrition and dislodgement of clays or fillers while the cyclone effects an average size separation at 70 μm.

The vibrating or rotating dewaterer accomplishes dewatering of the sand and thus improves its cleanliness and handling. This type of machine often has a self-regulating water feed through partial recycling of the cyclone's overflow waters.

The water from the dewatering operation proper is of course recycled. The total quantity of the water is controlled by the cyclone.

2.6.6 EQUIPMENT FOR SPECIAL TREATMENT FOR REMOVING IMPURITIES; WOOD, SHELLS, SCHISTS

The equipment for separation of impurities are usually classifiers used in ore-dressing operations. Their main function is to differentiate the particles, already classed into size fractions, on the basis of their density and shape, by recourse either to their equivalence in a zone of flow or decantation in a fluid.

This classification by gravimetry is accomplished by equipment wherein flow is interrupted by impediments to the flux of pulp.

Table XII.2: Separation equipment based on decantation and draining

Equipment	Working principle	Spent wash waters Permissible limits	Particle size separation, in microns	Moisture content of sands produced	Permissible tonnage per machine	Water consumption rate, m³/t	Power, kW/t	Wear
Hydro-separators	Static decantation	Negligible clay	80 to 150	30 to 40%	0 to 200	1 to 1.5	0	Localised at points
Pumping tube	Separation by surface current (with evacuation at the top)	High clay content	100 to 300					
Wheel with buckets	Decantation and separation by surface current	Negligible clay	100 to 160	18 to 25%	0 to 200	1.5 to 2.5	0.04 to 0.08	Very little
Screw	Mechanical extraction of sands	Negligible clay	80 to 160	18 to 25%	0 to 200	3 to 6	0.1 to 0.2	Localised at blades
Cyclones	Decantation and separation	High clay content	10 to 150	30 to 50%	5 to 200	Minimum 3	0.4	Localised at points
Tubiflux	By centrifugal current	High clay content	60 to 150	30 to 50%	100 to 500	3 to 10		Very little
Vibratory dewaterers	Liquid-solid separation Drainage accelerated by vibration and suction	Negligible	No separation	10 to 20%	0 to 300	0.3 to 0.5	0.05 to 0.1	Little in case of grids made of synthetic materials
Dewaterers	Equipment invariably fed downstream Hydroseparator, screw, wheel, hydrocyclone	Clean		4 to 15%		0.3	0.1 to 1	Localised at screen

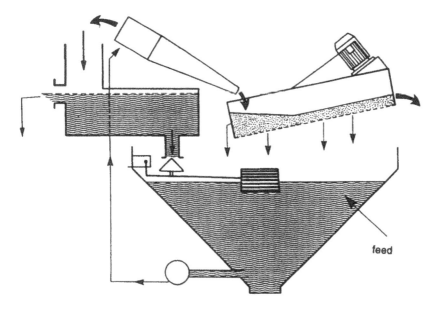

Fig. XII.21: Washing-cum-dewatering unit for sands.

The LD type straight Lavodune is an equipment for hydraulic treatment based on gravimetry which selectively removes particles placed in a flow inside an inclined tube, taking advantage of the difference in their speed of entrainment. Under these conditions the speed of entrainment varies proportionate to the square root of the density of the materials in the support liquid. It is practically independent of the size of the grain above 2 mm and varies also with respect to the diameter of the particles between 2 mm and 0.1 mm. Upon introducing into the inclined separator tube the polluted product (sands or gravels) to be treated, the lighter particles are entrained upwards by the rising water stream, which conveys them over draining equipment (Fig. XII.22). The heavier particles descend countercurrent forming a dune, up to a constriction which stabilises the dune to preclude the straying particles from escaping and rising upwards. When the dune is sufficiently built up with heavier particles, it is discharged into the lower section of the equipment from where it is evacuated through a pipe for sands or a mechanical extractor for aggregates, as the case may be, towards a draining system.

3. CLASSIFICATION OF SANDS

3.1 Definition

The separation of sands into single sized aggregates has long been an imperative in industrial applications. In practice, most deposits contain materials of wide grading varying during exploitation, which in turn entails equipment which are

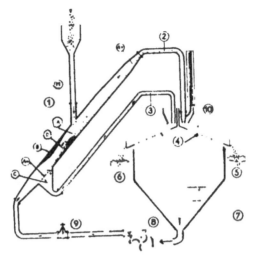

1.   "Lavodune" tube
2.   Pipe for evacuating lighter particles
3.   Pipe for evacuating heavier particles
4.   Drainers
5.   Devices for evacuating lighter products
6.   Devices for evacuating heavier products
7.   Pumping basin
8.   Circulation pump
9.   Gate for regulating wash water flow
10.   Control nozzle and manometer

Fig. XII.22: "Lavodune"

adapted to supplying sands of a very high cleanliness, with rigorous and consisten particle size composition.

The choice of the equipment is made with regard to the following factors:
— Particle size characteristics of the sand to be treated.
— Particle size characteristics of the sand desired.
— Intrinsic qualities of the classifying equipment in view of the given task efficiency, wear resistance, operational cost. If need be, trials are conducted to determine for each task the quality of separation or efficiency and the yield by weight, all of which are intimately interlinked in determining the profitability of exploiting a deposit.

3.2 Recapitulation of Some Methods of Control in Size Grading of Sands

This control seeks to verify not only the results obtained on the sands, but also the quality of separation of the equipment utilised. Several methods are employed for this purpose.

• *Efficiency of separation*

This efficiency in respect to classifiers is given by the formula:

$$E = 10^4 \frac{(p - a)(a - r)}{a(100 - a)(p - r)}$$

where a, p and r are the respective proportions, expressed in percentage, of the weights of grains smaller in size than m of the feed, overflow and underflow; m is the separation mesh of the sieve 95% of whose passing fraction is in the overflow.

This efficiency generally lies between 40 and 60% and for the most part gives the ratio (in percentage) of the classed product in the overflow to the classable product in the feed.

In practice, this figure depends on various parameters (curve of the raw material, mesh of separation effected), which makes its use difficult for comparison of classifying equipment between them.

• *Accuracy of separation g 35/65*

Like other related coefficients $\frac{d_{25}}{d_{75}}$ or $\log \frac{d_{75}}{d_{25}}$ this coefficient is derived from the separation curve which gives the probability of passage of different grains into the coarse fraction (Fig. XII.23).

The coefficient g 35/65 is the ratio of the diameters of the grains corresponding to the probabilities of 35 and 65% passing. The definition of $\frac{d_{25}}{d_{75}}$ is similar.

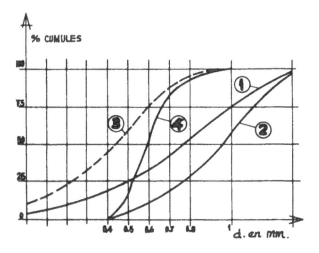

1. Raw sand
2. Coarse sand
3. Fine sand

4. Separation curve

Imperfection: $I = \dfrac{d_{75} - d_{25}}{2d_{50}}$

Fig. XII.23: Separation curve

Log $\dfrac{d_{75}}{d_{25}}$ is a direct function of the separation curve's slope. The other coefficient are related to it. The slope constitutes an excellent criterion for the separating power of the various classifiers, the perfect equipment giving a vertical line.

The drawback of these various formulae, especially the first, is that they take into account only the median portion of the separation curve and ignore both ends. Furthermore, these coefficients are not wholly independent of the separation diameter d50 (corresponding to the probability of 50% passing).

The separation notion should be related to a notion of imperfection (Fig XII.24).

• *Imperfection*

This coefficient, also derived from the separation curve, is defined by:

$$I = \frac{d_{75} - d_{25}}{2d_{50}}$$

Compared to the earlier mentioned methods, it has the merit of eliminating th influence of separation diameters (Fig. XII.24).

In certain cases (if the water current of the underflow traverses the equipment) a reference should be made to the reduced imperfection derived from the reduced separation curve. This curve is obtained by multiplying the ordinates of the separa tion curve by the ratio $\dfrac{100}{100-s}$, where s is the proportion of water entrained in th underflow.

Though its use is less convenient than the E.J. Ivers method, this coefficien at least overrides all other methods in defining a given classifier for its norma range of use.

A simplified idea of imperfection, often employed in respect of cyclones, i the percentage of materials recovered in the underflow at the mentioned d50.

All the equipment can be employed outside their nominal application rang but to the detriment of either the efficiency or the output. Investigation into th accuracy becomes essential given the fact that the characteristics of the sand utilised in public works, and more particularly in hydraulic concrete are deter mined in accordance with civil engineering specifications, which commend the us of sands whose particle size distribution lies within the limits of overall gradin; curves.

All plant managers seek to ensure a production of sands falling within re quired grading envelopes. Nor should it be forgotten that additional requirement are stipulated in some cases for certain construction works, as happens with th current tendency amongst engineers to be insistent on the quality and consistency of the sands used. This leads one to suppose that the machinery meant for treatin; sands should be capable to give several particle size fractions in order to have better control of the grading.

The ideal case would be to separate the sands into:

— coarse sand 3(5 mm) to 1 or 1.5 mm,
— medium size sand 1 or 1.5 to 0.4 or 0.7 mm,
— fine sand 0.4 or 0.7 to 0.04 or 0.08 mm.

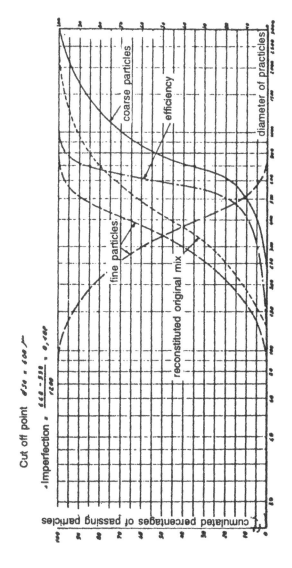

Fig. XII.24: Coefficient of imperfection

However, many plants supply satisfactory sands with a fewer number of par ticle size fractions. Yet, it may be observed that in a number of alluvial deposit the fraction smaller than 0.4 mm generally lacking, a low cut-off enables correctin; this category of sand by the addition of scarce fine elements.

3.3 Function of Sand-classifying Equipment

The function of hydraulic classifiers consists in volumetrically classifying the par ticles by differential decantation. Particles smaller than the separation mesh ar entrained in the overflow water; the larger ones settle down, the classification ii this case taking place by surface current; when the current is from the bottom topwards in the tank and overflows by the pipe at the top, the process is termec 'classification by rising current'.

The path followed by the particles during decantation is the product of the resultant of two forces: the force of gravity and the force of the pulp current whicl passes through the classifier. Gravity causes the particles to fall at a speed whicl depends on the resistance offered by the density and viscosity of the pulp t(the descending movement. The current entrains the particles towards the overflov weir.

In the treatment of aggregates, the pulps are highly diluted and offer only feeble resistance to settling of solid particles; hence the resultant path in decidec by the speed of the current, which explains the usual designation 'classification b; speed of flow'.

By virtue of their principle of working, classifiers separate two fraction: of sand with a overlapping of the granulometric curves, which results fron the particle size distribution of the raw sands and the equipment's separatioi point.

For assessing the qualities of a classifying machine, the definitions of efficienc) and imperfection usually employed in ore processing have been adopted.

3.4 Classifying Equipment

Different types of 'classifiers' working on different principles are available for car rying out separation of intermediate particle size fractions in sands.

3.4.1 CONVENTIONAL SCREENS AND SCREENS WITH VERTICAL SIEVES

When the sands received are mixed with a considerable volume of water, thes(screens can classify them down to 1 or 2 mm; their yield depends essentially o; the granulometric distribution near the cut-off point. The more the curve slopes the greater the number of particles close to the mesh opening size, which make: the separation more difficult to achieve.

3.4.2 STATIC SIFTERS

Curved or inclined grids as well as sifting panels are eminently suited for clas sifications in the 1.6 to 0.1 mm range. The sand should be clean at feed and th(particle size range wide, 0 to 3 or 0 to 6 mm (Figs. XII.25 and XII.26).

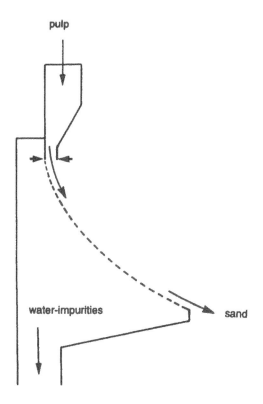

Fig. XII.25. Curved grid

3.4.3 SURFACE CURRENT CLASSIFIERS

These enable a selective decantation of the grains in hoppers disposed in the path of flow of a liquid, wherefrom the designation 'V-bottom tank' or 'Sand Sort' is derived. The equipment is eminently suited for extracting excess fine sand fractions when their grading gives convex curves (Figs. XII.27, XII.28, XII.29, XII.30).

Spiral screws are employed for classification of fine sands in the particle size range 0.4 to 0.08 mm; they enable removal of excess fine and very fine sands and extraction of graded sands that are almost fully drained.

3.4.4 RISING CURRENT HYDRAULIC CLASSIFIERS

These have a fairly wide operational range; their working principle is based on the selective entrainment of the particles by countercurrent in a rising vertical flow, completed in some models by a lifting inclined flow.

3.4.5 CENTRIFUGAL CURRENT CLASSIFIERS

Hydrocyclones are used in classifiers below 0.150 mm and in thickener-separators for sands of larger particle sizes (see Fig. XII.18). they are very commonly employed in extractive industry undertakings:

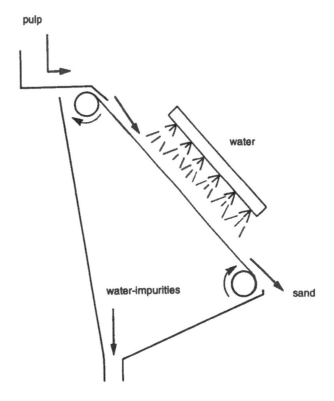

Fig. XII.26. Sifting panel

— Recovery of fine sands most often carried away in wash waters.

— Thickening of all sands, coarse and fine, leaving a classifier.

— Removal of clayey waters when the pollution of a deposit starts to increas since they concomitantly prevent loss of fine sands while affording the possibilit of handling voluminous flows of wash water when the space available is limitec

Coarse sands graded by screens, static sifters and screws are extracted with residual water content below 22%. Fine sands graded by this equipment as als all sands obtained from other classifiers should be drained and dewatered b hydrocyclones, screws, dewaterers or filters.

The capacities of each installation are necessarily decided according to th imperatives of putting up plants for recycling the waters used for classificatio and the stocking plan of sands. The wear in all this equipment is marginal o highly localised.

4. TREATMENT OF WATERS AND STORING OF MUDS

Quarry managers often find it necessary to submit the materials quarried to washing for scrubbing out and dislodging the clays extracted along with th

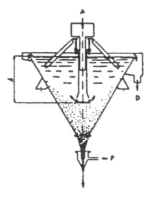

Fig. XII.27. Classifying cone with additional water source.
A—Feed; Δ—Overflow; P^m—Additional water source.

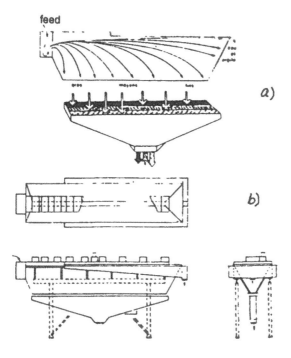

Fig. XII.28. Multiple-tank sizer
(a) Principle of equipment; b) Practical realisation

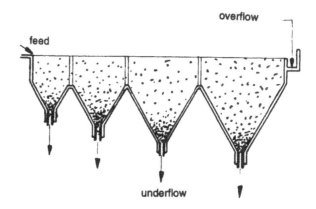

Fig. XII.29. Surface current classifier

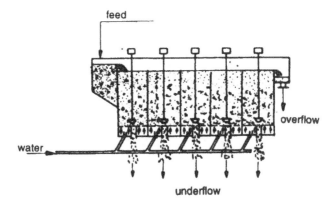

Fig. XII.30. Surface current classifier

natural products, and also rinsing and dedusting aggregates after crushing and grinding.

The water, mostly pumped off into rivers, carry fine elements and clays when released from the plants. Many plant managements have laid out settling ponds which retain the deposits of fine elements and clays. These arrangements, which are not elaborate, can provide a sufficient safeguard when the quantities of water and suspended materials are not considerable, provided that a frequent maintenance of the tanks and the feeding channels is ensured. In any case, accidental or planned discharges into rivers of water effluents after washing materials affect the growth of aquatic flora and fauna by turning the rivers turbid and forming sediments in the zones of placid flow. It is possible to mitigate and even totally eliminate this problem of mud in quarries equipped with a plant for washing with water and which have a sewage farm or a mined pit which can serve as a dumping

Table XII.3: Classification of equipment

Equipment	Principle of classification	Service range for classification of sands, in mm					Water, m³/t	Power, kW/t	Imperfection, in mm
		5–1.6	1.6–0.4	0.4–0.8	0.08–0.04	0.04–0.2			
Screen	Vibratory sifting with watering and under water	+ 1.6					1 to 2	0.7 to 0.2	0.15 to 0.30
Curved grids	Static sifting on inclined fabrics and grids		1.6 to 0.1				2 to 4		0.15 to 0.25
Sifting panel	Static sifting		1.6 to 0.1				1 to 1.5		0.10 to 0.20
V-bottom tank (Sand Sort)	Surface current Decantation			0.4 to 0.08			1 to 8		0.4 to 0.7
Screw (spiral)	Surface current Decantation Mechanical extraction			0.4 to 0.08			1 to 5	0.06 to 0.20	0.4 to 0.5
Rheax-Floattex hydrosizer	Rising current			1.6 to 0.8			3 to 5		0.15 to 0.20
Lavodune-Lavoflux	Rising current Current in inclined pipe		1.6 to 0.04				3 to 5		0.10 to 0.20
Turbiflux hydrocyclone	Centrifugal current	In separator-thickener 5 to 0.150			In classification 0.150 to 0.020		3 to 10		0.3

place for concentrated muds. The waters, after clarification, can be recycled for washing the materials, which can reduce the need of freshwater requirement to one-tenth the total circulating water for washing.

4.1 Natural Decantation

This method, often resorted to in installations that have a large area for ponds or dead riverbeds at their disposal, is based on the principle of letting the particles suspended in the water settle to form sediment. To be adequately effective, that is, to ensure proper decantation of the particles, the basins should be large in area and sufficient in depth to hold the sediments. In practice, the standing time of the waters, before being let out into the river or recycled into the washing circuit of the materials, should be long enough, of the order of several hours, between 8 and 24 hours.

Where basins have to be created with embankments, their permissible height should not exceed 12 m with due regard to statutory regulations. Furthermore, construction works should be carried out with the approval and under the control of the Commission on Large Dams.

For a 12-m height, a width of 64 m is needed which, for a given area of water, considerably increases the total surface of the basins.

The polluted waters are sent to one end of the pond where a sediment of the heaviest grains forms rapidly; a portion of the finest elements can be held in suspension by the water currents and the convections caused by wind, heat exchange and movements of sediment bodies.

The gradual settling of deposits can give rise to preferential flows of water, headed towards the pumping point, thus using up a tiny portion of the settling basin. The standing time gets shortened, entailing a direct transfer of the finest elements entrained by the water without decantation.

This procedure allows the use of only 70% of the total volume of a settling basin so as to keep the sedimentation zone away from the surface currents.

4.2 Decantation of Flocculated Waters

Flocculation of effluent waters can be treated in two ways:
— By injecting the flocculant into the pipe which conveys the waters to the basin of natural decantation. This device allows the finest particles to agglomerate for accelerating the settling rate of the wastes and thus cuts down the standing time of the waters in a basin.

In this manner a slight improvement in the system described in Sec. 4.1 is achieved but the volume of water circulating in the basin being large in relation to the surface and the volume, often disrupts the formation of sediments by preferential currents which are invariably powerful and troublesome.

The volume of the sediments can only be marginally improved, if at all.
— By injecting the flocculant into the treated water from a clarifying machine, dewatering flight conveyor or mud bed decanter.

This equipment, designed for separating clarified water from a volume of residual mud, permit immediate recycling of 80 to 90% of the water for washing the materials, depending on the nature and amount of the suspended substance.

The dewatering flight conveyors comprise a perfectly still hydraulic circuit in a basin of dimensions matched with the flow of the water treatment: 0.25 m2 per m3/h water.

• *Scheme of installation: principle of working (Fig. XII.31)*

The effluent waters arrive in a feed tank into which the flocculant has been injected. Using a nozzle, the water is directed to the centre of the clarifier. A central ring impels channelling of the water towards the peripheral overflow. In the course of this travel the flocs settle to the bottom of the basin.

A flight conveyor brings the muds to the centre of the clarifier in a cone of concentration. A central 'log washer' homogenises the muds to improve their thickening and to prevent leakages during pumping.

For the muds to be extracted at maximum density, a mud pump (obligatorily volumetric) should be placed as near as possible to the cone of concentration.

Automatic control by a programmable mechanism of Flocculation through turbidimetry as well as of the densities of concentrated muds, enables extraction of the latter at the rate of 600 to 1100 g/l dry material depending on the kind of solids treated. The muds are then sent to the drying basin.

Use of programmable automechanisms permits running this type of installation without manual supervision.

Mud-bed decanters (Fig. XII.32), whose cross-sections are five times smaller, have hydraulic circuits of higher speed to ensure mixing of the flocculants and filtration of the water through the mud bed maintained in suspension. The muds are renewed by the fresh inflow of clays and are evacuated through the underflow of the mud bed at the bottom of the equipment. The areas of sections are of the order of 0.04 m2 per m3 of treated water.

Thus it is established that these systems facilitate reducing the total volume of the wastes. Given the good flocculation of the fines achieved by this equipment, a concentrated mud of between 150 and 300 g/l decants in the settling tanks under the effect of the flocculants. This mud settles down and forms a sediment whose density on a dry basis increases progressively; the water remains clear on the surface and can be discarded or recycled without difficulty. Because of the passage of a smaller volume of polluted water, the preferential currents are weaker than in the system described in Sec. 4.1, which imparts a better stability to the deposits.

4.3 Thickening the Mud

After treatment of the effluent waters in a clarifying machine, evacuation of the muddy waters can be reduced by sedimenters or filters (such as a centrifugal pump) which enable to obtain a mud that has the consistency of a concrete mortar, whose concentration can attain 800 to 1200 g/l (Fig. XII.33).

Belt filters are also used and make it possible to obtain muds with 25–35% water content, which can be shovelled and transported (Fig. XII.34).

Thus the volume of the waste mud is reduced by 3 to 6 times depending on the nature of the wastes. The muds can then be dewatered on a platform or in chambers; they even restore the interstitial water on drying for some days. They can then be recovered as fairly solid lumps of clay containing about 30% water.

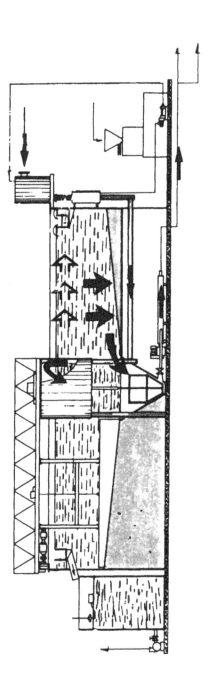

Fig. XII.31: Decanter with a sludge scraper

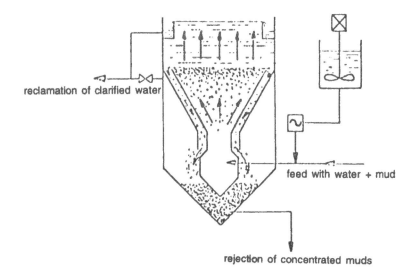

reclamation of clarified water

feed with water + mud

rejection of concentrated muds

Fig. XII.32. Mud-bed decanter

Fig. XII.33. Battery of silos (sedimenters) for drying mud

The density of the sediment on a dry basis is about 1500 kg/m3, which yields an optimal residual volume and a near-total recovery of the water.

Mud thickeners on the other hand consume large quantities of flocculants.

The muds can be transported by truck and dumped into hollows, quarried pits or utilised for improvement and reclamation of lands.

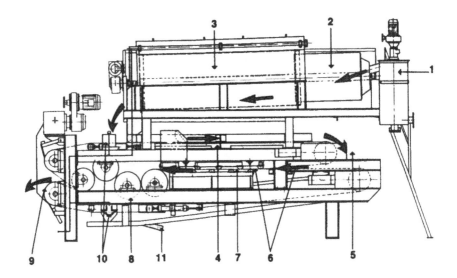

1.	Mixing zone	6.	Pressing zone
2.	Homogenising zone	7.	Progressive pressing zone
3.	Preconcentration zone	8.	High-pressure zone
4.	Dewatering zone	9.	Extraction of muds
5.	Overturning of cake	10.	Zone for washing belt
		11.	Exit of filtrate liquor.

Fig. XII.34. Belt filter

Some clays or calcareous fines can find use for specific applications.

A self-contained plant for treating effluent waters and thickening muds would appear too costly in terms of investment and operational expenses given the size-able consumption of flocculants.

However, the quarry manager stands to gain from such a plant on the following scores:

— Water consumption.

— Reduction in land area for ponding and investment in acquisition.

— Elimination of embankments for settling basins and investment in their layout and digging.

— Avoidance of accidental pollution of the environment (rivers, water table, lands).

— Reduction in volume of wastes and drying which allows their transport in trucks or on conveyor belts, and reclamation of natural soils.

— Avoidance of rehandling of muddy sediments by shovels and draglines, the cost of which is very high, as well as elimination of problems in transport such as deterioration of machinery and pollution of public roads.

The balance sheet of investments and overhead reveals a very real short-term gain and secures the environment as much as future expansion in the exploitation of a deposit.

CONCLUSION

Land areas available for opening up quarries have continued to shrink, with environmental constraints, thereby forcing quarry operators to turn more and more to contaminated deposits.

Concomitantly economic imperatives to exploit all deposits to the fullest and the obligation to respect standards of cleanliness make it necessary to wash mined materials. This washing entails substantial investments:

— Firstly, to upgrade the cleanliness of the aggregates, an objective sought by all quarry managers.

— Secondly, to clarify the effluent waters and the waste muds which account for considerable amortisation and operational costs. Their repercussion on the prices of the aggregates should be acknowledged by all aggregate producers and users, who strive in every way to obtain materials of quality while conserving the natural environment at the quarrying operations.

— Exploitation of alluvial and fluviatile deposits, already equipped for washing for the most part, can easily evolve towards improved or more comprehensive equipment. On the contrary, the trend towards washing for dry quarrying operations means radical changes in design and exploitation. It is therefore essential to ascertain in this case the entire range of new benefits washing can offer as well as the concomitant constraints.

REFERENCES

E. Condolios. 1975. Les décanteurs centrifuges statiques pour le dessablage des eaux chargées. La Houille Blanche N. 5/6.

J. Costes. 1968. Matériels d'extraction et de préparation des minéraux, carrières, ballastières, gravières, sablières et minières. Ed: Eyrolles. pp. 160–184.

P. Galabru. 1962. Traité de procédés généraux de construction. Volume II. 2e partie. Chapitre IX. pp. 137–157.

L. Svarovsky. 1984. Hydrocyclones. Ed: Holt, Rinehart and Wiston.

XIII

Dedusting

Robert Campanac, Jean Allombert and Thevan Delobelle

1. INTRODUCTION

The French National Standards Association defines atmospheric pollution as 'the presence of impurities in air which can notably affect health and comfort, or cause damage to property'.

Dust present in the atmosphere in the form of solid particles therefore constitutes a form of pollution when its concentration reaches a certain threshold value.

This threshold value can differ according to the particular consideration:

— health, in which case one should take into account the intrinsic toxicity of the dust on the living organism;

— damage to machinery;

— environment—creation of a dust haze constituting a hindrance to residents of the locality.

Dusts incidental to the processing of aggregates are extremely fine particles. These fine particles can adhere to the body of larger-size elements while in motion or at rest in the manufacturing chain, can settle upon the ground or remain suspended in the air due to various factors, in which case atmospheric pollution results.

Dedusting consists of capturing the dust particles suspended in the air in order to abate or remove the ill effects. This operation should be distinguished from fines removal, which also addresses fine particles adhering to aggregates and which can be removed therefrom to ameliorate the quality of the products (for instance, improvement of the surface cleanliness of coarse aggregates).

We shall examine in the following paragraphs:

— the objectives: why dedust? where and when to dedust?

— the means: how to dedust?

— the costs: how much does the dedusting operation cost?

This study is confined solely to the technological aspect of dedusting in aggregate processing plants; we shall deal with the fines removal aspect only briefly.

2. WHY DEDUST?

2.1 How the Need Arises

Most of the aggregates used in building construction and public works are 20 mm or less in size. To arrive at this size, it is necessary to reduce the size of raw materials whose volume can sometimes reach several cubic metres. This size reduction operation entails a whole range of sizes, from the smallest to the largest element. In other words, each size reduction operation is accompanied by production of fine elements, hence dust particles.

What is more, interruptions in the process of handling the crushed products in the various circuits upset the equilibrium and favour suspension of the finest elements in the air.

The emission of dust particles by an aggregate processing plant is quite clearly proportional to the volume of its output. Furthermore, by and large the capacity of these plants is steadily increasing.

It is not uncommon to see incursions by urbanisation and advance of residential buildings towards quarries, which is responsible for the stiffening of environmental constraints. Even when the plants are situated in a rural zone, farmers tend to organise themselves into a resistance group armed with legal or powerful political backing. In addition, the machinery operating personnel are increasingly being protected against occupational hazards, and industrial legislation empowers the various competent agencies to insist upon a continuous improvement in working conditions.

Lastly, even the machinery in aggregate manufacturing plants are undergoing improvements, often at the cost of ruggedness. Since they are expected to offer the highest reliability, the imperative condition for quality production, there is need to effectively protect their vital components against dust.

For these diverse reasons the battle against dust constitutes the purifying function of dedusting.

Since the technical properties demanded of the aggregates are also constantly being upgraded, it is sometimes necessary to dedust the product to make it conform to the stipulated quality requirements. This is the fines removal function of dedusting.

2.2 Toxicity of Dusts

2.2.1 CATEGORISATION OF DUSTS

Air invariably contains suspended particulate matter. This is instantly evident if one observes a ray of sunlight passing through a dark room. The suspended particulate matter is known by the generic term aerosols.

In Gibbs' classification of industrial aerosols, the term dust is reserved for particles below 1.0 µm in diameter, which float in still air at a uniform acceleration and are not susceptible to diffusion through a porous wall.

There are other classifications as well:

— Lahet: vegetable, animal, organic and radioactive dusts.

— Heim de Balzac: active dusts (toxic, caustic, infectious), inert dusts.

But in common parlance, the term 'dust' applies only to inert particles and excludes all living organisms.

2.2.2 CONCENTRATION OF DUSTS AS A FUNCTION OF THEIR SIZES

Only those dusts which remain floating in the air, hence in suspension, are harmful. Therefore it is useful to know the capacity of retention of this material by air; as shown in Table XIII.1, it is inversely proportional to the dimensions of the dust.

Table XIII.1: Rate of fall of spherical particles of density 1 in still air at 20°C

Dimension of particles (µm)	Falling rate, cm/s
5000	875
1000	395
500	277
100	29.6
50	7.4
10	0.296
5	0.074
1	0.0035 (12.6 cm/h)
0.1	0.000035 (0.126 cm/h)

Consequently, for a given quantum of dust emission, the concentration of the dust particles in the ambient air increases with the fineness of the particles.

2.2.3 PHYSIOLOGICAL EFFECTS OF DUSTS

Dusts mainly affect the respiratory tracts. Particles 3 to 5 µm in size lodge in the upper respiratory tract. They are expelled by expectoration and are practically harmless.

The smallest particles penetrate to the air-cells of the lungs, wherein they are phagocytised. Depending on their nature, they produce various effects.

Silica particles provoke agglomeration of the cellular protoplasm around themselves, then death and destruction of the cells which have absorbed them.

Calcite particles dissolve inside the cells. The products of dissolution are the originators of inflammatory reactions called conioses with the appearance of nodules. The nodulous lung has to adapt its air-blood exchange mechanism and becomes more sensitive to infections than a normal lung.

It should be noted that dusts which penetrate into the lungs, that is, those with dimensions smaller than 3 µm, are selectively retained; retention is maximum for those around 1 µm in size; their harmfulness is a function of their morphology.

2.2.4 PHYSICAL EFFECTS OF DUSTS

As mentioned above, the finest dusts remain suspended in the air longer. If their concentration is high, this causes a reduction in visibility which can give rise to accidents. Lack of visibility also affects monitoring of operating machinery.

Dust particles can likewise readily infiltrate electrical instruments and affect their proper functioning.

As for the coarsest dust particles, their abrasiveness becomes the paramount factor for machine components; this abrasiveness can in turn be exacerbated by the mineralogical nature of the particles (silica, dolomite etc.).

In the various operations of aggregate manufacture, unchecked dust emissions can lead to serious concentrations and dislodgement of dust particles, affecting the dimensional consistency of the products.

Lastly—another aspect of the question—it would be imprudent to let useful dust particles escape, particles for which much precious energy has been expended; this is the case with sands rich in fines.

2.3 Measurement of Dust Loading

Two methods are available for determining the level of dust loading at a work place.

The first method, highly approximate, measures the degree of opacity of the cloud by means of a photoelectric cell device.

The second method is based on samplings of suspended particulate matter.

3. WHERE AND WHEN TO DEDUST

Production of dusts is nearly non-existent in sandpits; however, some gravel pits carry out a major size reduction by crushing and grinding of alluvial sands; starting from a certain stage in the processing of the material (generally at the level of tertiary crushing and grinding), the problems of dedusting approximate those encountered in a quarry.

After listing the sites of emission, we shall examine the parameters which influence the quantity of dusts actually entering the atmosphere in a suspended state:

— grading of the material,
— moisture status of the material,
— type of stacking mechanism,
— atmospheric conditions.

The following distinction needs to be made:

— sheltered plants, case generally of permanent installations wherein the emission sites are localised in one or more buildings;

— open-air plants, case frequently of temporary installations wherein the emission sites are rather scattered. In such installations dust emissions are greatly influenced by atmospheric conditions.

3.1 Dust Emission Points

With no intention of presenting an exhaustive enumeration, we shall now examine the processing operations which give rise to dusts or favour their dispersion in the air (Fig. XIII.1).

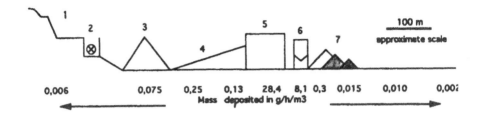

Fig. XIII.1

3.1.1 SIZE REDUCTION

• **Blasting:** This is done in the quarry by means of explosives which are generally introduced into vertical holes of some 15 m depth and about 100 mm diameter. The boring of these holes is accomplished by a tool impacting upon the rock, which is crushed, and the broken rock pieces are then brought up to the surface by a stream of compressed air. About 20% of the volume of the rock thus drilled is converted into dust particles.

• **Firing:** At the moment of explosion a considerable fragmentation of the rock takes place within a very short time; hence a voluminous dust cloud arises which, however, disperses quickly. It should be noted that this cloud is mostly constituted of dust particles kicked up from the quarry floor and sent back into suspension by the air blast from the explosion.

• **Crushing:** Crushers bring about a reduction in size of materials through compression or by impacts. Quite obviously, this operation generates fine particles.

Jaw crushers and gyratory crushers effect fragmentation by crushing the rock between a moving jaw and a fixed jaw; the speed of the moving parts is relatively low. These machines generate little dust.

Impact crushers effect fragmentation by impacts, owing to which the moving parts have a high speed which favours dispersion of the fines in the air current induced by the machine's engine; this dispersion is particularly high at start-up and switch-off since the machine serves as a 'ventilator' and disperses the fines accumulated inside.

Rod mills are machines manufacturing products of small particle size and hence producing a high proportion of fine elements. By virtue of their design, they are relatively dust-tight.

3.1.2 SIZE GRADING OF AGGREGATES

Prior to marketing, and during manufacture, the materials are graded by size; this grading operation is effected by screens.

Separation with reference to a certain dimension of the elements constituting a material is accomplished on screens subjected to vibration. This causes dispersion of the fines; furthermore, the friction of the material particles between one another entails an additional production of the fine elements, which becomes higher the softer the materials in question.

Dispersion of dust particles into the atmosphere due to use of vibrating screens depends on the state of fineness of the product being screened; the higher the fines content, the greater the dispersion. It is during screening operations that the largest dispersion of dusts arising from mechanical treatment of the material occurs.

3.1.3 TRANSPORT AND STORAGE OF AGGREGATES

There are three types of transport (see Chapter XIV):

Loose transport, in which each of the elements is entirely free to move with respect to the rest; it is mostly transport by free fall through the air.

Contiguous transport, in which each of the elements, so to speak, rests on a moving support so that relative displacement between the aggregates becomes impossible. This is the case of transport by conveyor belts, for instance.

Lastly, compact transport, wherein the elements remain packed against each other.

Dispersion of fines into the atmosphere can be considered a special kind of segregation and we have taken up this distinction in the different modes of transport of aggregates for assessing their respective influence on the dispersion of dusts.

• **Loose transport** is obviously the one which disperses fines most into the atmosphere: it is to be found in all the stages of manufacturing:
— at the time of loading trucks in the quarry,
— at the inlet and outlet of crushers and screens,
— at the discharge end of belt conveyors,
— in the stacking and loading of the waste or marketable products.

The extent of dispersion of the dust particles depends on the free-fall distance of the materials through the air.

• **In contiguous transport**, transport by conveyor belt, dust emission is due essentially to the return belt during its passage on the idlers.

• **Compact transport** is essentially the transport of products in highway trucks for marketing or for delivery of fragmented rocks to the processing plant, starting from the quarry. It generates few fines.

3.1.4 DUSTS KICKED UP FROM THE GROUND

This aspect is of great importance for the environment and concerns the surroundings of the processing plant, open stacking yards and pit head, especially the haulage tracks between the latter and the plant.

This return of particulate matter into suspension is due to atmospheric turbulence caused by the air blast from the explosion at the time of blasting, movement of vehicles and most of all, wind.

Atmospheric conditions play a dominant role in this case.

3.2 Quantities of Dusts Emitted

The Quantities of dust emitted in a plant depend on several factors, of which the foremost are:
— state of fineness of the material,

— moisture content,
— free-fall heights,
— atmospheric conditions.

• State of fineness of the material

Starting from a massive rock, a quarry installation puts out a production generally including 40 to 50% sand, that is, elements of maximum size below 5 mm obtained by crushing and grinding. As the materials progress through the production stages the quantity of fines they contain increases.

• Moisture content of the material

It is a well-known fact from everyone's experience that to agglomerate fine particles, it suffices to moisten them, whether the purpose be sand pâte or clay mouldings.

Yet the mechanism of this agglomeration differs markedly in these two examples.

In the first case the bonds ensuring cohesion of the sand pâte are due to surface tensions between the water and the interfaces of the minerals; a surplus of water causes the pâte to break down by liquefaction.

In the second case the bonds due to surface tensions matter less since there exist electrostatic bonds between the clay minerals and the water particles, which confer a special property upon clays: cohesion.

It is thus seen that the moisture content of the material plays a dominant role in the emission of dusts.

— Free-fall heights

The material is sometimes in free fall; during this trajectory in the air, a large quantity of fines escape from the material and find their way into the ambient air. This quantity is directly influenced by the fall height, which may reach:

— some 20 cm while being transferred from one means of transport to another and while feeding the screens;

— some 50 cm to 1 m at the inlet and outlet of crushers and at the exit from screens;

— some 10 m and occasionally more at the time of stacking the products.

• Atmospheric conditions

Atmospheric conditions intervene mainly in two ways:

— rains wet the material, especially if it has already undergone blasting, and cause the airborne dust to settle around the working face and the processing plant;

— wind has the opposite effect, drying the material and redispersing the fines that have settled on the ground into the air through squalls. Its action is quite pronounced in open-air installations, which permit a direct uptake from the materials in free fall or agitated on a screen or even on transporters. Dispersion of dusts will be that much higher according to the degree these major points of fugitive dust emission are not protected by an enclosure.

3.3 Where and When to Dedust in an Installation

A programme of monitoring the atmospheric pollution caused by dusts was carried out at an industrial site; it covered a period of three years and the total duration of the measurements was around 12 months.

The values of fall out have been shown in Figure XIII.1 and represent the results obtained. Examination of this date together with that of measurements of transport of dust particles as a function of climatic conditions (not presented here) illustrates four points:

— relative impact of the different emission sites;

— the fact that depositions of dust are generally highly localised around the working face;

— at the periphery of this site, the masses deposited by diffusion are relatively less and their magnitude is the result of an extended process of build-up;

— under some meteorological conditions (fairly high wind speed, low humidity), transport of dust takes place: it is directional and can be substantial; it happens to freshly emitted dusts but equally (and this is the main source) to dusts previously settled on the pit head or escaped from stacks of fine products.

Hence the external nuisance at the working site will depend on the following, leaving aside the dedusting measures described later:

— emplacement of wind shelters for buildings or equipment and stocks of finished products;

— upkeep (cleanliness) of the site.

4. HOW TO DEDUST?

4.1 Parameters of Dedusting

When a decision has been taken to install a dedusting device, it is first of all necessary to circumscribe the problem by clearly defining all its aspects:

— Nature of the dust, particularly its abrasiveness, could rule out certain methods.

— Concentration of the dust has an effect on the rate of suction and sometimes necessitates multistage systems.

— The efficiency demanded is a function of the nature and concentration of the dust, its physiological and physical harmfulness, and the situation of the zones to be dedusted (inside or outside a building, whether frequented or not by personnel). Often it depends, more and more, on the dust sensitivity of the environment.

— In an already existing plant, retrofitting the type of equipment ideally suited to the situation is not always possible if, for instance, the space required by it is incompatible with the room actually available.

— The energy and ultimately the water resources available would need to be assessed. For example, for a high-capacity installation, it is pointless to think of a hydrocyclone if the water has to be ferried through tankers.

— Side by side, one should attend to the removal of sludges or dry dusts, materials which are not easy to handle and transport without special equipment, and whose storage may pose ticklish problems.

If for economic reasons it is not possible to install the 'ideal' solution in one go, one should adopt a step-by-step approach, namely removal of the coarsest fraction of the dust in the first instance, reserving the possibility of completing the process, as need be, through installation of dedusting subsequently.

Once the foregoing criteria have been weighed, an intensive study needs to be conducted on the investment and maintenance costs, bearing in mind the fact that while dedusting brings in some benefits, a major part of its outcome does not directly contribute to productivity. So, it is prudent to tailor the solution to the problem posed, taking into account the scenario of probable development of constraints and selecting from amongst the methods in practice, the one of appropriate efficiency.

Lastly, one should not forget 'the art of the quarryman' whose knowledge of the equipment and the material, not to mention his ingenuity, can materialise simple yet efficacious devices, of extract the maximum advantage out of existing dedusting equipment.

4.2 Dedusting Techniques

There are two kinds of techniques. The first seeks to prevent dust from diffusing into the atmosphere, the second employs methods for capturing the dust particles by sucking in the surrounding air, then separating the former from the latter.

4.2.1 NEUTRALISATION OF DUST PARTICLES WITHOUT CAPTURE

The simplest means is to encapsulate the equipment emitting dust to ensure dust-tightness (see example of hood placed over vibrating screen; Fig. XIII.2) To be truly effective, this method should, of course, be applicable to all equipment. A study soon reveals, however, that this is not feasible. There will always be leaks somewhere, rendering this solution far from effective.

Another method consists of bringing the dusts down by moistening them with water jets or a spray of atomised water, or a spray of wetting water. If pure water is used, the large quantities needed alter the moisture content of the material and aggravate the risks of screen clogging or chute clogging.

When 'wetting' water is to be used, a very sophisticated installation has to be envisaged. The risks of clogging are thereby eliminated. However, the surface-active agents used in this process could alter the behaviour of the aggregates (especially sands) vis-à-vis hydraulic or bituminous binders.

It should be noted that the fine dusts of some limestones are not readily wettable and hence higher doses of surface-active agents are required, which minimises the advantages offered by this method.

Lastly, the risks of freezing during the winter season should not be overlooked.

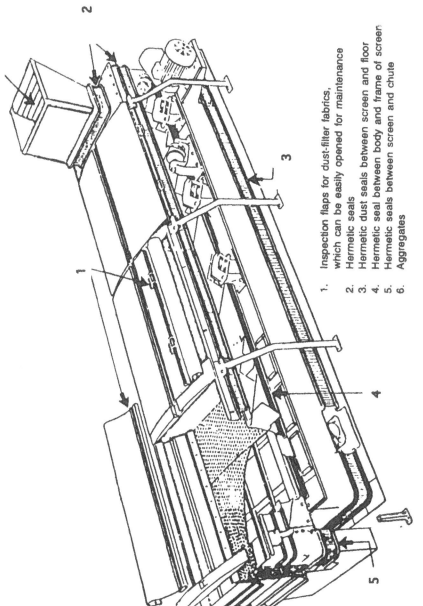

1. Inspection flaps for dust-filter fabrics, which can be easily opened for maintenance
2. Hermetic seals
3. Hermetic dust seals between screen and floor
4. Hermetic seal between body and frame of screen
5. Hermetic seals between screen and chute
6. Aggregates

Fig. XIII.2. Hood over screen. Anti-dust protection system

4.2.2 CAPTURE OF DUSTS BY EXHAUST SUCTION

• *Ventilation of buildings*

It is possible to replace the polluted air in buildings by exhaust suction of the former followed by induction of fresh air. However, only the finest dust particle will be sucked out in this process; the coarser particles will fall and be deposited before exhaustion can be completed. To increase the airflow rate, with the expectation that this would exhaust the coarser particles, is a false notion; an increment in airflow would create turbulence whereby the coarser dusts would be brought back into suspension.

So, this method protects the personnel and the equipment inside the building only to a very limited extent. Dust-tightness of the building, very costly indeed is still necessary.

• *Single capture*

This consists of encapsulating each device in a dust-tight chamber and then exhausting the dust-laden air from the enclosure. This method has a relatively lower power demand. But a great deal of care is needed in fabricating the enclosures which are not always compatible with proper monitoring of the functioning of the equipment, the mechanical components of which are otherwise not protected from dust.

• *Emission: point-wise capture*

As seen earlier, the preferential zones of dust emission are situated at very well defined points.

So, for an expedient dedusting, it suffices to create a negative pressure in the immediate vicinity of these emission points. The negative pressure can be effected by means of a hood, but the large air inlets at the periphery of the hood call for high suction rates. Hence this method is employed only when encapsulation of the equipment is not possible or when such an arrangement would interfere with the smooth operation or maintenance of the installation.

The most commonly employed method consists of a hood localised over the zone to be protected. Only the dust-emitting zone is encapsulated within an enclosure and not the entire device, as in the case of single collection (Figs. XIII.3 and XIII.4).

In this case, it is not possible to realise a perfect dust-tightness without affecting the smooth function of the moving parts. Hence the suction rate has necessarily to be higher than the case with single dedusting or by general ventilation of the building. Despite this drawback, this method is the one adopted most often, given its ease of execution, its minimal space requirement, its efficiency and the minimum disturbance it occasions upon the good functioning of the equipment.

4.3 Dedusting during Drilling

During drilling the ultimate purpose of dedusting devices is protection of the personnel engaged in this operation. Three techniques are employed.

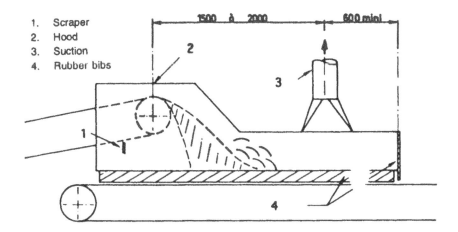

1. Scraper
2. Hood
3. Suction
4. Rubber bibs

Fig. XIII.3. Hood over discharge end of a conveyors

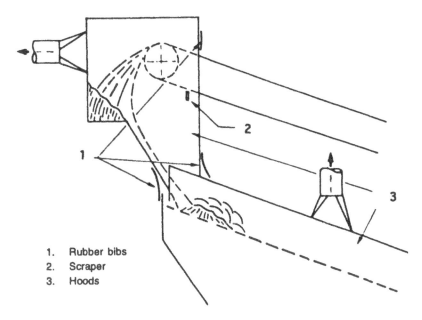

1. Rubber bibs
2. Scraper
3. Hoods

Fig. XIII.4. Capture of dusts over screen

• *Capture and deflection of dusts*

The apparatus based on this principle essentially comprise:

— a conical cap placed on the ground across the hole being drilled, the drill-bits passing through the cap;

— a suction tube welded to the cone into which compressed air is tangentially injected and, as need be, water for suction or expulsion;

— a delivery tube of some 15 m.

This system is the least cumbersome, relatively economical (compressed air consumption: 1000 l/min, water: 100 l/location and per device) and also offers the advantage of permitting observation and sampling fly rock.

• *Capture and separation of dusts*

The apparatus employing this principle essentially comprise:

— a capturing head which may be either mobile and positioned on the ground, or affixed to the drilling machine itself, having a dust-tight gasket around the drill-bit;

— sucking hosepipes;

— a cyclone and filters assembly (these devices are described below in Sec. 4.4).

The dust particles are collected by the capturing head above the hole being bored and recovered by the cyclone-filter unit. These devices are highly effective but cumbersome, rather heavy and costly. Consumption of compressed air is relatively high (4000 l/min for the caterpillar-mounted, heavy-duty models).

• *Wetting of dust particles as they form*

This method, commonly used in underground workings, calls for devices provided with a water injection device alongside the drilling apparatus. It has the advantages of being least cumbersome and relatively more effective, provided the large quantities of water formed can be removed.

All the same, mention may be made of a solution which seems to give attractive results and which consists of introducing a mist of atomised water into the compressed air injected at the bottom of the borehole for forcing out the bits of fly rock.

4.4 Separation of Captured Dust Particles

The dust particles captured by entrainment in an airstream must now be disposed of. One could of course vent this air into the outside atmosphere. This would solve the problem of dedusting the working areas inside, no doubt, but the dust load of the outside air would be aggravated—which is not at all desirable. So the question is how to extract the dusts contained in this air so that only air practically freed of suspended dust particles is vented. A number of methods of varying efficiency are available for successful accomplishment of this task. We shall mention here only the main ones, the various manufacturing firms supplying variants derived from these archetypes, from amongst which the user can select the one that best resolves his specific problems.

Dust-extracting apparatus can be divided into six categories:

— devices for expansion of sucked air;

— dry separators, upstream of suction;

— exhausters—dry dedusters;

— wet separators;

— filters;

— electrostatic precipitators (mainly used for dedusting furnace gases, industrial boilers).

4.4.1 DEVICES FOR EXPANSION OF SUCKED AIR

These consist of a sheet-metal container for holding water (Fig. XIII.5). The dust-laden air enters and exits the container at the top portion through pipes whose cross-section is far smaller than that of the container. Therefore an expansion of the air takes place which becalms it and allows a portion of its dusts to decant, especially the coarsest ones, which are later recovered in the form of slimes at the container's bottom. A median baffle is generally an additional feature for forcing the airstream to graze the water surface, thus improving the efficiency. But if this baffle is faulty in dimensions, it can give rise to eddies retaining the dusts in suspension.

In brief, this highly simple and very economical device is mediocre in efficiency.

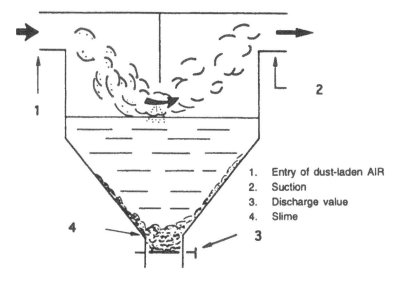

1. Entry of dust-laden AIR
2. Suction
3. Discharge value
4. Slime

Fig. XIII.5. Water pool expansion chamber

4.4.2 DRY SEPARATORS, UPSTREAM OF SUCTION

• *Cyclones (Fig. XIII.6)*

These devices comprises a sheet-metal cylindro-conical container. The dust-laden air enters tangentially at the top of the cylindrical portion. The centrifugal force thus created precipitates the solid particles against the wall from where they slide

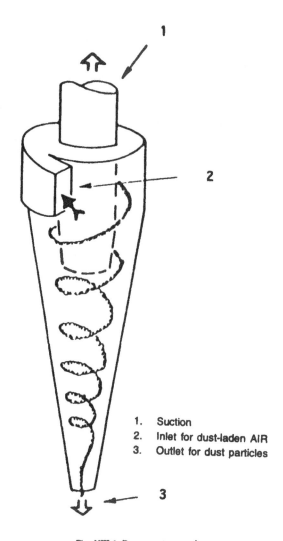

1. Suction
2. Inlet for dust-laden AIR
3. Outlet for dust particles

Fig. XIII.6. Preseparatory cyclone

down by gravity to the tip of the cone. The air, freed of the coarser particles, escapes through the upper orifice.

• *Radial cyclones (Fig. XIII.7)*

These devices work on the same principle as cyclones. The tangential suction produces a centrifugal force impinging the dust particles upon the wall from where they slide down by gravity. For equal space occupied, the output is higher for radial cyclones for a comparable efficiency.

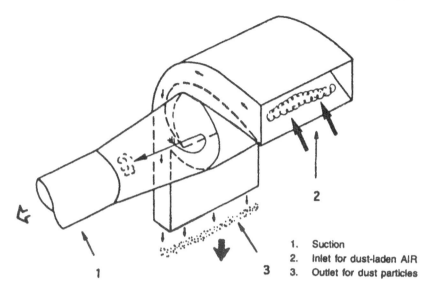

1. Suction
2. Inlet for dust-laden AIR
3. Outlet for dust particles

Fig. XIII.7. Radial cyclone

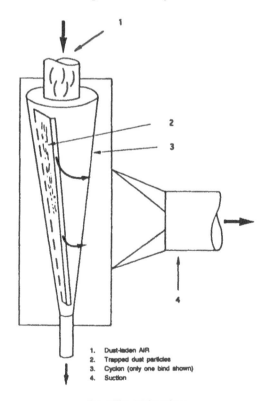

1. Dust-laden AIR
2. Trapped dust particles
3. Cyclon (only one bind shown)
4. Suction

Fig. XIII.8. Multicyclone

• *Multicyclones (Fig. XIII.8)*

A number of cones made of sheet metal fashioned in the shape of venetian blind: along the generatrices are disposed side by side inside an enclosure subjected to lower pressure. The dust-laden air penetrates the cone through its base (situated at the top portion), hitting the venetian blinds at a very low angle, and deposit here a fraction of its dust particles. The coarsest particles, trapped by the blinds then fall by gravity to the cone's tip from where they are collected.

4.4.3 DRY DEDUSTERS (Fig. XIII.9)

A centrifugal turbine exhauster sucks in the dust-laden air. The dust first land: on the blades of the turbine and then slides along them by centrifugal force to later be discharged at the lower portion of the device.

The great speed attained by the particles along the blades causes wear. Hence recourse to this equipment is limited to dusts of low abrasiveness.

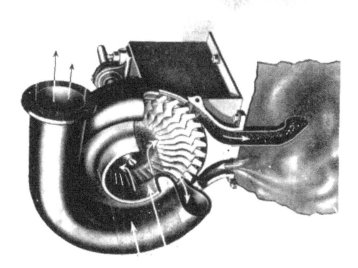

Fig. XIII.9: Exhauster

4.4.4 WET DUST SEPARATORS

• *Water-film deduster (Fig. XIII.10)*

This is a metallic chamber partitioned by a double-wall curved in the shape of ar 'S', the impeller. The lower portion is filled with water. The polluted air travel: through the impeller dragging along with it a thick water film. The speed of the liquid filaments is such that the water film laden with dust particles, after having assumed the profile of the lower part of the impeller, is thrown against its soffe and then ejected into the second compartment. The difference in the water leve

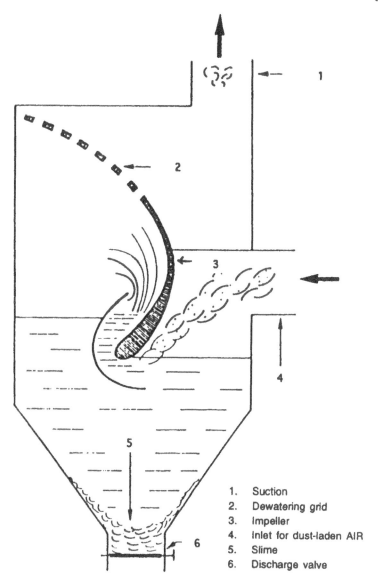

1. Suction
2. Dewatering grid
3. Impeller
4. Inlet for dust-laden AIR
5. Slime
6. Discharge valve

Fig. XIII.10. Water-film deduster

between the two compartments corresponds to the loss in the dust load of the air across the impeller. The air which is sucked and freed of its dust particles then passes through a grid which is an extension of the impeller's profile. This grid retains by impacts the entrained water droplets. The dusts are entrained in the form of slimes at the bottom of the chamber from where they can be discharged by an ejector provided with scrapers.

Purification with this device is highly satisfactory.

• *Wet-wall cyclone*

This is similar to the cyclone shown in Figure XIII.6 except that the central nozzle for evacuation of air carries a perforated tube along its axis for injection of water. The coarse particles slide down the cone wall, as in the case of dry dedusting. The fine particles remaining in suspension in the vortex of air are moistened by the mist of water and fall down as slime to the tip of the cone.

The purification efficiency of this device is excellent.

• *Air washers*

These devices are identical to the dry suction apparatus of Figure XIII.9. Water nozzles moisten the surface of the rotor blades. The air admitted axially undergoes an initial washing as it passes through the mist of water produced by the nozzles. In the rotor's action zone, the air and the water are vigorously agitated and the intimate contact frees the air of almost all dust particles.

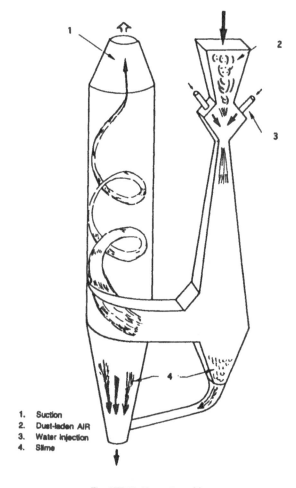

1. Suction
2. Dust-laden AIR
3. Water injection
4. Slime

Fig. XIII.11. Venturi scrubber

• *Venturi scrubbers* (Fig. XIII.11)

The Venturi sucks in the polluted air by the water blast principle. At the funnel exit, this polluted air is intimately mixed with water atomised by the nozzles of the Venturi. The major portion of the slimes settle at the base of the venturi cone. The wet air is then sucked in by a tangential admission cyclone in which the water-spouts, still carrying fine dust particles, arrive to impinge upon the wall by centrifugal force, from where they flow downwards to the tip of the cyclone for discharge.

The equipment ensures a very good purification.

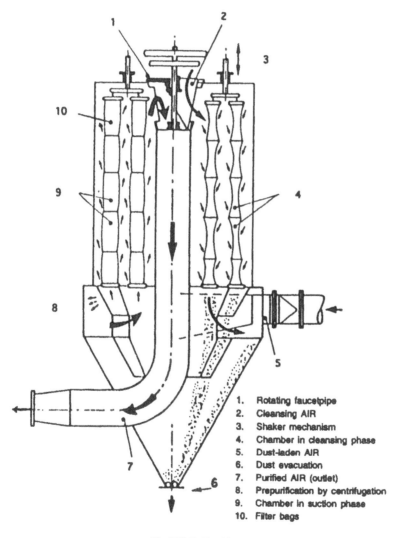

1.	Rotating faucetpipe
2.	Cleansing AIR
3.	Shaker mechanism
4.	Chamber in cleansing phase
5.	Dust-laden AIR
6.	Dust evacuation
7.	Purified AIR (outlet)
8.	Prepurification by centrifugation
9.	Chamber in suction phase
10.	Filter bags

Fig. XIII.12. Bag filter

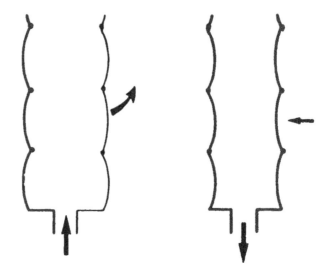

Fig. XIII.13: Pneumatic cleansing

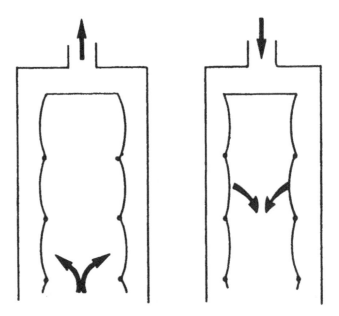

Fig. XIII.14: Counter-current cleansing

4.4.5 DEDUSTERS BASED ON FILTERING

There are a number of filter systems but for dedusting as applied to aggregates manufacture, only special fabric filters are used. These work on the principle of a domestic vacuum cleaner in which the polluted air passes through a fabric bag whose texture is capable of retaining suspended particles (cotton, felts, terylene, glass fibre etc.). The bag units are split into a certain number of cells which are pressed into service one by one, disconnected, and finally cleansed by automatic devices. At the time of cleaning the bags, a pulse of fresh air for cleansing is passed through the fabric in the opposite direction.

The automatic cleansing device can be:

— mechanical or electromechanical (Fig. XIII.12);
— pneumatic (Fig. XIII.13);
— a counter-current type (Fig. XIII.14).

The efficiency of these devices is excellent and dust recovery efficiency can reach 99.9%.

In view of the very high cost of installation, electrostatic precipitators are generally employed for dedusting huge quantities of gases laden with more or less harmful impurities: hot rotary cement kiln gases, smokes from thermal power stations.

4.5 Choice of Separating Equipment

This depends on the concentration of dust particles in the air sucked in during the dedusting operation. It can also be dictated by their toxicity and special environmental constraints. Sometimes it is advantageous to provide for multiple types of equipment in cascade, for instance a dry cyclone preseparator to catch the coarse particles, followed by bag filters intended to trap the finest particles, so that the latter will not be prematurely clogged by an excess of large-size dusts.

Table XIII.2 synoptically presents the various types of apparatus described above with respect to their efficiency vis-à-vis dusts of different size ranges.

4.6 Art of the Quarryman

One should not lose sight of the fact that simple solutions adopted as such, or as adjuncts to particular commercial devices, can often very effectively upgrade the fight against diffusion on dusts, among which mention may be made of:

— a judicious orientation of the buildings, stocks, with reference to the prevailing winds;
— exhaust chambers at the outlet of screens, crushers...;
— hoses of loading fine elements onto trucks;
— hoods or screens at sensitive points.

To illustrate our idea, we adduce here a device which can be installed over an impact crusher; we have already seen (Sec. 3.1.1) that this device lowers pressure at the inlet and raises it at the outlet. Hence an astute arrangement would be to connect the outlet hood to the feed chamber through a bifurcated pipe in which the slopes of the two arms are such that the dust particles entrained in the airstream can be evacuated by sliding down the lower wall (Fig. XIII.15).

Table XIII.2: Effective operational range of various dedusters

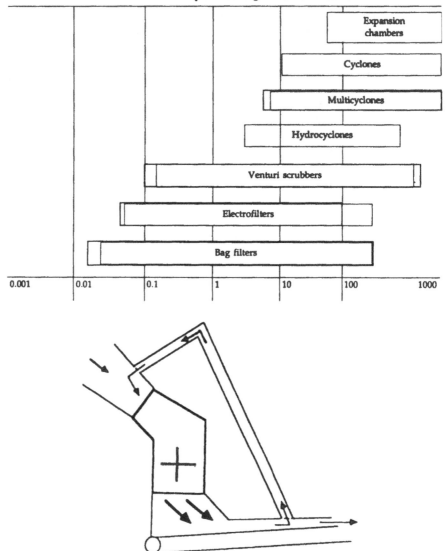

					Expansion chambers	
				Cyclones		
			Multicyclones			
			Hydrocyclones			
	Venturi scrubbers					
	Electrofilters					
Bag filters						
0.001	0.01	0.1	1	10	100	1000

Fig. XIII.15. Equalisation of air pressures in an impact crusher.

5. COST OF DEDUSTING

5.1 Introduction

The scope of this section does not extend to details of equipment sizing, which calls for equations governing air flow. This would, in effect, go into the theory of fans, pressure drop inside the ducts, influence of the geometry of cyclones on collection efficiency etc. Even a voluminous treatise would not suffice so we shall confine ourselves to simple mentioning some general principles that will give some idea of the cost for dedusting, proportionate to the overall cost of investment and working costs of a crushing-screening installation.

Let it be noted in this respect that the best is the enemy of the good!

For instance, it is tempting to think a priori that stepping up the rate of suction and the speed of air would ipso facto lead to a higher efficiency of dedusting. Such is not the case, however; a very high suction rate entrains in the sucked air dusts which would otherwise have remained within the flow of the materials. In addition, too high a speed inside the ducts exacerbates the phenomenon of wear by abrasion, which in turn inflates maintenance costs. Lastly, since power consumption varies with flow rate, the operational costs increase correspondingly. It may be noted in this regard that the pressure which occurs is proportional to the square of the flow rate and the power consumed is proportional to the cube of this flow rate.

5.2 Some Simple Rules

For an initial estimation, one need make use only of the following rules:

— Categorise the various divisions of the installation into a few simple points: primary crushing, secondary crushing and screening, tertiary crushing and screening.

— For each point, based on the nominal capacity in t/h, estimate the production of particles smaller than 100 μm in t/h.

— For good conditions of hooding at the dust emission points, and for average conditions of particle size, temperature and humidity of air, take the proportion of suspended fines to be equal to 10% of the total fines for each of the points considered.

The suction rate is obtained by limiting the mass of particles in suspension to between about 10 and 20 g per cubic metre of air sucked.

The suction speed is limited to 2 to 3 m/s at the section heads and between 25 and 30 m/s in the ducts.

All these elements enable sizing the various components that go into a dedusting set-up. Further, it is known that the power consumed for dedusting varies from 0.5 to 1 kW per t/h of material produced, depending on the case.

In accordance with the distances between the various points of the crushing-screening installations, as defined above, it often becomes necessary to provide for not one but several dedusting units to avoid wastage of energy through pressure drops in ducts that are too long. This envisages diversification of the designs of dedusting between primary crushing and secondary and tertiary crushing-screening, the degree of sophistication increasing from upstream to downstream operations.

5.3 Cost of Dedusting

Since absolute costs have no meaning, we shall indicate costs as percentages of the total investment for aggregate manufacture:

— installed power: 10 to 15%;

— investment: 3 to 5% (excluding land cost);

— operating cost (amortisation, consumables, maintenance) 2 to 3%.

In the case of an old plant, when dedusting is retrofitted, these figures would obviously be higher.

6. PNEUMATIC SEPARATION AND UTILISATION OF FINES

Pneumatic processes are widely employed for effecting a size separation, especially for obtaining lower particle-size fractions, for which purpose the conventional screens are not suitable, and in such cases wherein hydraulic classification is ruled out.

Special mention may be made of:

— control of the maximum particle size of products, which is the case in cement plants for closed circuit operation of a crusher;

— classification of the fine elements into categories such as for sale, inputs for animal feed industry, paints etc.;

— partial or total removal of ultrafine particles, dedusting, fines removal from sands (especially crushed sand for hydraulic cement).

The materials should have a grading suitable for their intended utilisation. In the field of civil engineering and building construction, according to the type and quantity of binder going into a composite material, this concerns the normal production of the crushing equipment:

— less rich in fines (sands for hydraulic cement);

— more rich in fines (sands for bituminous cement, asphaltic mastics);

— very rich in fines (make-up fillers). The fines removal operation consists of extracting a certain quantity of fine elements form a sand. This can be accomplished by the hydraulic separators described in Chapter XII or by means of pneumatic separators.

6.1 Pneumatic separators

These separators can be divided into three categories:

— separators utilising vertical air current,

— separators using centrifugal force,

— separators incorporating both these actions.

• *Vertical air-current Separators*

— *With selection chamber with rising currents* (Fig. XIII.16)

The material for separation is introduced at the base of the device into a current of rising air. The coarsest elements fall to the bottom while the material carried along the current undergoes a second classification in the outer conical envelope

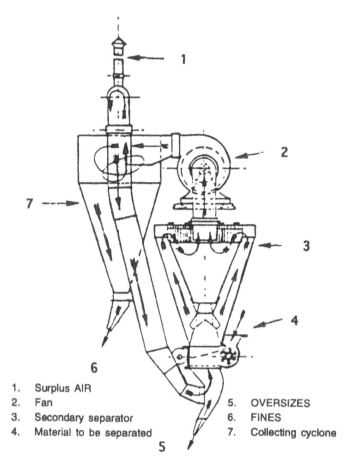

1. Surplus AIR
2. Fan
3. Secondary separator
4. Material to be separated

5. OVERSIZES
6. FINES
7. Collecting cyclone

Fig. XIII.16. Pneumatic separator with selection chamber

where the speed of the air slows down, then a third while passing through the central conical chamber after passing through a propeller annulus. The particles deposited slide down the walls of the two envelopes and are evacuated at the base. The finest particles alone are forced up by the fan into the collecting cyclone wherein they settle down.

Several variants of this device are available, incorporating a combination of selection chamber, fans, cyclones, fresh air injection into the circuit and in the upper portion of the selection chamber, and one or more mobile devices for improving the efficiency of separation.

— *With zigzag chutes*

In the set-up shown in Figure XIII.17 the material to be treated is introduced at the base of the device upon a perforated vibrating table from under which air is

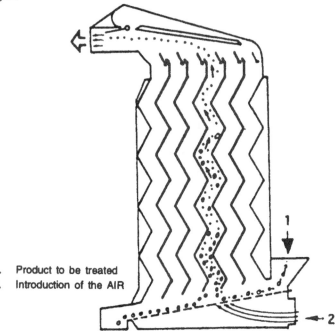

1.   Product to be treated
2.   Introduction of the AIR

Fig. XIII.17. Pneumatic separator with zigzag chute

introduced, which effects a first separation by carrying along the medium and small particles while the larger ones are evacuated by the vibrating table.

In their upward movement the particles meet with the counterslopes of the parallel zigzag chutes in which a second separation takes place; the particles which are stopped fall to the bottom of the device. The fine elements are sent into a collecting cyclone.

• *Centrifugal separators*

In the separator shown in Figure XIII.18 a cylindrical chamber is divided into two by a propeller annulus. The material to be classified is introduced tangentially. The larger particles are classified at the periphery and evacuated by an Archimedes' screw. The air carrying the fines is sucked to the centre by a fan.

• *Separators combining both actions*

These generally consist of a cylindro-conical container inside which a fan rotates. The products to be treated are distributed upon a rotating table which disperses them into the air current created by the fan. The finest fractions are carried along in the air current while the larger ones fall to the bottom from where they are recovered.

Various set-ups have been conceived by manufacturing firms: diffusion table, horizontal diaphragms and circular shutters for varying the path of air filaments, auxiliary turbine for dispersion, lateral feeding, enhancement of the sedimentation column for:

— expediting the regulating operations;

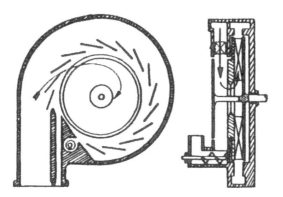

Fig. XIII.18. Centrifugal pneumatic separator

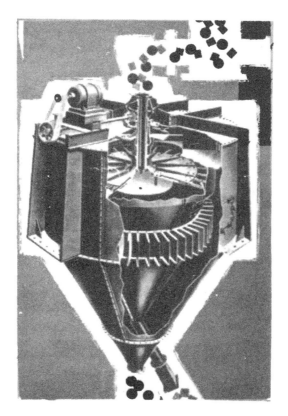

Fig. XIII.19. Pneumatic separator with rising currents and centrifuges

— increasing efficiency and yield;
— reducing wear and the space requirement of the equipment;
— multiplying the cyclonic effects which sometimes become preponderant and thus enable precise classifications with very small particle sizes.

• *Field of application of air separators (Table XIII.3)*

Compared to the hydraulic-based system, these devices have the following drawbacks:
— low processing rate,
— high investment.

Table XIII.3: Utilisation range of air separators

Types of selector		Particle size range of separation (μm)
— with selection chamber:	zigzag rising currents	80 – 120
	with centrifugal	40 – 100
	superstructures	5 – 80
— centrifuges		3 – 40
— combining both actions		30 – 150
— suited for extreme fineness		8 – 15

6.2 Fines Removal from Crushed Sands

Sands obtained by crushing have a rather high fines content (about 15% passing through a 0.3 mm).

Though they may be of a proper cleanliness, these fines should be partly removed so as to obtain a performing sand for use in hydraulic concretes.

Two technological solutions can be utilised:
— hydraulic treatment (see Chapter XII),
— pneumatic treatment.

The solution adopted in the latter case is dry treatment, with the help of a classifier which effects a fraction cut-off close to 80 μm while maintaining a controlled proportion of the fines (about 9%), and retaining the 0.1–0.4 mm elements imperative for the making of a quality concrete. The purpose of the equipment is not to effect a rigorous classification conforming to a particular particle size, but to produce a sand which is poorer in fines and uniform. This solution has the advantages of retaining a portion of the 'clean' fines beneficial in a hydraulic concrete and giving dry products, which are easier to handle and batch. The devices used most often are air selectors that combine the actions of separation by rising airstreams and centrifugal force.

To obtain a proper output, the sands for processing should be free from clay particles and have fairly low moisture content. Moisture conditions are very strict: a good efficiency cannot be obtained from the device if the moisture content exceeds 1%, however, preheating the air reduces water contents exceeding 1% and also very strikingly increases the throughout of the equipment by reducing this content to 0.5%.

Trials conducted on a sand containing 17% fines gave the following outputs:

— water content: 1%—30 t/h of sand treated.
— water content: 0.5%—50 t/h of sand treated.

The ingenuity of manufacturers has led to the introduction of simple but effective solutions for ensuring the removal of fines from sands. The device in question is one of forced suction, wherein the stream carrying the material to be treated is led to a bag filter. Suction is effected at that place in the stream of material to be treated where segregation takes place: the finer elements remaining longer in suspension, a preferential trapping results.

The act of bringing into suspension the fine elements may be either intrinsic to a device in the manufacturing chain—at the middle of a vertical screen, above a screen—or artificially caused. Figure XIII.20 shows the chamber of a screen equipped with pipes in which the lower part is perforated and subjected to a lower pressure; bringing the finest elements under the pipe into suspension ensures selective trapping.

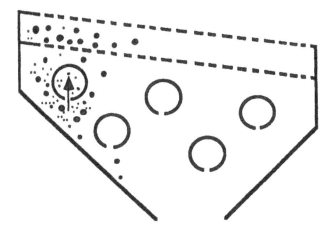

Fig. XIII.20. Suction of dusts from a screen

6.3 Utilisation of Recovered Fines

The problem of utilisation of fines is essentially faced on crushing massive rocks, and this to an extent which becomes that much more acute as the production of sand meant for making quality hydraulic concrete is subject to a fines removal operation. What is the use of calcareous fines? Producers are exploring every possible use for this by-product but the current outlets invariably cannot absorb the totality of the fines recovered, which amounts to 2% of the total production from crushing.

The current uses are as follows:
— incorporation in compositions rich in fines (asphalt mortars and bituminous mixtures);
— supply of calcareous raw materials to cement plants;
— agricultural improvements;
— various fillers;

— incorporation in quarry by-products (prescreening material);
— storing.

REFERENCES*

Ann. Mines. Special number on dusts, Jan–Feb. 1978.

⁺Bauer, W.G. 1963. Design trends in mechanical air separators. *Pit and Quarry*, Dec. 1963.

Blanc, E.C. 1951. Classification of materials using air. *Equipment mecanique*, Jan. 1951.

⁺Buchhi, R. and E. Pescali. 1965. Untersuchungen über die arbeitweise des zyklon—Umluftsichters in der vermahlung von zement. *Zement Kalk Gips*, Nov. 1965.

Bur. Int. Trav. Geneva. Guide for the prevention and suppression of dusts in mines and quarries.

Detrie, J.P. 1969. Atmospheric pollution. Dunod, Paris.

⁺Janich, H.J. 1969. Der turbo-windsicher Typ JSU, ein neuer umluftsichter für hohe leistungen. *Zement Kalk Gips*, Dec. 1969.

⁺Kayser, F. 1962. Der Zickzack-sichter, ein windsichter nach neuer prinzip. 1st European Symposium on Fragmentation, Frankfurt, 1962.

⁺Kayser, W. 1962. Neuentwicklungen auf dem gebiet der streu-windsichter. *Zement Kalk Gips*, Nov. 1962.

⁺Martin, C. 1926. Researches on the theory of fine grinding. Chapters 4, 5, 6, and 7. British Institution of Chemical Engineers, Henley and London.

Muhlrad, W. and A. Rebours. 1969. Dedusting. *Techn. Ing.*, Paris.

Quesnel, C. 1963. Pneumatic classification, Part A 908, *Techn. Ing.*, Paris.

⁺Schauer, S. 1968. Die Streuring-sichter: ein neuer umluftsichter. *Zement Kalk Gips*, Sept. 1968.

⁺Taggart, A.F., 1945. *Handbook of Mineral Dressing*. Sec. 9: Air sizing and dust collection. McGraw Hill, New York-London.

⁺Wesssel, J. 1966. Schwerkraft und flichkraftsichter. *Aufbereitungs Technik*, March 1966.

⁺Wessel, J. 1967. Grundlagen des siebens und sichtens. *Aufbereitungs Technik*, Feb. 1967.

*All entries in French except those marked with a cross (+)—General Editor.

XIV

Handling, Stockpiling, Mixing, Weighing, Dispatch and Transport of Aggregates

Michel Yernaux and Alain Maldonado

1. INTRODUCTION

In the scheme of an aggregate processing plant, materials-handling operations account for up to 65% of the total investment.

The portion of production costs chargeable to materials handling varies between 20 and 40%.

But then we are discussing matters of investment in an industry, namely that of exploitation of quarries and sand-pits, which is obliged to bear very heavy costs.

In fact, the initial investments on setting up a quarry exploitation are 2 to 3 times the future annual turnover and should therefore be amortised over a period extending well beyond a decade.

Whether it be a simple matter of linkage between different machines, or the final point in a plant, equipment handling materials are of very great importance, albeit they contribute little to improving the latter as these have already been produced and graded. Nevertheless such equipment constitute a substantial part of the total investment and hence should be as reliable as, say, a crusher or screen, so as not to precipitate a stoppage of production.

Furthermore, this equipment represents a large percentage of the total equipment needed in a quarry or sand-pit plant and even one out-of-kilter can bring production to a sudden halt.

There are two opposing trends. The first is traceable to the constraint of transport costs and dictates that quarries should be situated as close as possible to the consumption centres.

In France there are at present nearly 5000 quarry or sand-pit enterprises or, on average, 60 per administrative district.

The second trend is consequent to the shift from alluvial exploitation to quarrying massive rocks, often located much farther from the centres of consumption.

The problems dealt with in this chapter—handling of materials, their storage, mixing, weighing, loading and dispatch, transport and unloading—can be subscribed under a technique newly termed 'bulk transit'. According to the documents issued by the Ministry of Industry, this might be defined as: 'All transit operations of the products necessarily appertaining to a complete and automated industrial process.'

The objective is to limit the quantity of indispensable stocks to the barest minimum and to organise their management, hence movement, in such a way that the entire handling operations studied herein are as economical as possible.

Therefore, lowering the cost price in a domain in which extension is anticipated and one which already accounts for almost one-third of the total production cost of aggregates, is a matter of great import.

We shall examine this matter according to the following scheme.

After defining the term 'materials handling', the pertinent equipment, i.e., belt conveyors with all components are described and the problems arising in the wake of using such equipment examined from the viewpoint of safety and precautions in this regard suggested.

A summary examination of the rules which enable calculation of the dimensions of equipment handling materials seemed imperative to us. We have done so in discussing the problem of stockpiling, first and foremost stockpiles of primary products.

We then take up the interface between crushing and screening machinery and the effect this has on the last point of the manufacturing plant, namely stockpiling of the processed products.

Obviously these stockpiles have to be considered in terms of how they were mixed but, more importantly, weighed. The various types of weighing equipment are discussed in Section 8.

Problems of loading and dispatch have to be solved when the processed materials are ready to leave the plant. Aggregates are not utilised at the place of their production. They are transported to users mainly by road, say 90% of the total mass, but also by railway and waterway. We shall attempt a harmonious presentation of the machinery involved, the operating conditions and every aspect concerning movement of the raw material within the processing plant and its transport to the user.

2. MATERIALS-HANDLING EQUIPMENT

2.1 Definition of Continuous Handling

In the context of the plants with which we are concerned, i.e., the production of aggregates meant for job sites of public works and buildings, continuous handling is that operation which ensures bulk movement of products, free from disruption, from one point to another by means of mechanical implements.

Considering the complexity of the problems and the implication of the results, we believe that studies on materials handling by conveyor should always be entrusted to specialists in the field.

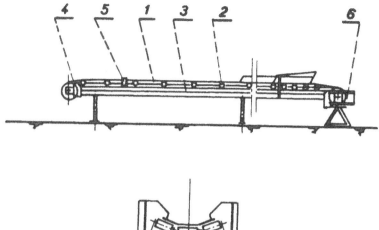

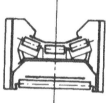

Fig. XIV.1. The belt

2.2 Belt Conveyor

Handling of materials in a quarry, sand-pit or gravel-pit is today invariably accomplished by belt conveyor set-ups (Fig. XIV.1).

We shall adopt the term 'belt conveyor', which more vividly highlights the active role of the equipment, and which we prefer to 'convoy' since this denotes 'to travel with' or 'escort' and appears to us palpably inadequate in the context of quarries and sand-pits.

2.2.1 THE BELT

The main element of a material conveyor is the band or the belt. Needless to say, every attempt has been made to strengthen its properties and to lighten its supports.

Progress in the capabilities of belt conveyors has considerably extended their range of utilisation in the field of materials handling. Synthetic textile carcasses and especially steel-wire carcasses have augmented strength and enabled use of belts several kilometres long, flow rates of several thousand tonnes per hour, and incorporation of driving motors of several thousand hp. Furthermore, the strength of the rubbers utilised and their perfect adhesion to the belt structure, whether it be made of polyester or polyamide fibre or consist of layers of metallic cables, have led to equipment that can withstand impacts, punches and wear in general.

This applies equally to materials of large dimension or high density (Figs. XIV.2 and XIV.3).

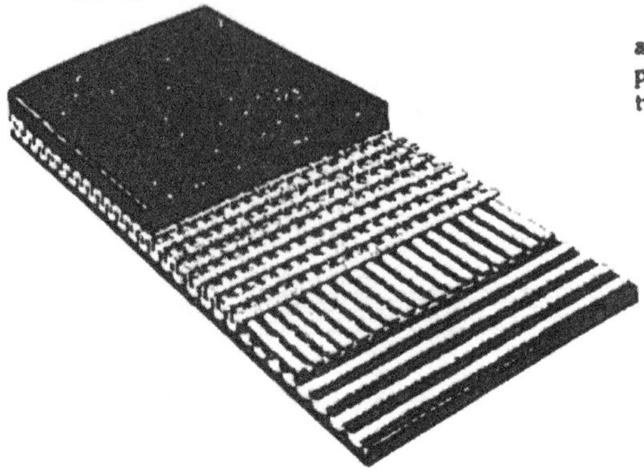

a) Single-ply made up of a combination of polyester cord layer in warp contained between two polyamide cord layers in filler.

b) Double-ply of interwoven tissue of polyester warp and polyamide filler fused to an elastic rubber sublayer.

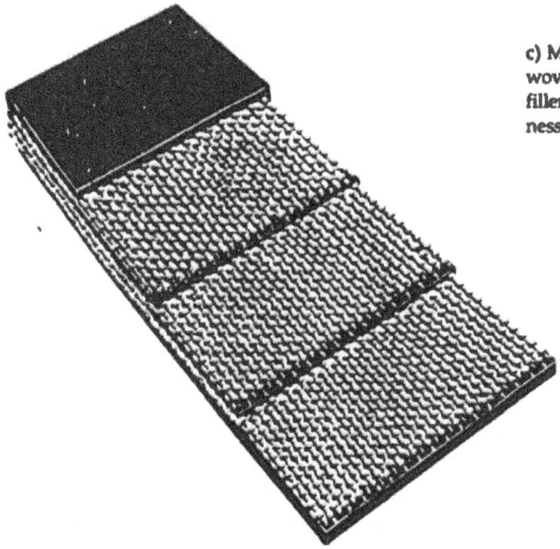

c) Multi-plies made of three plies of interwoven tissue (polyester warp, polyamide filler) with rubber interfaces whose thickness varies with the nominal tension.

Fig. XIV.2.

a) Three plies of interwoven tissue of polyester warp and polyamide filler contained between two transverse stiffeners (polyamide) joined to a protective rubber layer.

b) Three plies of interwoven tissue of polyamide warp and filler jointed to a shock-resistant cover made up of a polyamide breaker and a shock-absorbing underlayer.

c) Assembly of three layers of elastic metal cables impregnated in a rubber body:
• In the warp direction a layer whose diameter and pitches of wires vary with the nominal tension.
• In the filler direction two layers disposed on either side of the chain.

Fig. XIV.3.

Impact-resistant covers and shock-absorbing underlayers allow one to envisage handling huge blocks or even replacing the metallic decks of feeders with rubber-lined conveyor belts.

2.2.2 IDLERS AND IDLER SUPPORTS

Next only to the belt conveyor, these constitute an important element of the continuous materials-handling equipment. Considerable advances have been made in this field, especially with regard to dustproofing and hence durability of the mechanical parts of idlers, without entailing unrealistic conditions of supervision and maintenance.

In fact, in conveyors which can now reach tens of kilometres in length, supervision and greasing of the supporting idlers numbering three to five per metre every two metres or so, is inconceivable and hardly necessary if the tension is very high and the risk of 'scalloping' thereby reduced. Thus idlers are lubricated for their service life.

Furthermore, in most cases the belt speed reaches 8 m/s; there is no question of idlers becoming even slightly unbalanced and hence subject to intense vibrations at this speed.

Given the number of idlers which an extra-long belt conveyor sets rolling under the impulse of the driving systems, it is necessary that each idler have as low a resistance to rotation as possible; this is incompatible with the search for methods that rightly increase the resistance to rotation when the number of idlers is increased.

Lastly, the idlers should have good resistance to friction and corrosion; it is to be noted that the angle imparted to the lateral idlers to effect recentring of the belt conveyor increases the frictional wear of the carcass of the support rollers, especially when a layer of sand or abrasive dust finds its way between the belt and the lateral idlers.

The increase in power consumed due to pinching of the lateral support idlers of the upper or lower side obeys the following formula: $\Delta f = 0.66 \, \varepsilon \sin \lambda \cos \lambda$, ε being the pinch, λ the lift angle of the trough-shaped belt.

Moreover, this value of Δf changes with the conveyor profile (horizontal or rising). Consequently, one should not hesitate to impart a good pinch angle to the lateral idlers in order to assure good centring (see Sec. 2.2.5).

The drawback due to increase in the resistant torque and wear on the side of the belt not carrying materials is greatly offset by the advantages of continuous belt realignment.

When the conveyors are equipped with ilders and idler supports of the Limberollers type (Fig. XIV.4), an attractive solution in some cases, it should be borne in mind that the Limberollers assembly will be entrained by the belt and come into a decentring position, but certainly not in an automatic recentring position since the pinch angle of the lateral idlers will be negative.

So, it is imperative that the belt conveyor reach the tail pulley perfectly centred, that the head pulley, even when moving, always be strictly perpendicular to the conveyor's main axis and that the load of materials always be perfectly discharged along the axis at the middle of the trough. If these conditions are observed, the belt will remain centred upon the Limberollers. The belt conveyor should receive its load only when trough-shaped. The skirt boards should be adequately safe-

Fig. XIV.4. Conveyors with idlers in position before passage of the belts, NORDBERG Manufacturing

guarded so that the materials land in the trough-shaped central zone of the belt and not on them.

The idler supports in the loading zone ought advantageously to be of the shock-absorber type, that is, made of rubber. In difficult cases these rubber idler supports should additionally be cushioned on rubber blocks.

A relatively recent technique has introduced a set of rubber strips with a shock-absorbing part within the thickness and a sliding surface for the belt (Fig. XIV.5).

2.2.3 FRAMEWORK FOR THE CONVEYORS

We did not deal with this part of the conveyor at the outset because it is altogether static and were it possible to reduce it to a very simple exposition of technique, all the manufacturers would have thought of that.

In the case of short conveyors this framework should withstand the compression engendered between the head member and the tail member; in the case of longer conveyors where the head member and the tail member are anchored, the framework needs to support the weight of the idler supports, the belt and most of all the materials conveyed.

Accordingly, all solutions have been envisaged for the purpose of simplification in fabrication:
- framework made from light alloys: aluminium case under pressure etc.;
- flexible framework of stretched wire ropes;
- framework made from sections, heavy but economical nonetheless;
- framework made of simple tubes;
- framework of composite beam.

Fig. XIV.5. Impact bars in loading zone. Blocking of boards which are perpendicular to the belt

In the case of conveyors erected on the ground a compromise is always sought between spacing of the support points and minimum weight to the running metre.

The problem of cableways for conveyors which have to cross over natural obstacles, such as rivers or valleys, or which ought to have very large overhangs for making voluminous stocks, belongs to structural engineering rather than the discipline of materials-handling equipment.

Portable conveyors are ground conveyors whose lateral movement and realignment are facilitated by a semi-flexible link, such as a rail. Good stability of the belt at the mid-point of the framework will be all the better as the idler roller supports will be strictly perpendicular to the conveyor's axis and the lateral idler rollers will have a good pinch angle.

If the axis is not a straight line but a gentle curve at places, the belt will readily train itself on this curved line.

Lastly, it seems worthwhile to us to give some idea of the maximum possible inclines for conveyors to handle the various types of products, which we have done in a table (Table XIV.1) in the Appendix. The apparent density in tonnes per cubic metre for each of these products is also given in this table.

2.2.4 ENTRAINMENT

To achieve instantaneous flow of several tens of thousand tonnes per hour and covering distances of several kilometres from the head member to the tail member, it was necessary to design on the one hand, the driving heads of several thousand

Table XIV.1: Open stockpiling

H_m	L_m	VT for a cone		V_p with 1 evacuating passage		V_p with 2 evacuating passages		M_m	VT for a prism	
		m^3	t	m^3	t	m^3	t		m^3	t
1.5	3.6	5	7.5	1	1.5	1.3	2	0.40	2.7	4
2	4.8	11.9	17.5	2.3	3.5	3.2	4.8	0.53	4.8	7.15
2.5	6	23.2	54.8	4.6	6.8	6.2	9.3	0.66	7.4	11.2
3	7.2	40.2	60.3	7.5	11.8	10.8	16.2	0.79	10.7	16
3.5	8.3	63.8	95.7	12.5	18.8	17.1	25.7	0.93	14.6	21.9
4	9.5	95.2	142.8	18.7	28.1	25.6	38.3	1.06	15.1	28.6
4.5	10.7	135.5	203.3	26.6	39.9	36.4	54.6	1.19	24.1	36.2
5	11.9	185.9	278.9	36.5	54.8	50	74.9	1.32	29.8	44.7
5.5	13.1	247.5	371.2	48.6	73	66.5	100	1.46	36.1	54.1
6	14.3	321.3	481.9	63.1	94.8	86.3	129.4	1.59	42.9	64.4
6.5	15.5	408.5	612.7	80.3	120.4	110	164.5	1.72	50.4	75.5
7	16.7	510.1	765.2	100.2	150.4	137	205.4	1.85	58.4	87.6
7.5	17.9	627.5	541.2	123.3	185	168.5	252.7	1.99	67	100.6
8	19.1	761.5	1142.3	149.6	224.5	204.5	306.7	2.12	76.3	114.4
8.5	20.3	913.4	1370.1	179.5	269.2	245.2	367.9	2.25	86.1	129.2
9	21.5	1084.2	1626.4	213	319.6	291.1	436.7	2.38	96.5	144.8
9.5	22.6	1275.2	1912.8	250.6	375.9	342.4	513.6	2.51	107.6	161.3
10	23.8	1487.3	2231	292.2	438.4	399.3	599	2.65	119.2	178.8
10.5	25	1721.7	2582.6	338.3	507.5	462.3	693.4	2.78	131.4	197.1
11	26.2	1979.7	2969.4	389	583.5	531.5	797.3	2.91	144.2	216.3
11.5	27.4	2262	3393	444.5	666.7	607.3	911	3.05	157.6	236.4
12	28.6	2570.1	3855.1	505	757.6	690.1	1036.1	3.18	171.6	257.4
12.5	29.8	2904.9	4357.4	570.8	856.2	780	1170	3.31	186.2	273.3
13	31	3267.6	4901.4	642.1	963.1	877.4	1316	3.44	201.4	302.1
13.5	32.2	3659.3	5489	719.1	1078.6	962.5	1474	3.58	217.2	325.8
14	33.4	4081.2	6121.8	802	1202.9	1095.8	1643.7	3.71	233.6	350.4
14.5	34.6	4534.3	6801.4	891	1336.5	1217.5	1826.2	3.84	250.6	375.8
15	35.8	5019.7	7529.6	986.4	1479.6	1347.8	2021.7	3.97	268.1	402.2
15.5	37	5538.6	8307.8	1088.3	1632.6	1487.1	2230.7	4.11	286.3	429.5
16	38.1	6092	9138	1197.1	1795.6	1635.7	2453.6	4.24	305.1	457.6
16.5	39.3	6681.2	10021.8	1312.9	1969.3	1793.9	2691	4.37	324.5	486.7
17	40.5	7307.2	10960.7	1435.9	2153.8	1962	2943	4.51	344.4	516.6
17.5	41.7	7971.1	11956.6	1666.3	2349.5	2140.2	3210.4	4.64	365	547.5
18	42.9	8674	13011	1704	2556.7	2329	3493.6	4.77	386.1	379.2
18.5	44.1	9417.1	14125.6	1850.5	2775.7	2528.5	3792.7	4.90	407.9	611.8
19	45.3	10201.5	15302.2	2004.6	3006.9	2739.1	4108.7	5.03	430.2	645.3
19.5	46.5	11028.2	16542.3	2167	3250.6	2961.1	4441.6	5.17	453.2	679.8
20	47.7	11898.5	17847.7	2338.1	3507.1	3194.7	4792.1	5.30	476.7	715.1
20.5	48.9	12813.4	19220	2517.8	3776.7	3440.4	5160.6	5.43	500.8	751.2
21	50.1	13774	20661	2706.6	4059.9	3698.3	5547.6	5.56	525.6	788.3
21.5	51.2	14781.4	22172.2	2904.5	4366.8	3968.8	5953.2	5.70	550.9	826.3
22	52.4	15836.9	23755.3	3112	4667.9	4252.2	6378.3	5.83	576.8	865.2
22.5	53.6	16941.4	25412.1	3329	4993	4548.8	6823.1	5.96	603.3	905
23	54.8	18096.1	27144.2	3555.9	5333.8	4858.8	7288.2	6.09	630.4	945.7
23.5	56	19302.1	28963.2	3732.9	5689.3	5182.6	7773.9	6.23	668.1	987.2
24	57	20560.6	30840.9	4040.2	6060.2	5520.5	8280.8	6.36	686.4	1023.7
24.5	58.4	21872.6	32808.9	4238	6447	5872.8	8809.2	6.49	716.3	1073
25	59.6	23239.2	34858.6	4566.5	6849.8	6239.7	9359.6	6.62	744.8	1117.3
25.5	60.8	24661.6	36982.5	4846	7269	6621.6	9932.5	6.76	774.9	1162.4

For materials of density 1.5

VT = total volume

V_p = reclaimable volume

kilowatts, and on the other, elements for transmitting such intense forces to the rubber-lined conveyor belts.

For conveyors of short length and especially the mobile ones, use of driving pulleys is an attractive solution, given the reduction in weight of the unit and suppression of the lateral overhang. The compact unit incorporates inside the pulley the motor and the reduction drive. At present, the thoroughness in fabrication of these devices makes it possible to consider that the risks of breakdown are practically non-existent.

A very wide range of hollow-shaft reduction drives induces adoption of this elegant solution for conveyor equipment up to relatively high power ratings.

However, the most traditional drive system, which is also the best suited to powerful performing outfits, comprises an electric motor and a speed reducing system, the motor and reducer being connected either by flexible coupling or by hydraulic torque converter and that between the reducer and the driving pulley by torque converter (Fig. XIV.6).

Links by trapezoidal belts between the motor and the reducer enable subsequent manipulation of speed, but their efficiency is not above doubt and they constitute a hazard (see Sec. 2.3).

Fig. XIV.6. Conveyor with motor, torque converter and reducer assembly

Fig. XIV.7.

The drive pulleys these days are almost all rubber-lined so as to enhance the grip between pulleys and belt conveyors (Fig. XIV.7). According to various tests, the best grip is obtained between rubber and neoprene. There are other attractive linings, such as ceramic ones. The rubber linings are of great interest, especially for high speeds or for service in humid atmosphere, through provision of rafter or rhombus-shaped reinforcements, or sometimes by being well-glued to an inter-calating material.

Hydraulic converters help minimise the belt's instantaneous tension at the moment of start-up. The margin of such reduction is around 1.4, that is, the maximum starting power at the connecting point with the conveyor belt is at most 1.4 times the nominal power in normal working. These converters may be of the 'bailer type', which permits the electrical motor to be started up practically with no load, the oil then gradually entring the converter. A safety plung ensures absolute safety: if the oil temperature in the converter shoots up due to an abnormal functioning, the safety plug melts and the oil flows, of course arresting the machine's operation. With the 'bailer type' converter, the maximum power can be limited to 1.2 times the normal power, and starting of a wide belt with full load is possible. If this has to be done quite often, a cooling system must be provided for the converter oil.

The conveyors should be able to work without supervision and conveyor installations are now made with centralised control.

Therefore, various safety devices are to be provided, especially rotation monitors (which enable detection of slip in the drive pulleys which, in the case of high-power conveyors, can spell major disasters quite fast through rise in temperature), monitors of belt behaviour etc.

The head pulley on large conveyors is equipped with strain gauges which continuously register the value of the tension on the belt's upper and underside.

2.2.5 BELT TRACKING AND CENTRING

With regard to tracking and centring the belt conveyor, incorporation of lateral idlers with a 3 or 4° pinch is very helpful. The belt, in the event of becoming misaligned, comes into contact over a larger width with the inclined idler, which tends to push it back towards the centre. The pinch imparted to the idler support becomes that much more necessary as the belt width becomes smaller, since the slight shift between the end fibres and the central neutral axis removes a certain rigidity which is natural to a narrow belt. The same principle of centring can be applied to the underside provided two idlers are used in place of one for supporting it. Each of these two idlers has a pinch so as to serve the same purpose of recentring as the upper idlers.

We saw in Sec. 2.2.2 the influence of the pinch angle of lateral idlers on wear and the power consumed in regard to recentring the belt.

Theoretically, from the moment and the idler supports, whether on the upper or the underside, are perfectly perpendicular to the conveyor's axis, and as long as there is no external influence, such as wind or difference in moisture content between right and left, the belt conveyor follows its track of the tail or head pulley, without moving away from the ideal axis. The same holds true for the underside.

Sad to say, matters are altogether different in practice.

Firstly, outside factors intervene, such as wind, or a load lies eccentrically on the belt and quite often, the idler supports, and hence the idlers themselves, are positioned with a generous 'approximation'. In which case, at starting up, a 'resetting' is carried out.

The idler supports on one side of the conveyor are given blows with a hammer, alternately forward and backward, and at each blow the belt responds until a balance is established, which again will hold, for example, only as long as the moisture content on the idlers to the right and those to the left remains in equilibrium.

As a matter of fact, a conveyor should be started up as meticulously as any manufacturing apparatus.

All the central idlers or the bottom idlers, when they are single, should be perpendicular to the conveyor's axis, or in the case of curved conveyors, conform to a pre-determined axis.

The solution here is to employ a tee which adapts to the conveyor frame width, that is, rests against members which represent the conveyor's axis.

Concomitantly, at each station the lateral idlers should have, with respect to the conveyor's axis, an equal angle to the right as well as the left, and an angle which is theoretically sufficient if it is of the order of 1.5° but still more effective if between 3 and 4°.

In respect of the underside, the solution is to have a station not with a single idler for every 3 or 4 stations, but a station with two idlers, each having a pinch angle of 3 or 4° with respect to the perpendicular of the conveyor's axis.

Spacing of the carrying posts of the upper or underside should be calculated with due regard, of course, to the fact that if the belt 'scallops' too much, the power consumed increases. One should also take into account the fact that a reasonable spacing of the idlers has an economic influence on the conveyor's cost price on the one hand, and an influence on driving torque transmitted to the idler on the other.

When this idler is fitted with due regard to safety rules, which we shall discuss in Sec. 2.3, there is every reason for its start-up to be 'powerful'.

2.2.6 TENSIONING SYSTEM

The problem of tensioning the belts becomes more complex with increase in length of the conveyors but has been greatly simplified in recent years with the advent of belts with metallic carcass and hence low relative elongation. Compared to some belts with synthetic textile carcass, the choice of a steel carcass has indeed helped to bring down the magnitude of relative belt elongation from 2 to 0.5%.

For the rest, while making the belts endless at the time of vulcanisation in hot or cold, an additional precaution is taken by way of keeping the belt in tension for a few hours prior to its final cutting and vulcanisation. By this means the need for allowing substantial movement to the tensioning system is precluded, which we shall now study.

Tensioning can be achieved very simply by moving the tail pulley through manipulating two lateral screws (Fig. XIV.8). But then this tension is neither automatic

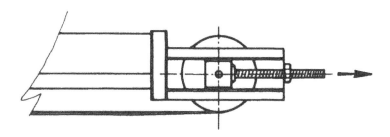

Fig. XIV.8. Tensioning by screw

nor sensitive to the conveyor's operating conditions. In starting up, the tension on the upper side close to the driving pulley rises instantly while that on the underside drops immediately after the driving pulley. This is one reason why it is preferable to locate the automatic tensioning system as close as possible to this pulley. This automatic tension is achieved by movement of a winding pulley which may be either the tail one for interaxes distances which are not too long, or an additional pulley moving in a predetermined plane, most often vertical, the belt passing before and after this counterweight pulley upon two inflection pulleys (Fig. XIV.9). Movement of the counterweight pulley is automatic since it is connected either to a counterweight by a system of cables, or a dynamometer which is also connected through metallic cables, or lastly to a set of jacks and compressed air reservoir (oil-pneumatic assembly). Therefore there is constant equilibrium between the belt's strength and the influence of a constant tension upon the mobile pulley along a predetermined plane. It is to be noted that to avoid any misalignment of the belt with respect to this counterweight pulley, the latter should move along a given plane, remaining strictly parallel to itself during these movements. Furthermore, it is recommended that the inflection pulleys, i.e., those in contact with the conveyor's bearing surface, be lined with relatively soft rubber (Shore's hardness below 45) so as to preclude their clogging, which would result in belt misalignment and risk of degradation.

The ballast boxes should be placed at a low height from the ground, but with sufficient clearance however, and in a protected-access zone.

Fig. XIV.9. Tensioning by traction on the tail pulley

2.2.7 BELT CLEANING

This question of possible clogging of the pulleys straightaway brings us to the problem of cleaning the belts. It is a very ticklish problem and challenges a perfect solution.

Mention may be made of the following amongst the solutions adopted to date:

— Scraping systems incorporating a rubber blade, for which purpose some manufacturers supply highly abrasion-resistant rubbers unless it be a conveyor belt whose cover has undergone damage. This conventional device is of the articulated type with a counterweight whose position is adjustable so as to exert upon the belt a predetermined and adequate pressure for scraping the adhering materials. In some cases shock absorbers are fitted to these articulated scrapers to prevent their vibrations from a beating phenomenon. Some scrapers are fitted with steel blades which yield very good results but require that the belt be in perfect condition and free from incipient tears (Fig. XIV.10a).

— Rollers with spiral blades too, which may be made of metal or rubber, assure a certain decrusting of the belt but it is necessary to increase their number since each leaves an area untouched and, what is more, they can cause the belt to shift (Fig. XIV.10b).

— Brush manufacturers have made excellent advances and the wear is sufficiently reduced so that this method of cleaning conveyor belts, which is altogether attractive in many respects, may be adopted in a number of cases (Fig. XIV.10c).

— Another excellent method is to shake the belt rather than scrape it. In this case use is made of drums with bars or vibrators entraining a group of three idler supports (Fig. XIV.10d).

— When possible, cleaning with water jets gives excellent results but then dewatering becomes necessary; this can be done by passing the belt between many sets of rubber-lined rollers (Shore's hardness less than 50). Experiments have also been conducted on using compressed air judiciously directed against the belt at a predetermined angle (Fig. XIV.10e).

— A piano wire held at predetermined distance from the belt can also dislodge the materials adhering to it but here, again, it is essential that the belt surface be in perfect condition.

In the case of conveyors of fairly large length, the most efficient and consequently the most commonly used method is based on a system of belt reversal. Immediately after the head pulley and when the conveyor has completed its work, the belt is made to turn by 180° in such a way that the proper side of the belt is the one which is going to come into contact with the return idlers, and shortly before its passage on the tail pulley the belt is once again reversed in the reverse direction.

2.2.8 CONVEYOR MAINTENANCE

Materials-handling installations based on belt conveyors no longer demand constant supervision.

This does not preclude the need for a programmed maintenance of the equipment, however.

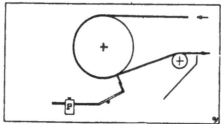

a) Belt cleaning by scraper

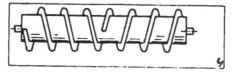

b) Decrusting roller with spiral blades

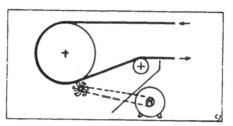

c) Belt cleaning by rotating brush

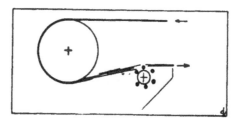

d) Belt shaking;

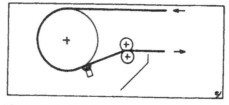

e) Belt washing and rinsing.

Fig. XIV.10. Conveyor belt cleaning devices

This consists of restoring the surface condition of the conveyor belt by hot or cold applications of rubber, if need be, with the prior repair of the textile or steel carcass.

All the idlers are checked for their free rotation and those that have stopped for one reason or another and bear a major flattening on their cylindrical lining are replaced.

The free clearance of the automatic tensioning systems must be checked.

Lastly, we shall mention in passing the conventional but crucial check-ups on oil levels and the general condition of the driving head's mechanical parts.

2.2.9 CURVED CONVEYORS

The first belt conveyor made of 'leather' dates back to 1795.

The belt conveyor from then on was necessarily of flat configuration in a horizontal plane.

From 1980, it appeared that the conveyor could not only have an upward and downward incline, kilometre after kilometre, depending on the needs of its operating situation, but could also negotiate curves to either the right or the left.

Therefore the belt conveyor has been made self-centring by horizontal and vertical positioning of the idler supports, whereas the problems of the driving head or of the tail pulley remain the same as those of conventional conveyors, i.e., the flat ones.

A systematic study, greatly aided by computer techniques, enables determination at each location where the belt conveyor is supported of the various forces at play in different situations, viz.: start-up, stopping, belt under loading, belt without materials, etc. Based on this data positioning of the idlers for the upper and underside can be established.

This is one amongst the most attractive solutions to the problems of conveying materials over long distances, in the field of mines as much as public works, as demonstrated by the job site for Djakarta airport layout.

France has thus become a leading specialist in this area (Fig. XIV.11).

2.3 Safety

Belt conveyors have numerous points that involve winding the belt on pulleys and its pressure on the idlers at this time results in zones of drag called 're-entrant points'.

It is these points which cause accidents that are often serious. The victims are seized by the re-entrant points when they intervene while the belt is in motion to set its working right or to clean the product.

Therefore the design and commissioning of belts should be such as to make their working as smooth as possible to obviate the need for such interventions.

2.4 Design Calculations

We have been called upon to look at a large number of plants with very costly crushing and grinding machines whose commissioning had been delayed due to materials-handling problems, or whose optimum output was far from being

Fig. XIV.11. Curvoduct in New Caledonia (REI Manufacturing)

achieved due to such problems. Had belt conveyors been installed, such problems would not have arisen. No doubt it is a question of theoretical knowledge, but more so of experience and commonsense.

Readers interested in the design of conveyors are referred to the Tables given in the Appendix at the end of this chapter.

However, design of a heavy-duty conveyor is a matter for specialists. Often the design is done jointly by the conveyor belt manufacturer and the supporting framework manufacturer in consultation with the project mechanical engineer.

2.5 The Chute

2.5.1 THE CHUTE—A CONTINUOUS CONVEYING DEVICE

The belt conveyor is not the sole engine of 'continuous transport' since the chute can fulfil this function.

However, it should be noted that in this case the material's own weight has to be used as the motive force and, consequently, the widespread use of chutes necessitates laying out an installation with substantial differences in levels.

Furthermore, the abrasive action of aggregates on the metal sheets of the chute causes notable wear which, it is true, can be mitigated by lining them with wear-resistant materials, such as rubber, or using the very aggregates as false lining (Fig. XIV.12). Care is to be taken to provide for lateral protection by setting in relief the layer of protective materials in the shape of small lateral triangles.

Fig. XIV.12. Protection of chutes by the materials themselves

2.5.2 CHUTE AS ACCESSORY TO BELT CONVEYOR

Quite often, good performance of a conveyor depends on how exactly the materials are discharged upon it.

Never should the materials arrive countercurrent or at right angle to the belt's axis, nor should they fall from a great height as this causes bouncing and escape of materials to either side of the belt.

In quarries and sand-pits the most judicious way of discharging materials in most cases consists of deposition in a pebble box or row of such boxes close to the belt.

A device with a triangular aperture (Fig. XIV.13) facilitates excellent centring of the materials in continuum despite possible variation in the feed rate to the conveyor.

The chute has to be retractable when a crusher is to be fed. Replacement of the wearing components of the size-reduction machinery indeed requires the space situated below the feeding zone to be uncluttered.

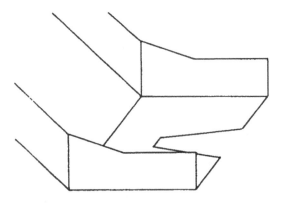

Fig. XIV.13. Chute which discharges materials centrally upon conveyors

3. MATERIALS-HANDLING OPERATIONS—PRIMARY STOCKPILE

3.1 Conveyance of Materials from Primary Crusher to Primary Stockpile

3.1.1 MATERIALS CONVEYED

As the dimension of the materials to be conveyed becomes smaller, handling entrails correspondingly fewer problems. Quarrying massive rocks, contrarily, poses several.

In the exploitation of massive rocks it is the primary conveyors, those which receive the primary materials, which have to be designed with the greatest care and incorporate certain devices, or at least conform to certain rules with regard to their disposition.

3.1.2 SETTING MATERIALS IN MOTION

This is a common problem but the consequences are more serious and swifter if not solved in the case of conveyors meant for large-size materials.

Under a primary crusher of large size and wide opening (200 to 300 mm), the materials, at their largest, can far exceed 250 kg. Under these conditions it would be highly dangerous if they do not land on the belt upon initial velocity and, of course, even more so if they do in the direction opposite to the belt's movement.

To make a protective cover on the rubber belt, based on the finest particles, manufacturers have sometimes provided their feed chutes with grates. Unfortunately, quite often the grates, even when trapezoidal in cross-section and disposed divergently, become clogged and the flow of materials is totally disturbed.

It is also possible to have a buffer-hopper under the primary crusher, a hopper with an extractor at the exit, especially a vibratory extractor, either electromagnetic or with counterweights. This arrangement is highly satisfactory since it does not impart to the materials a velocity of their own; on the contrary, it permits discharge nearly level with the conveyor belt.

3.1.3 INTERMEDIATE CONVEYORS

Sometimes a solution is adopted which comprises introducing an intermediate conveyor between the primary crusher and the actual conveyor, that is, a conveyor belt which takes on most of the wear and mitigates that of the main conveyor belt.

The arguments in favour of this arrangement stem from the fact that the main belt is thereby protected and should there still be some accidental wear and tear, for instance as a result of the presence of a sharp sheet, only a short-length belt is damaged, which means less expenditure and stoppage.

Let us detail a particular situation.

Considering the size of the materials at this stage in a plant, the belt in any case will be of larger width and the conveyor cannot have a very high speed. So much so, there is no chance of speed being imparted to the materials, as is the case with conveyors operating at 5 to 6 m/s.

The wear caused mainly by the impact of the materials upon the belt and their friction while landing, will be dispersed over the entire surface of the belt and, consequently, the rate of conveyor belt replacement will be inversely proportional to its length provided there is no accident.

However, there is an intrinsic merit in favour of the intermediate conveyor under a primary crusher; it safeguards against accidental damage. Hence it is essential to examine the possible chances of sharp elements passing on this spot, which like a razor can cut a conveyor belt width-wise in two.

3.1.4 RAPID OFFTAKE OF MATERIALS

Let us emphasise once again the fact that handling of materials by conveyor belts becomes increasingly simpler as the size of the materials reduces.

The cost price per linear metre also comes down since at higher speed width lessens and the rubber covers can be less in thickness without compromising safety.

Increase in belt speed also has an immediate effect on service tension, which again leads to a reduction in the equipment price.

That is why there is every need in a plant to find methods of very fast offtake of materials, either by providing 'secondary' size-reduction machinery at the primary station of receipt of the materials, after the primary crusher but before the primary stockpiling with, for instance, a link by gravity; or by using equipment that assure the highest reduction ratios possible.

In the case of calcareous materials having only a low percentage of abrasive silica, i.e., free silica, percussion machinery are advantageous; in the case of hard and abrasive materials, primary gyratory crushers present an advantage over primary jaw crushers in this regard.

3.2 Primary Stockpile

3.2.1 INDISPENSABLE ROLE OF BUFFER STOCKS

Fluctuations are inevitable at the primary crushing stage, regardless of whether the primary crusher is fed by dumpers or directly by mechanical shovels at the foot of the working face, if the crusher is mobile and moves along the working face.

Furthermore, it often happens that the primary size-reduction machinery is purposely oversized to enable acceptance of huge blocks of rocks.

Revamping, starting with the secondary size-reduction post, is thus possible as well as necessary for achieving an operation with steady output, which explains the need for primary stockpiles.

3.2.2 NECESSARY VOLUME

The principle of computation is as follows.
   • *Quarrying the deposit* proceeds intermittently, especially dictated by the exigencies of maintenance of the primary crusher. Hence stoppages occur at the primary station (or in quarrying in the case of exploitation of massive rock deposits), whose duration D, equal to the work stoppage time, should be determined.

Let us assume that all through this stoppage in the upstream part of the exploitation the processing plant continues to work to maximum capacity Q. Let it be noted that this maximum capacity is not the instantaneous nominal output for which this processing plant was designed, but the actual output calculated over a relatively long period. Therefore a factor of reduction should be applied to the instantaneous nominal capacity which could range from 0.6 to 0.9 depending on the country and the organisation of the job site.

To preclude stoppage in the downstream section of the plant, the primary stockpile should be at least DQ.
   • *The same reasoning* should be applied to possible stoppage in the processing plant. However, it may be noted that total stoppage of the plant is hardly likely since several of most of the machinery work in parallel. But the case of the processing plant working to minimum output versus the primary station delivering maximum output should be kept in view. In this case there is a surplus, say q, in the upstream flow vis-à-vis the downstream. If the duration of the incident (or the maintenance operation) in the processing plant is t, the volume of the stock should be at least qt.
   • *It is evident* that the larger of the two volumes DQ and qt is to be taken into consideration for computing the volume of the primary stockpile.

3.2.3 COVERING THE PRIMARY STOCKPILE

Is it necessary to cover the primary stockpile? This is a major problem in some cases since covering accounts for a higher investment than handling prior to and after stockpiling (Fig. XIV.14).

Fig. XIV.14. Covering of a primary stockpile

In the open, the reclamation cone of the materials constitutes a perfect funnel which will ensure the maximum collection of rain-water over a large surface. Since the fines in the precrushed materials, howsoever small in quantity, will be entrained by the waters seeping in the first stage through the interstices of the dead cone, the funnel in question will rapidly become watertight, unless it is periodically reclaimed by loader to be conveyed to the general plant.

The primary stockpile can be left uncovered if the material originates from a wide-open regulated machine.

In effect, in a 0/250 for example, there will hardly be any fine elements if those smaller than 60 mm have been removed before the primary crusher.

But then it is these fine elements of large specific surface which can entrain moisture and upset the functioning of the screens.

Also, if for instance elements of up to 20 or 30 mm are removed at the exit of the primary crusher for stockpiling or processing separately, the rest of the primary crusher's output can be stockpiled without covering.

In most other cases it is preferable and sometimes imperative to shelter the materials for smooth functioning of the rest of the plant.

3.2.4 RECLAIMING FROM THE PRIMARY STOCKPILE

Considering the invariably huge magnitude of this stockpile, open stockpiling has increasingly been adopted, which obviously is the most economical.

In the case of traditional quarry materials the natural angle of repose can be considered to be around 40° to the horizontal. Under these conditions the volume stockpiled at a single point, expressed in cubic metres, is very nearly equal to the cube of the pile's height in metres. When this stockpile is reclaimed by a front loader or shovel, the volume reclaimed can be equal to the volume stockpiled. But this is not the case when reclamation is done from the bottom of the pile, for instance by a conveyor inside a tunnel (most frequent case); a sizeable portion of dead stock will remain in place unreclaimed. Indeed, with a single extracting hole at the centre of the pile it is possible to reclaim only one-fourth of the total volume of materials. If there are two extracting points equidistant from the boundaries of the pile, the volume of materials reclaimed will be equal to 8/17 of the total volume, ignoring a minor approximation (Table XIV.1.).

At the moment of reclamation from under the pile sizeable flow rates (equal to the capacity of the processing plant) are often demanded. Placed inside a tunnel, the conveyor can simply be fed by one opening in the roof of the tunnel. Blocking of this outlet for flow can be secured either by swivelling plates, by trap-doors, or by regulators, unless the evacuating passage has a certain slope and is equipped with a liftable spout at its end.

This operation may be manual, electrical, through an electrical jack or a motor unit with reducer, hydraulic or pneumatic. The last mode is the most rational and the most economical.

3.2.5 UTILISATION OF INFRASTRUCTURAL FEATURES

The site layout of some plants offers advantages under unique conditions.

For instance, in the case of a limestone quarry a huge 'reservoir' is dug into the homogeneous rock body; it offers an ideal storage space sheltered from bad weather.

Of course, the entire mass of limestone rock dug out has already been passed on to the crushing plant and put to use. It is the more valuable part coming from the primary crushing station which is therefore stockpiled and also sheltered.

This reservoir, semi-natural, has open space on either side and so additional stockpiling can be done.

Reclaiming materials is done in the conventional manner, followed by handling of the products by conveyors.

Fig. XIV.15. Storage under reclamation tunnels • REI Manufacturing. Road construction site of TAI LAM DTP Doc.Bouygue

4. LIAISON BETWEEN MACHINERY IN THE PROCESSING PLANT

4.1 Feeding Conveyors

To obtain quality hydraulic or bituminous concrete, and particularly in the case of the latter, aggregates of high compressive strength or resistance to attrition should be used. There are some aggregates which have little free silica, such as diorites, but in some countries where the choice is limited, the foremost available materials essentially contain free silica:

— quartzite: 95% silica,
— granite: 55% silica, and so forth.

Wear in all metallic components is consequently excessively fast, especially in sheet parts which need to be replaced often, constituting one of the most important aspects of maintenance. Use of a rubber belt conveyor, though more expensive initially, markedly reduces maintenance outlay and quite often improves the conditions of the machinery.

At the exit of a screen of large surface area and with three decks, the exit box ahead calls for very frequent intervention to maintain it in good condition. This is also the case with the hopper for fines located below the screen and meant to recover elements which have traversed the entire screening surface. Provision of a series of small evacuating conveyors across and ahead of the screen and additionally of a large-width conveyor below the screen for its entire width, greatly reduce the number of sheet parts and thus the hours of intervention for maintaining them in good condition.

4.2 Spill Prevention at the Foot of Conveyors

Plants which before long find themselves half cluttered with cubic metres of materials that have fallen from the loading zone of the conveyors are rather rare.

Prevention of spill is imperative, especially from the viewpoint of safety; to effect this, certain precautions must be taken.

Conveyors shall be of sufficient width; they shall be fed in the direction of materials movement; they shall be provided with guiding passages for materials of sufficient length since the loading zone should extend over several metres to help the materials settle down upon the belt if the conveyor operates at high speed.

When dedusting is taken up in the plant as a whole, the conveyors shall be enclosed over several metres and this chute, terminated by a flexible damper, shall be kept down. There is every need to sufficiently raise the sides of the conveyor and provide spill-prevention skirts beyond the strip of materials.

The feeding chute of the conveyor shall be placed in the zone where the band is trough-shaped and never in the intermediate zone.

4.3 Dispersion on Screen

The high flow rates that have to be achieved in the scheme of materials processing concomitantly entail the fabrication and commissioning of screening equipment of very large surface area, reaching up to 30 sq m, and consequently large width. But such widths cease to be productive if the equipment is not properly fed, that is, materials should be uniformly spread over their whole width. Some solutions based on distributor chutes exist but they also entail rapid wear and consequently frequent intervention. More expensive at the stage of investment, the solution by way of a large-width conveyor is more radical and permits a saving in height, which has repercussions for the length between axes and the size of the buildings.

5. STORAGE OF FINISHED PRODUCTS: MATERIALS-HANDLING OPERATIONS ASSOCIATED WITH STOCKS

5.1 Size of Finished Product Stocks

We have already seen what should be the volume of the buffer stocks for maintaining smooth production operations. The stocks of finished products should enable continuation of production even with a disruption in evacuation of materials.

For example, a plant might cater to more than one destination and hence work both day and night, but loading of trucks is concentrated during certain hours of the day. When the quarry is connected to a railway network, the trains will be scheduled to fill time slots between passenger trains and other goods trains. Lastly, in the case of plants dispatching materials by boat, some irregularity in boat arrivals is inevitable.

Then there are the seasonal variations; the larger construction works being concentrated in the fair-weather period, there is inevitably a period of the year when the plants producing the materials have to stockpile them.

In the case of some huge job sites, such as construction of dams or nuclear power stations, the Contractor-in-charge often insists on a reserve stock against disruption in supply, which is apt to interrupt the progress of civil works.

5.2 Storage in Bins or Silos

It is customary to call the storage structure a bin when it is square or rectangular in cross-section and a silo when it is circular.

5.2.1 CLOGGING

Storage in a silo or bin cannot be discussed without reference to clogging.

The fall of a material inside a silo or bin is accompanied by compaction and adherence of the material to the walls in the case of 0/D sands and all-ins having a certain grading curve and a certain moisture content.

When the silo or bin is required to be emptied, the natural slope of the material is much higher than the slope of the walls and a substantial mass, if not the whole, may remain blocked.

Storage in a silo or bin cannot even be considered for certain materials whose properties are particularly favourable for compaction. In less problematical cases a few precautions should be taken to preclude lump formation here and there. (Fig XIV.16).

• *Smooth sliding* on walls of sufficient slope, that is 3 to 5° above the natural repose angle of the material, will not be hindered by untoward features, such as the presence of a metallic piece or flange (often the precipitator of clogging).

• *The discharge gate* should be of sufficient size and so located as to be perfectly flush with the sloping walls.

• *Varying slopes* and generally any asymmetry will disfavour the formation of arches and hence preclude jamming of materials during flow.

• *Some paint coatings* and some plastic linings markedly facilitate sliding.

Fig. XIV.16. Storage in vertical silos with asymmetrical double conical exits to avoid clogging. PMO Manufacturing.

• *An injection of air* at a judicious location engenders fluidity when the stored material comprises powdery elements.

• *Vibrating* the bin or the silo or a part there of should be considered only after a thorough study since in some cases vibration is more harmful than favourable to the product's flow. When this solution is adopted, only the lowermost portion of the silo or the bin is vibrated when clogging is first observed through a slackening in flow. But if the vibration does not yield an immediate result and has to be continued beyond the time limit, compaction of the material is accelerated. This means it stays trapped inside the silo or the bin until intervention under hazardous conditions for breaking up the formed mass is arranged.

• *Incorporation of inflatable walls* made up of rubber cushions into which a volume of air is pumped has the effect of demolishing the formation of arches and thereby aids the flow of materials.

5.2.2 BIN OR SILO?

In the case of bins of a capacity above 40 m^3 and up to 200 m^3 per unit, the quantity of metal generally required is 140 to 150 kg per m^3 material stored, and this for bins fabricated from 5-, 6- and 8-mm thick sheets. These different thicknesses of sheet are distributed respectively between the upper vertical sections, the middle sections and the inclined sections.

Silos, with their circular cross-section, have the advantage of requiring less metal for fabrication.

As a matter of fact, for capacities of the order of 150 m^3, the weight of metal needed for 1 m^3 material stored is of the order of 65 to 75 kg. However, these silos are favoured less when it comes to locating them above those of the plants for treating materials.

As in the case of bins, reinforced concrete construction may be envisaged for silos too, but this option is sometimes discarded due to the impossibility of shifting it later or even a modification in the storage complex.

Lastly, mention may be made of a configuration which consists in doing away with the truncated cone portion, which is the costlier one, by extending the cylindrical portion down to the roof level of the collecting tunnel. Under these conditions the material itself forms the bottom of the receiver by flowing spontaneously towards the central hole.

The construction is decidedly less costly; the weight of metal per m^3 stored is of the order of 40 to 50 kg per m^3 for silos of a capacity of the order of 500 m^3, with due allowance for the dead stock at the lower portion.

5.3 Open Storage

We underscored in Sec. 3.2.1 the interest this solution holds in respect of primary stockpiles. The advantage is less evident in respect of stockpiles of finished products since it is imperative to avoid intermixing the various products on the one hand, and their contamination on the other, either by the dust pervading the quarry-head if dust-control measures are not scrupulously observed, or by the ground if it is not properly managed, or by vehicles. Therefore, this type of storage presumes meticulous care in the preparation of the storage areas.

To preclude pollution, the ground surface shall be scraped and a layer of intermediate materials laid and carefully compacted; the best solution, of course, is to lay a concrete or bituminous course, especially if the materials have to be reclaimed in full by loader.

When stockpiles of different particle sizes are made side by side, some overlapping between them cannot be helped and hence a zone in which a mixture of two adjacent particle size grades results. If reclamation is automatic through openings across a tunnel and done by extractors and loading conveyors, the overlapping zone of adjacent stockpiles should not be close to the reclamation openings; this results in some unusable dead stock (Fig. XIV.17). But if it is desired to reclaim the materials in full by loader, boundary walls should be raised, bearing in mind that one pile may be fully packed, with the adjoining zone altogether devoid of materials, which leads to considerable resultant forces.

Fig. XIV.17. Open storage

5.3.1 LOADING OF STORED FINISHED PRODUCTS

The type of loading adopted obviously depends on the nature of storage. When in bins or silos, generally on higher level ground to allow trucks or wagons situated lower to be loaded by gravity flow through an opening or trap-door, two options are available for loading: if substantial level differences prevail (due, say, to a favourable topography), the stocks are loaded by gravity through chutes starting from the screens situated above. If such differences in level are not available, recourse is had to conveyor belts.

5.3.2 BELT CONVEYORS

These permit stockage of the materials by discharging at a fixed point, whether it be open storage, or in bins or silos. But often, starting from a single loading point, a link with several receiving points needs to be ensured. Different solutions are available for the purpose, such as portable conveyors or travelling trippers.

• *Portable conveyors* are either oriented by pivoting on a vertical axis or placed over a roller-track. The latter solution is probably the simplest as it permits use of absolutely standard equipment, and with the advantage of being able to make the belt move in both directions, it is possible to cover a large surface. The portable conveyor pivoting on a vertical axis, an equipment very similar to the conventional belt truck loader often used in mobile crusher-screening plants, has an advantage over the former in requiring only a small infrastructure.

• *The travelling tripper* permits mobile discharge points, the conveyor itself remaining stationary. The principle of the travelling tripper is simple: it imprints an 'S' on the belt, resulting in discharge onto a chute and stacking over the entire length of the conveyor along which the tripper can move.

A par excellent achievement in southern France is complementary storage in bins and open storage with automated reclamation (Fig. XIV.18).

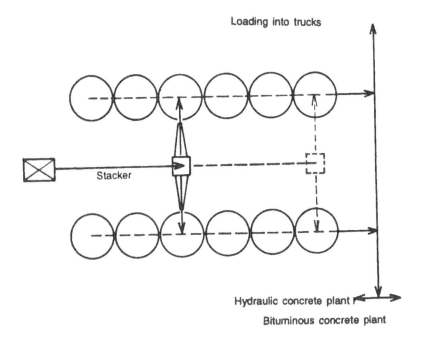

Loading into trucks

Stacker

Hydraulic concrete plant

Bituminous concrete plant

Fig. XIV.18. Storage installation with automated reclamations.

The main plant comprises 12 buffer bins whose stock status is monitored by an automatic monitoring outfit.

This challenge lies in managing the storage of the materials in step with their production upstream and the possibility of their offtake downstream.

The status is monitored by a set of ultrasonic sensors in the bins for the crushed material: inductive sensors for the positioning of the trap-doors and capacity sensors for the level of the material in some intermediate bins.

The objective is timely emptying of materials from each bin before it becomes full and transferring them to the open storage zone reserved for them, then managing the stock in open storage with regard to its possible withdrawal for loading into trucks to feed the various batching and mixing plants.

5.4 Segregation

5.4.1 DEFINITION

We consider segregation as covering all actions caused either by machinery or personnel which separate the elements from one another by gathering together the largest ones on the one hand, and the finest ones on the other, thereby making a particle size distribution which was originally homogeneous, heterogeneous.

5.4.2 THEORY OF SEGREGATION UNDER EFFECT OF TRANSPORT

• *Different types of transport:* A transport of aggregates wherein each of the elements is totally free to move relative to others is called 'loose transport'. For example, at the end of a stacker the elements move over a certain distance in the air before touching the surface of the stack, which obviously is a loose transport.

On the other hand, a transport of aggregates wherein the elements remain closely packed against each other is called 'compact transport'. The closer the packing, the lesser the segregation to be feared from such transport. But the moment the material is less compact, segregation can take place under the effect of vibration.

In a third type of transport, 'contiguous transport', elements of disparate particle size are placed on a conveyor belt. So long as they remain on this belt, there is no relative movement between the particles and thus this type of transport obviously cannot lead to segregation.

The effect of grading differs according to whether transport is of the loose or the compact type with vibration. Very weak in loose transport, the effect is truly considerable in compact. Segregation in this case tends to manifest less in a well-graded material than in a gap-graded material. This tendency is readily explained. A well-graded material is more compact so that the frictional forces are more powerful and better able to resist segregation.

If the material contains much sand, the coarse elements are, as it were, located in a medium at once fluid but with high viscosity. Under the influence of jolts in a compact transport subject to vibration, the coarse elements, as though the viscosity of the medium had been reduced, move under the influence of bulk forces (weight).

In a material which contains much gravel, the phenomenon is quite different. The coarse elements cohere with each other so well that they can barely move under the effect of vibration; yet substantial voids poorly filled with fines occur in-between and such fines can easily move into these voids under the influence of bulk forces. In the most common case, wherein the sole bulk force is weight, these fines move downwards and the lower portion of the material becomes richer in fine elements.

The foregoing account makes it clear that in compact transport with vibration, quite the most common type of transport, aggregates rich in gravels segregate by downward movement of the fines (supposing weight to be the only bulk force), the highly sandy aggregates segregate by downward movement of the coarse elements and well-graded aggregates segregate little.

The more workable a material, the less its angle of internal friction and hence the smaller the forces of friction which resist segregation. Therefore angularity will resist segregation.

Further, since angularity has a greater influence on workability in the case of sands than in coarse elements, it should be expected that the 0/D with angular sands will be less prone to segregation than materials containing rounded sands.

• *Role of water content:* Added in small quantities to an aggregate containing fine elements, water plays the role of a binder owing to the forces engendered by the interfacial tensions (water films creates bonding bridges between articles). Added in larger quantities, water plays the role of a lubricant, i.e., it reduces friction between grains.

In the first case the water creates forces which resist segregation, the fine elements adhering to the coarse aggregates, which has been aptly described in a culinary metaphor: 'each bit carries its sauce'.

In the second case, on the contrary, the decrease in friction between grains minimises the forces which resist segregation exactly in the same manner as vibration.

Inside a bin meant to store either 0/d material or d/D aggregate, segregation is produced: firstly, at a right angle to the chute, a cone is produced inside the bin and everyone knows that the cone (or dihedron) resulting from emptying is an important aspect of segregation (sliding of the aggregates along the sides of the cone or the dihedron is a form of loose transport). Accordingly, the coarse elements gather at the walls. When the trap at the lower portion of the bin is opened, the materials present in the centre are emptied first and only thereafter those along the walls. Therefore, emptying the bin does not totally suppress the segregation produced inside it.

5.4.3 HOW TO AVOID SEGREGATION

The problem of segregation being so monumental, we could only study certain aspects. As seen above, a close scrutiny of the classification of materials is one step towards its mitigation.

As a matter of fact, from the moment production takes place in the form 0/2–2/4–4/6.3–6.3/10–10/14–14/20 and from the moment it is mixed again, the problem simplifies greatly.

However, to reduce the risks of segregation when one should, for example, store a 0/31.5, the solution of a pivotal conveyor with variable inclination is very attractive.

Sensors stop the conveyor apprise the automat connected to the stacking equipment.

The equipment thus effects superposition of layers and the low height of drop limits segregation.

As the material requires adequate moisture content with its being placed under compaction, a water sprinkling on top will check dispersion of dusts and also segregation, and prepare the material for removal and transport to roadwork undertakings.

6. RECLAMATION OF STOCKS

6.1 Reclamation by Loaders

Reclamation by loaders is the most common solution in large work sites abroad and in mobile aggregate-producing plants.

This kind of reclamation for loading trucks poses no special problems.

6.2 Reclamation by Chutes

In the case of stationary plants with open storage, we have already seen that the simplest solution for reclamation of the materials is the tunnel with conveyor; here we have only to regularise withdrawal of the materials in storage.

When the question is one of batching these materials or extraction at regulated rates, a belt feeder is the only solution and we shall discuss this in Sec. 7 (Mixing).

On the contrary, when simply reclamation at a constant rate is desired, the intermediate means of chutes suffices.

6.3 Extractors and Collecting Belts

The collecting belt will pass below different feeders or different openings.

Its lateral lift may be at 40° or even at 45°, especially since no system for guiding the materials need be provided, considering that there is a succession of dropping points.

In some cases the collecting belt has a provision for sample collection by means of a device which does not require stopping the belt to effect sampling of the materials on it over a length, say, of one metre. Thus sampling is truly representative of the bulk loading in progress.

7. MIXING

7.1 Purpose and Objective

To realise the supplies demanded by clients, it is often necessary to mix in well-defined proportions certain of the aggregate sizes produced. It might seem curious to classify products [into different size fractions—Technical Editor] only to mix them again later, but the intrinsic curve of a product of even 0/D denomination might vary depending on the end use. Thus a 0/6 sand meant for base courses can comprise on average 50% or 72% of 0/2 sand according to whether it will form part of a gravel-slag or gravel-cement mixtures. Mixing makes it possible to stock the supplies meant for all the layers of pavements from a limited number of well-defined aggregate sizes, for example, for highway applications (fines for mixing or addition, 0/2, 2/4, 4/6, 6/10, 10/14, 14/20, 20/31.5). So much so, these supplies can be made with no modification in the screening line or change in setting in the crushing line. It may be recalled that sand meant for bituminous concrete is delivered in 0/2 or 0/4 mm size depending on the stipulated uniformity in the desired application: sand meant for base courses or foundations is delivered in 0/6 mm designation.

Further, a mixing line enables compensation for a possible excess of one aggregate size since control of the blended material's granulometric curve, which should be situated within a control zone, allows a certain modification in the percentage of one aggregate size while remaining within the specifications. For example, it is possible to change a 2/4 fraction by 40 to 60% in a 2/6 fraction while conforming to the specifications generally prescribed for a 2/6 fraction (percentage of misclassed fractions of 2 and 6 mm sieves less than 10% and percentage passing 4 mm between 30 and 60%).

It may be added that the mixing line is further justified when loads have to be programmed and automated.

As a matter of fact, belt feeders have a regulated flow rate. When mixing has to be done, a number of them are started up at different speeds, but when a single aggregate size has to be taken, only one feeder is operated at maximum speed and its working duration determines the quantity of materials extracted, conveyed and loaded. (Fig. XIII.19).

Fig. XIV.19. Feeder in a tunnel. Guadeloupe sand pits. NORDBERG apparatus. Automation AES.

7.2 Machinery Used

Mixing of aggregate is accomplished by the use of feeders which deliver upon a collector. Each feeder, whose working is controlled, fulfils several functions:

— Extraction from the stock of a certain volume of materials per unit time, this volume differing according to the percentage of the constituents desired.

— Uniformity in the weight of the materials delivered in a given time. When volumetric control becomes vague or imprecise (variable density of the aggregate), weight control is resorted to. In fact, a feeder should invariably have a self-regulating device for its working. Furthermore, the feeders should be equipped with devices which stop the mixing when uniformity in delivery is no longer assured, for example the silo feeding them has been emptied.

A distinction can be made then between feeders which rely on different procedures to ensure these two functions. Extraction depends on the 'flowability' or fluidity of the materials, which itself is a function of the water content and the particle size of the product for batching.

7.2.1 BELT FEEDERS

These feeders, also called 'rubber-lined volumetric extractors', comprise elements which constitute:

— a conveyor with two short plow arms;

— a stay-bolted chassis supporting the hoppers and the drive assembly (motor, reducer, tachometric dynamo);

— a screw system ensuring tension of the belt through a reversing drum with hubs located on top;

— a smooth belt conveyor supported by idlers;

— a drive machanism.

Belt feeders can be installed in two different ways:

The hopper discharge bears directly upon the feeder: The ratio H_o/L is less than $\tan \phi$, where ϕ is the angle of internal friction of the aggregates, and the feeder works in 'shear' (see Fig. XIV.20). Incorporation of a shutter may be necessary in some cases to enable repairs below large-capacity silos. The maximum extraction speed of these feeders reaches 33 m/min.

When the ratio H_o/L is less than 1, the materials situated upstream of the feeder flow first (Fig. XIV.20 and XIV.21).

The hopper discharge bears upon the rear wall of the hopper (Fig. XIV.21). In this case the feeder works in 'flow'. The ratio H_o/L is greater than $\tan \phi$.

The maximum rate of the feeder is determined by the natural flow of the product through the chute's opening; the feeder's disconnection is found facilitated. The ratio H_o/L determined earlier is greater than 1.

In fact the hopper is provided with a spout ABC in such a fashion that the point of contact of the aggregates upon the feeder is upstream of the vertical of the shutter's lower edge (L is then exceedingly small, almost zero).

The speed V of these feeders should not be so high as to not let the height H of the materials vary. It is often less than 20 m/min.

The materials preferably flow along a path of slip. As a general rule, a study of the materials flow should be conducted so as to prevent clogging of the silo and to decide on the location of the vibrators.

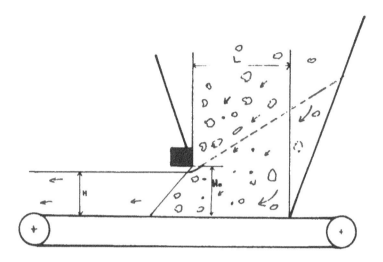

Fig. XIV.20. Installation of a feeder working in shear (H_o barely differs from H).
H_o = height of trap; H = height of strip of materials equal to H_o − D; D = size of largest element;
L = length of hopper exit opening

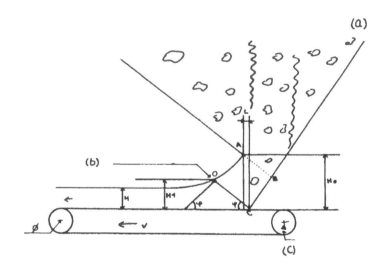

Fig. XIV.21. Installation of a feeder working in flow
(a) Path of slip (b) Compulsory passage point (c) Possible articulation point

Calculation of feed rate: In the absence of slip between the driving drum, the belt and the aggregates, the feed rate of a volumetric belt feeder is calculated by means of the following formulae:

— Rate in m³/h

$$Q = V \cdot l \cdot H \cdot 60$$

where V = speed of the belt in m/min; l = width in m of strip of materials between the guiding edges; H = height in m of the strip of material.

— Rate in t/h

$$D = Q \cdot \rho$$

where D = rate in t/h; Q (see above); ρ = density of materials generally around 1.7 t/m³ in quarry.

Variation in rate: The feeder's maximal rate is reached at the instant of preliminary adjustment by varying the opening of a gate. This opening should be equal to at least 5 times the size of the largest grain of the materials to be fed, so as to check the drawing-out effect which tends to reduce the thickness of the strip of materials. Consequently, different rates are obtained by changing the belt's travel speed.

To this end, use is made of motors whose rotation speed can be varied.

• *DC motors* have been the most appropriate choice since the origin of electrotechniques because at constant excitation, their speed varies with the voltage applied to the armature.

A set of thyristors is used even with a rectifier bridge to vary this voltage, the thyristors being controlled to let only a fraction pass: adjustments of alternations.

Excitation of the motor, independent of the main current, is effected by coil windings or permanent magnets.

Various procedures (auxiliary commutation poles) enable minimisation of the tendency towards instability in the speed arising from the variation in the flux created by the armature reaction.

To work at constant torque until the nominal speed is reached, the automatic excitation function is controlled through a second thyristor bridge: for the motor, with separate excitation (see diagrams and curves).

• *AC motors* can now be used at variable speed since they are no longer associated with the slip due to the grid's frequency.

In effect, the slip in question is raised by increasing the rotor resistance in the stable operating zone. Further, there are two types of converters which make it possible to vary the speed of asynchronous motors: the phase angle or wave train dimmers and especially the frequency converters made up of a rectifier stage and a thyratron inverter stage.

To improve the starting torque, strong squirrel-cage motors with dimmers and motors with double squirrel-cage or with deep slots and torque converters are used.

Which then to use for mixing-plant feeders? The speed variation for DC motors covers a 100-fold range and even more while that for AC motors is limited to 10, or say 15-fold. Therefore, *a priori*, only DC motors are suitable for granulometric mixing in a quarry, except when introducing less than 10% of a component in a mixture is inadmissible, which fact should be made known to the user.

For the same power, a DC motor is the costlier, especially in the top-grade brands which permit working at a very low speed for extended periods and without external ventilation. They are amenable to use in a dusty or highly humid atmosphere. The price of a dimmer is equivalent to that of a DC variator. That of a converter is higher.

Once set, belt feeders should operate at constant speed. For this, every motor should have a closed loop regulator in addition to the speed variation control.

There is a voltage corresponding to the speed selected and it should be displayed. There is then a measured voltage corresponding to the actual speed, this voltage coming from a tachnometric dynamo. The regulator serves to maintain a balance between these two voltages. For this, a speed amplifier receives and matches the data and supplies a voltage which is transformed in the switching device for switching on the thyristors. This voltage increases with the difference in speed. The result of this is a longer response time for switching of the thyristors and a stronger current to the motor, which accelerates its speed.

All these occur several times per second and the response time is very brief, resulting in a perfect agreement between the actual speed and the selected value which is displayed.

Display on the status of mixing is effected by a potentiometer connected as a voltage divider for the reference voltage. The display can be analogue or digital.

When the flow of the material is irregular, which happens with products rich in fines or variable in moisture content, incorporation of weight control is imperative. The speed of extraction in then controlled by checking the weight passed on the extracting belt or on an additional belt. A correction is warranted for taking the moisture content into account.

Weight checking is effected through a weighing platform of rigid construction supporting the idlers. This platform rests upon a set of levers which transmit the force to an extensometer gauge or a sensor. This sensor is provided with a potential field detector fed by the output voltage of a tachnometric generator whose voltage is proportional to the belt's speed.

Hence the output signal of the load cell is decidedly proportional to P (weight) × V (speed) and hence the feed. It is worthwhile noting that the traditional fulcrums with bearings are replaced by deflection blocks or steel lamellae.

In quite a few cases it is the feeder proper which constitutes the weighing platform. In this case the feeder should be of natural flows, which partly separates the function of extraction from that of weighing. To separate these two functions completely, the weighing element which works at constant speed should be preceded by a feeder whose flow it will control.

Generally, the load should be recorded nearest from the feeding point of a feeder to avoid the effects of pumping. To minimise these effects, the weighing element is often connected facing the feeding point.

Any weighing outfit obviously comprises a device for automatic taring and a calibration device.

Belt feeders are very widely employed and their flow rate can reach 3000 t/h.

7.2.2 ELECTROMAGNETIC VIBRATORY FEEDERS

These feeders are made up of a trough whose inner surface is provided with a wear-resistant lining. This trough is actuated by means of an electromagnetic device (Fig. XIV.22 and XIV.23).

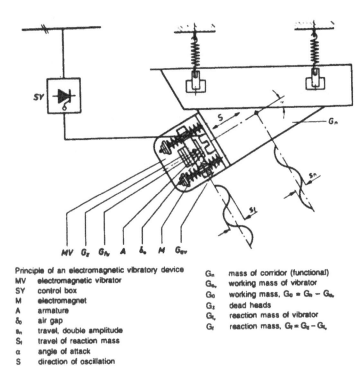

Principle of an electromagnetic vibratory device

MV	electromagnetic vibrator	G_n	mass of corridor (functional)
SY	control box	G_{s_v}	working mass of vibrator
M	electromagnet	G_o	working mass, $G_o = G_n - G_{s_v}$
A	armature	G_z	dead heads
δ_o	air gap	G_{t_v}	reaction mass of vibrator
s_n	travel, double amplitude	G_t	reaction mass, $G_t = G_z - G_{t_v}$
S_t	travel of reaction mass		
α	angle of attack		
S	direction of oscillation		

Fig. XIV.22. Electromagnetic vibratory feeder. Principle of an electromagnetic vibratory device

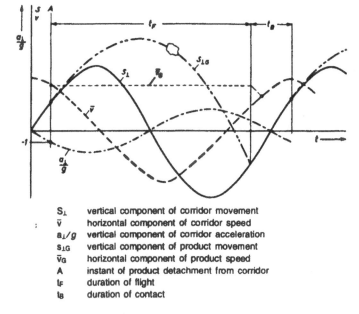

$S_\perp$	vertical component of corridor movement
$\bar{v}$	horizontal component of corridor speed
$a_\perp/g$	vertical component of corridor acceleration
$s_{\perp G}$	vertical component of product movement
$\bar{v}_G$	horizontal component of product speed
A	instant of product detachment from corridor
t_F	duration of flight
t_B	duration of contact

Fig. XIV.23. Path of product grains in a vibratory conveyor

The exciters are vibratory systems with two masses connected to each other by springs; one mass, constituted by the vibrator assembly including the protective cap, is the working mass. The second, called the reaction mass is inside the vibrator.

c = spring constant
G_a = working mass
G_f = reaction mass

G_f G_a

The assembly possesses a frequency of its own, $F_e = 5\dfrac{c}{G_r}$, where G_r is the resultant mass equal to

$$\frac{G_a \times G_f}{G_a + G_f}.$$

The useful amplitude is given by the force of attraction P, the working mass G_a, the vibrator assembly's own frequency F_e and the frequency of excitation F_a.

$$S_n = \frac{P \times F_e^2}{G_a (F_e^2 - F_a^2) \times F_a^2}.$$

The frequency of the electromagnet is proportional to the square of the magnetic flux, hence approximately to the square of the current or the voltage.

The particles inside the trough are subjected to an acceleration; they cease to hold together when this acceleration is greater than g (acceleration due to weight) and follow a ballistic path for a time t_F, fall back to the bottom and remain there for a duration t_B before the cycle starts all over again.

It is the speed of particles advance which conditions the flow rate of the equipment; it depends on the horizontal component of the corridor speed during the period of contact but never attains the maximal value since the separation (of the particles) overtakes it.

Figure XIV.24 shows that the speed of the particles movement depends on the vibration amplitude.

The conveying speed of the products is adjusted by changing the vibration amplitude.

Calculation of feed rate: The feed rate of an electromagnetic feeder is calculated like that of a belt feeder working in flow.

Volumetric feed rate $Q\ (\text{m}^3/\text{h}) = V \times 60 \times l \times H$, where V = speed of the product flow. The diagrams supplied by the feeder manufacturers give the values of this speed with respect to the vibration amplitude in m/min; l = breadth of the strip of materials in m; H = height of the strip of materials. This height is determined as in the case of belt feeders working in flow; H_o/L should be between 0.9 and 1.1. This determines a rule to be observed while installing electromagnetic vibratory feeders.

These feeders can handle up to 2500 t/h. Their use is widespread but one should nevertheless mark that the feed rates achieved depend on the moisture content of the materials, especially sands.

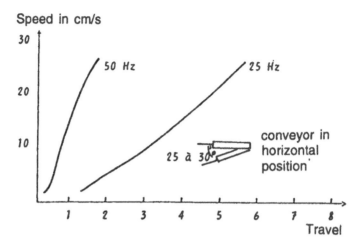

Fig. XIV.24. Curve of response speed V versus travel given by an equipment vibrating at 3000 vibrations/min and 1500 vibrations/min and for a 200-mm layer of materials

7.2.3 VISUAL DISPLAY OF MIXING

Once visual display of the mixing is done, the mixing operation proper can only start when the sum of the displayed voltages is equal to the reference voltage, which is called the 100%.

The feeders should only be started in a particular sequence, such that at any instant of feeding the material is correctly mixed. Any temporary stopping of the movement of the feeders is done simultaneously for all the feeders, as also a simultaneous restart. The automatic stopping at the end of the cycle is done in the same feeder sequence as at the time of starting.

After a prior calibration, depending on the product grading and the overall output of the mixing line, it is possible to display the desired quantities using a minuend electronic counter. There are counters which can display the quantities of each feed instalment and do the totalling.

Sensors have been installed on feeders to check the thickness of the strip of materials. Mixing lines are now installed in quarries which help lead trucks and wagons at rates upwards of 1000 t/h.

7.2.4 CONTROL POST OF A MIXING LINE

Generally, the control post of a mixing line is situated facing the feed point. The technician in charge of the dispatch, after having posted the desired mix proportion, can also control the manoeuvring of the transport vehicles. Often a synoptic table schematises the circuit of the materials and indicates the level of filling of the bins. Signals help visualise possible breakdowns. Further, the controls for starting up the whole installation are quite often integrated in the cabin of the mix control post.

7.3 Application of Mixing

Provided measures are taken against segregation, mixing permits the aggregate manufacturing enterprise to supply materials of measured 0/D or d/D to batching plants producing gravels stabilised with cement, bitumen or lime, which are sometimes situated straightaway at the end of the collector-conveyor.

The direct linking of a batching plant to the stocks of finished products in a quarry site offers several advantages:

— It avoids disruption in loading operations usually occasioned by successive loadings and discharges, the costs of which are by no means negligible.

— It confers upon the plant a certain degree of freedom in determining the relative percentages of the components in order to remain within the zones of specification and to minimise possible deficiencies in the particles size quality.

— It permits the plant to programme its manufacturing operations better and to have a wider range of finished products.

All the same, it is advisable to follow a proper policy of manufacture.

Special attention should be paid to the homogeneity in particle size of the different components.

The plant manager should then implement a quality assurance plan (matching of quality to the order from the phase of exploitation at the working faces to the processing phase).

It is also advisable to safeguard against the risks implicit in disruptions in stocks in the course of supplies contracted for large job sites. In this regard, it may be recalled that it is customary to stock 60% of the supplies before work is begun at such sites.

8. WEIGHING

8.1 Weighing Function

Today all quarry plants have weighing outfits appropriate to the modes of transport of the aggregates produced. The supplier should be able to provide a certificate on the weight dispatched and the relevant invoice within a short interval, whether it be transport by truck, train or boat.

The weighing technique adopted cannot disregard the means of transport employed.

Can one obtain the weight dispatched by difference in weights? Should the materials be loaded after weighing? Could the procedure adopted be other than a continuous process of loading? One should also take into account the elements related to the criticality of the job sites or, more simply, the constraints of the industrial environment.

Weighing is generally done at the point of loading. Nevertheless, if there are two or more loading points, weighing can be done at a common point on the routes of the trucks.

In all cases weighing of the materials supplied by the tonne is counterchecked on officially verified scales.

We shall return to these different points after defining the metrological characteristics peculiar to different weighing equipment.

8.2 Characterisation of Weighing Equipment

French standard specifications NFX 07 002 and NFX 07 001 define the technical terms relating to weighing equipment and the terminology of metrology. It is pertinent to refer to them here.

Weighing instruments are categorised according to three criteria:

- Nature of their working: —automated (a)
                               —non-automated (b)

- Class of display: —graduated (c)
                               —non-graduated

- Manner of obtaining equilibrium: —automated
                                       —semi-automated (d)
                                       —non-automated

(a) Calls for no manual intervention and starts an automatic process characteristic of the instrument.

(b) Requires intervention of an operator, especially for bringing or discharging the load upon the receiver as also for determining the result. Weigh-bridges come under this category.

(c) Has a scale graduated in mass units which enables direct reading of the results.

(d) The operator intervenes only beyond a certain range, called automatic indication, for restoring the possibility of automatic equilibrium whereas he should do so at all events if the equilibrium is non-automatic.

A weighing device generally comprises the following parts:

- Load receiver—part meant for receiving load.
- Load transmitter—transmitting the force from the receiver to the measuring system.
- Load measurer—part serving to measure the load by balancing with a measurable force the force transmitted and showing the mass corresponding to the value of the balancing force.

The different devices serve to receive the regulatory weights, balance the force resulting from the load, indicate its value, do the setting at zero at the time of balancing with nil load, counterbalance the tare (by adding or subtracting), block the whole or part of the mechanism and check only one or more of the principal devices.

The technical nomenclature (levers, scale-beams, knife-edges) is described in the French standard specification NFX 07 002 under the heading 'Elements of Weighing Instruments'.

All weighing instruments should have:

— A device for setting at zero when there is no load on the load-receiving part. Then general rules pertaining to the action or the manipulation of the devices used for setting at zero are applicable depending on the class of scales.

— A taring device which helps to bring the instrument display to zero when a load is placed upon the load receiver. The device used can be 'tare additive or subtractive', subject of course to rules concerning the sensitivity of the taring devices and their working range.

The quality of a weighing instrument is determined by features such as sensitivity, mobility, correctness, reliability, precision..., to which two more may be added:

— The weighing capacity is defined by:

• Upper limit; maximum weighing capacity with due allowance to the additive value of tare.

• Lower limit: value of load below which the weighings can be faulted by a very high relative error.

• Weighing range: interval between the upper and lower limits.

• Limit load: maximal static load which can be supported by the instrument without alteration in its metrological qualities.

— The scale, actual or conventional, of which one should know:

• The value: expressed in units of mass.

• The number: quotient of the upper limit divided by the graduation value:

$$n = \frac{Max}{d}$$

• The length: relative linear movement of the indicating part and the scale corresponding to the graduation value, this movement being measured on the base of the scale.

Based on the foregoing, it is possible to define the classes of precision of a weighing instrument.

Table XIV.2, relates to the usual instruments in public works and also gives the check value 'e' of the graduation.

All weighing equipment are then subjected to an inspection. This inspection comprises three broad phases:

Table XIV.2: Check value 'e' of the graduation

Precision	Upper limit, 'Max'	Graduation, d	Number of actual graduations, n	Check graduation, 'e'	Lower limit of lower range 'Min'
	Instruments graduated with automatic balancing				
	10 kg < < 100 t	10 g < < 10 kg	1000 < < 10,000	d	50 d
	15 t < < 100 t	20 kg < < 100 kg	750 < < 1000	Max/1000	1000 kg
III Average	Instruments graduated with automatic or semi-automatic balancing				
	25 kg < < 100 t	50 g < < 10 kg	500 < < 10,000	d	50 d
	15 t < < 1000 t	20 kg < < 100 kg	750 < < 10,000	d	1000 kg
	Instruments graduated with non-automatic balancing				
	4 t <	20 kg <	200 < < 400	Max/400	10 d
	8 T <	20 kg <	400 < < 1000	d	10 d
III Ordinary	Instruments graduated with automatic or semi-automatic balancing				
	4 t < 10 t	20 kg <	200 < < 1000	d	10 d

— *First inspection*, carried out on an equipment which is new, repaired or modified immediately upon its being fitted.

— *Periodic inspection*, aimed at checking stability over time depending on the usage of the weighing instrument.

— *Annual inspection*, depending on the local official regulation, carried out directly by the Metrology service (bureau of measuring instruments).

8.3 Static Weighing

This weighing, which permits the measurement after an equilibrium is established, quite specifically applies to weigh-bridges and weighing hoppers according as the pay-load is obtained by subtraction after the weighing of empty trucks, or directly by the weighing of a loaded hopper, which is then emptied into the truck after the former has been weighed following taring.

8.3.1 WEIGH-BRIDGE

Installed in a quarry or mobile in temporary job sites, weigh-bridges are characterised by their length and adaptation of their deck to the kind of vehicles to be weighed.

The infrastructure of a bridge and its cross section are depicted in Fig. XIV.25.

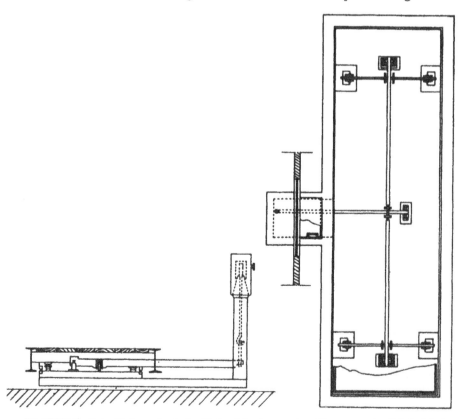

Fig. XIV.25. System with levers for deck with 4 points of support and its cross-section (left & right)

The weighing mechanism is:

— *Mechanical* with automatic or non-automatic gravimetric balancing. In this case the levers in action should be free from all torsion to guarantee exact weighing. Generally the deck rests on four or eight points of support depending on the bridge length.

— *Electronic.* A signal proportional to the load is given out by the load gauges. In this case the proving body is either a bar, a ring, a cylinder in compression, or a beam in bending. As for the sensors, the following list shows the technologies utilised by various load-sensor manufacturers:

- Metallic strain gauges with wires or pellicular screen called extensometer gauges
- Piezoresistive semi-conductor strain gauges
- Piezoelectric quartz crystal pressure gauge
- Sensors based on variation of mutual induction or differential transformer and mobile core
- Potentiometer sensors (SFIM)
- Vibrating wire sensors (Télémac)

Mounting of a load sensor on a moving support point in a pendulatory mount is depicted in Fig. XIV.26.

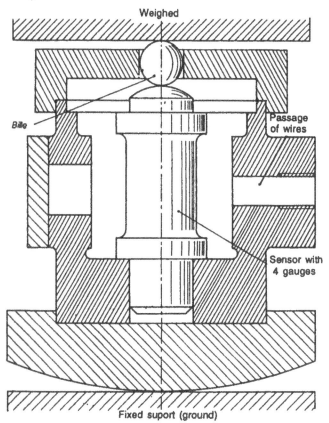

Fig. XIV.26. Mounting of a load sensor on a moving support point in a pendulatory mount
(Doc. Phillips)

In all cases an indicating head permits display of the weight and depending on the plant's options, arrangements for stamping the date and hour. Printers and monitoring computer can always be incorporated.

8.3.2 WEIGHING HOPPER

This is mounted on an electronic weighing system. By measuring its weight before and after loading of the vehicle positioned under the hopper, the weight of the load delivered is obtained; hence the weighing hopper effects weighing at the moment of loading. It should be filled before the transport vehicle comes into position. Often loading points have twin weighing hoppers, one of which gets filled while the other weighs and vice versa. This system makes it possible to carry out weighing almost continuously and at a high rate. Thus it is used in particular in large-scale dispatch (for example by rakes).

Depending on the case, the hoppers are mounted on tensile gauges (or sensors) of compression or bending gauges.

It may be mentioned that while using twin weighing hoppers mounted in parallel, use of a logic automatism is necessary to manage the flow of materials (Fig. XIV.27).

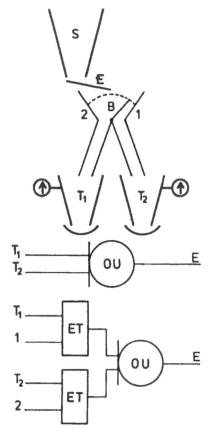

Fig. XIV.27. Loading scheme of a system with twin weighing hoppers

The extractor E under the silo S should feed two weighing hoppers T_1 and T_2; the first condition for normal working of the installation is that E should extract only when T_1 or T_2 indicates a zero weight. Hence E should be actuated by an element which makes sure that T_1 or T_2 is empty; this element is a 'logic module' called door 'OR'.

The second condition for normal working is to avoid feeding an already loaded hopper; therefore the bypass B should serve as a closing device and E able to extract only:

— if T_1 shows a zero weight AND if B is in position 1

OR

— if T_2 shows a zero weight AND if B is in position 2.

8.4 Dynamic Weighing

Dynamic weighing concerns determination of weight of a moving load.

Dynamic weighing can apply to determination of loads transported by trucks or wagons moving at 15 km/h.

The weigh-bridge used in this case is equipped with shock absorbers for filtering the oscillations of the deck and a special device for filtering the analogue signals of weighing.

As a general rule, dynamic weighing applies most particularly to weighing on the belt by means of an integrating scale, usually called the 'roller scale'.

8.4.1 PRINCIPLE

Weighing of a load transported upon the belt is effected by means of a weighing table comprising three parts:

— A rigid frame resting against which are the rollers meant to transmit the load of the belt to a load cell which will deliver:

  • An analogue signal in the case of an electronic table,
  • A mechanical lever movement in the case of integrating rollers (these techniques are almost out of vogue).

— A device which will also integrate the weight signal and the speed signal from the belt by a tachometer.

— A device for the display of the result from weighing which could be like those used in weigh-bridges.

The total mass transported is given by integration

$$Q = \int p \, v \, d \, t,$$

where p is the unit mass of the product which is variable upon a unit length vdt (Fig. XIV.28); v is the speed of the belt feeder.

8.4.2 PRECAUTIONS TO BE EXERCISED WHILE INSTALLING A WEIGHING TABLE

Unlike in a weigh-bridge, the data delivered by a weighing table can be disturbed by quite a number of factors and the list we have given is not exhaustive. These disturbances relate to:

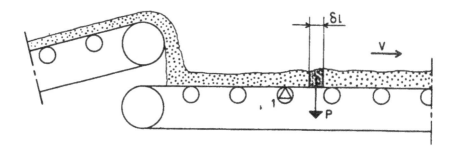

Fig. XIV.28. Roller scale.

• *Conveyor belt link* and load sensor. Variations in load can alter the tension of the belt whereupon it behaves as a data filter with changing characteristics.

If the band's rigidity is too high, its effect on weighing cannot be ignored.

The working under no-load and underload which might have to do with the contact between the belt and the weighing roller, should not disturb the taring operations. Deformation in the support frame or the shell might entrain a disturbance in the weighing.

The belt should be fairly flexible to ensure a permanent contact with the three rollers of the weighing point in the no-load state and should not be curved.

• *Characteristics of the conveyor* whose length should not disturb the taring which otherwise should come about automatically over a certain number of times.

The conveyor should not be subjected to vibrations which may be transmitted to the load cell and disturb the measurement. The angles of the conveyor's troughs, if very high, can accentuate the defect of alignment.

• *Measurement of the speed term* for the integral which gives the weight.

It is essential that no slipping occur between the techometer and the conveyor belt when loaded. This may happen if the rotation speed of the driving drum is the reference value.

Nor should there be any slipping between the load transported and the belt. This might happen if the slope of the conveyor is high and if the coefficient of friction of the materials is disturbed by variations in mixture content.

Accumulation of dust on the conveyor could give rise to a defect in measurement of the speed.

• *Measurement of the weight term* in the integral which gives the weight.

Integrating scales are highly accurate over a range of measurements between 20 and 100% of the conveyor's capacity. It is desirable that loading of the conveyor belt be as uniform as possible to minimise variation in thickness of the strip; the hoppers can be equipped with doors for adjusting this height (Fig. XIV.29).

• *Installation of weighing roller(s)*: The weighing roller should be so placed that the vibrations engendered by loading of the belt are not felt on the balance. Also, it should be located away from the driving drum (in principles more than 4 m) to avoid problems of variation in belt tension.

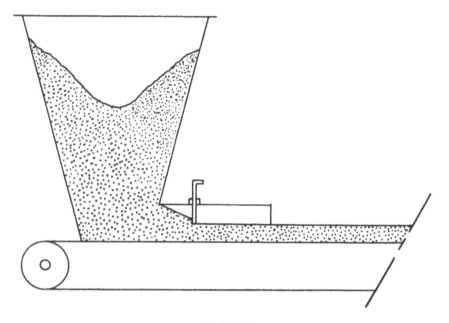

Fig. XIV.29.

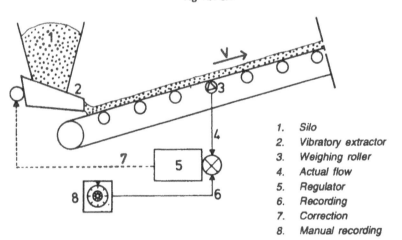

1.	Silo
2.	Vibratory extractor
3.	Weighing roller
4.	Actual flow
5.	Regulator
6.	Recording
7.	Correction
8.	Manual recording

Fig. XIV.30. Belt-mounted weighing outfit

8.4.3 APPLICATION OF DYNAMIC WEIGHING

Dynamic weighing can be resorted to as part of a process control scheme.

Sometimes it is profitable to feed a crusher at a relatively constant rate, for example, to improve its output. This is achieved by connecting a flow regulator to the belt mounted scale (Fig. XIV.30).

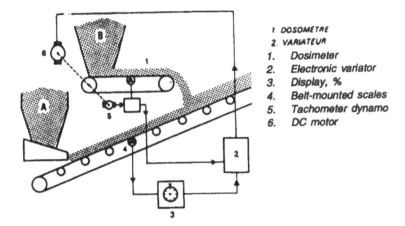

1 DOSOMETRE
2. VARIATEUR

1. Dosimeter
2. Electronic variator
3. Display, %
4. Belt-mounted scales
5. Tachometer dynamo
6. DC motor

Fig. XIV.31. Coupling of two feeders

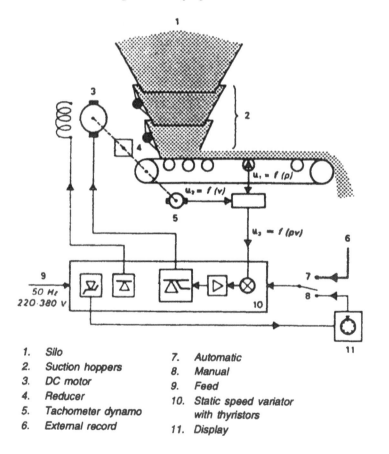

1. Silo
2. Suction hoppers
3. DC motor
4. Reducer
5. Tachometer dynamo
6. External record

7. Automatic
8. Manual
9. Feed
10. Static speed variator with thyristors
11. Display

Fig. XIV.32. Working principle of dosimeter

The regulator compares the actual flow indicated by the scale and the record of flow displayed on a potentiometer. The moment the two signals are not equal, the regulator transmits the necessary commands for correction to the conveyor belt's feeder (for instance, a vibratory extractor) so as to restore the equality. To prevent pumping, adjustable delays take into account the response time for extraction and the time for the material's travel from the extractor to the scale (Fig. XIV.31).

The simple and economical method makes it possible to obtain the *average flow rate* whereas an *instantaneous flow rate* still remains subject to minor variations.

The conveyor and the weighing element can be directly made part of a batching scheme. This assembly has been termed the dosimeter.

The dosimeter (Fig. XIV.32) is a continuous batching apparatus capable of delivering an *instantaneous flow* by weight which is constant and adjustable with precision within wide limits. It comprises a variable-speed belt extractor and an integrating scale. It is positioned directly under the silos, whatever their size may be.

The weigh-belt is driven by a DC motor fed by an electronic speed variator with controlled diodes.

The actual-flow signal weight × speed formed by the integrating scales is compared with the recording of flow displayed on a potentiometer or coming from an automatic general regulation.

In case of deviation due mainly to momentary variations in the density of the product to be batched, the resultant voltage is applied to the speed variator, which will instantly restore the equilibrium by acting on the speed of the DC motor.

The dosimeters can be equipped with weigh-belts of 500 to 2000 mm width for covering all the flow rates from 100 to 2000 t/h.

A combination of several dosimeters and weighing table permits mixing multiple aggregate sizes, as shown in Fig. XIV.33.

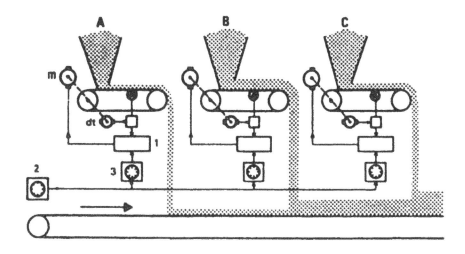

1—Variator; 2—Display or total; 3—Individual display.

Fig. XIV.33. Combination of several dosimeters

It appears that the tolerances on batching are of the order of 0.5% for securing the initial agreement. within the present scheme of use, this precision can reach 1%.

9. LOADING AND DISPATCH

9.1 Truck Loading

Upon entering the plant, the truck proceeds to a weigh-bridge. The driver receives a coded card which serves as his identity card for the duration of his stay inside the plant. He inserts it into a reading machine, which sets off weighing of the truck and enters the value in memory with the reference of the card.

The truck then goes to the loading point where another card-reading machine helps to enter the co-ordinates of the operation (product code, client's identity, supplier, transporter etc.) in memory.

While leaving the plant, the truck again passes over the weigh-bridge, the coded card sets off the weighing and recall of the data earlier entered in memory. A printer produces a delivery note: date, hour, co-ordinates of the operation, identity, weight at entry, weight at exit, net delivered or lifted.

If the total weight of the vehicle marginally exceeds the stipulated values in the Highway Code, the operation will be stopped and the delivery note will not be issued.

The vehicle therefore cannot leave the site without proceeding for a reduction in load.

9.2 Wagon Loading

Transport of aggregates by railway has long been practised in tip wagons of 20 to 40 t capacity, the unloading of which had to be done by crane. The crane's work had to be completed by cleaning the wagons, which perforce had to be done manually and constituted a major limitation, hardly compatible with the quest for productivity.

With a view to facilitating loading and unloading operations—and thereby cutting costs—railways have made available to users specially designed wagons, called hopper wagons, whose unloading can be done entirely by gravity, by means of trap-doors locked at the bottom of the wagons. This arrangement makes it possible to almost totally do away with manual cleaning of the wagons, unless they contain fine and moist sands, in which case a manual intervention becomes necessary for carrying out thorough emptying.

Further, the capacity of the wagons employed for transporting aggregates has been increased to 60 t, for a dead weight of 20 t. This has enabled reducing the number of wagons in a rake and shortening its total length.

The capacity of these wagons can go up to 70 t for the same dead weight, the load per each of the 4 axles of the 2 bogies being brought to 22 tonnes.

To improve the turnover of this rolling stock—which accounts for substantial investments—loading and unloading installations should be such as to accomplish these operations in the minimum time.

For convenience in loading the rakes, the aggregates plant should be provided with a special facility for their reception and an appropriate loading arrangement. The length of the tracks should be duly matched to the length of the rakes received into the siding; it should be adequate not only for accommodating the rake, but more so enable to wagons to position under a loading point. For instance, to be able to load a 1200 t rake whose length is about 300 m under a single loading point, 600 m of track have to be provided.

To line up the wagons under the loading point, power trolleys are customarily used. But if the track has no appreciable slope, recourse can also be had to a winch remote controlled from the loading point.

The loading point should also be a weighing point, which can be accomplished by:

— installing a weigh-bridge or a weighing hopper at the loading point,

— or even by fitting a conveyor scale or weighing roller on the loading belt conveyor (Fig. XIV.34).

Fig. XIV.34. Rake loading with distribution belt GOURAUDIERE quarry, automation AES)

It should be noted that regardless of the outfit used for loading the wagons, it should be so deployed that the load for each axle be less than the maximum permitted, which means that it is subject to the same constraint as applies in the case of loading trucks: the weight loaded should be optimal:

— neither too low, for reasons of profitability,

— nor too high, for reasons of safety.

The operations should be quite quick at all times, with a view to not holding up the rake of wagons. However, a few documents do have to be completed and present-day modern installations greatly expedite this matter, similar to the loading to trucks (several million tonnes per annum).

9.3 Loading Boats, Barges etc.

In the case of boats too one meets with problems already mentioned, say amongst others, the need for quick loading since here again craft hold-up coasts dearly, perfect distribution of the load along both main planes of the boat to ensure proper balance on the water and, of course, the need to mechanise and control loading.

Weighing the receiver of the load is out of the question here, as is having buffer hoppers; so the most appropriate solution is to weigh the materials upon the belt.

The process is invariably completed by a check based on the mobile scales. Conveyor scales permit perfect automation of loading onto the boats. The total weight of the loading to be made being displayed, the system will come to a stop or slow down as it approaches this displayed value.

9.4 Computerisation, Appropriate Solution

In the foregoing description of the operations of loading trucks and wagons, computerisation has been implicit.

Its use becomes inescapable in plants of a certain size and use of computers for management and accounting is as justified as using them for control of the operations.

A closer scrutiny of each of the cases presented above reveals that computerisation could expedite many tasks over and above operational ones; truck identification could be done on magnetic badges, client code and tallying his orders could be automated, drawing up delivery notes and preparation of invoices greatly simplified, not to mention stock inventory and constant update of a client's account.

Computerisation in the domain of quarries and sand-pits is dealt with in greater detail in chapters addressed to the engineering and design of modern large-scale plants, certainly needed on some job sites in some countries today.

10. TRANSPORT OF AGGREGATES

Transport of aggregates is effected by:
— road,
— railway,
— waterways.
In France, the relative share of these modes of transport hovers around:
— road 85%,
— railway 5%,
— waterways 10%.
If the fact that transports by railway and waterways have more often than not to be completed by transhipment by road is taken into account, the inevitable

conclusion is that the roadway mode almost singularly handles the transport of aggregates. Indeed it constitutes the only possible mode of transport for a great many quarry sites, which are connected neither to a railway nor a waterway. Moreover, it is the only means of regional surface transport of aggregates to highly scattered users.

10.1 Road Transport

Originally looked after by the simple tip-lorry, transport of aggregates is now done almost exclusively by the tractor-trailer unit with tipping arrangement. Indeed, by virtue of its higher capacity and lower cost price vis-à-vis all other vehicles, this means of transport has become a forced choice.

However, in some special cases where accessibility for delivery is limited, trucks, which occupy less space than a tractor-trailer, have necessarily to be used.

It is well known that transporters have to abide by French and European regulations as laid down in the Highway Code and the Social Regulations. Hence it would be appropriate to recall from them some of the essential provisions relating to the tractor-trailer unit: (France)

— total permissible weight of the unit when loaded not to exceed 44 t,

— load per axle not to exceed 13 t,

— length not to exceed 15 m,

— vehicular speed not to exceed 80 km/h on highways and 50 km/h in built-up areas.

The daily run and rest periods, very precisely regulated in the EEC, should conform to the following norms.

— Continuous run shall not exceed 4 h 30 min and shall be followed by a 45-min break.

— Total run shall not exceed 8 h per day, 48 h per week and 92 h in 14 days.

A 5-hour maintenance programme, executed in two spells, shall be observed once a week.

These provisions, conceived on the grounds of social considerations on the one hand, and for assuring better safety in road transport on the other, are progressively being implemented. Along with the Highway Code, they constitute a rigorous regime, undoubtedly opposed to the earlier 'laissez faire', which has become obligatory for all transporters.

Since the speed and total laden weight of vehicles have been limited, manufacturers have perforce striven to minimise the dead weight of their vehicles in order to increase the pay-load and thereby improve productivity. Plastic may be used for certain elements, such as cabins, and aluminium for others, such as buckets, and a dead weight of 11 tonnes for a total load of 44 tonnes thereby achieved provided an eddy-current braking system is fitted.

In practice, road transports can profitably be resorted to for a distance of 150 km. But then France is particularly rich in aggregate deposits and in most regions the users can draw supplies from quarries and ballast pits situated within 150 km, which explains the predominance of road transport over rail. It may be noted that beyond 150 km, transport by road—if undertaken by a public transporter—is subjected to the mandatory pricing envisaged by the unified provisions governing transports, in which case the transport cost is determined with due regard to:

— nature of the commodity,

— weight transported,

— ratable distance,

— category of the section over which transported.

The last of the foregoing, concerned with difficulty in transport over a given section, is determined by the category of destination locality among the four (A, B, C, D) under which all the localities of France have been classified. Upon the tariff price assessed, the transporter can obtain a maximum discount of up to 23%.

Road transporters typify the advantage of great flexibility since one can very easily adjust the number of vehicles to the tempo of work and the distance envisaged, as also a large delivery capacity.

10.2 Rail Transport

Transport by railway is generally resorted to for supplying regions inadequately bestowed with local resources of aggregates, and which have to draw from relatively far off sources, for example the Paris basin.

But railway is also used for transport over relatively short distances (for instance 100 km) of huge quantities of aggregates, whose dispatch by road would pose problems of safety and congestion of the roads and built-up areas traversed. This could be the case with large job sites, for instance construction of highways.

Executed once upon a time in tip-wagons of 20 t payload, railway transport has come a long way in the course of the last ten years.

In transporting heavy bulk materials, the railways found that the best way to remain competitive was to dispatch by full rake from the originating station to the final destination. This mode of transport permits use of traction engines in the most rational manner and achieves fast turnover of the rolling stock. It has become routine for a rake leaving a quarry site at the end of a day to arrive at its destination the following day, having traversed 400 or 600 km, indeed sometimes more.

The useful tonnage of the rakes is determined by the railway with due regard to the difficulties of transit. Generally it is between 1000 and 1500 t but can exceptionally touch 2500 or 3000 t. The transport cost mainly depends on the ratable distance between the starting and destination points, but also depends on the tonnage of the rakes moving in a given section, the heaviest rakes yielding the lowest costs.

The railway allows discounts on base cost in the case of regular dispatches, executed for instance at the rate of one rake per day for one month in a given section. The structure of the cost prices by rail comprises high fixed expenses and expenses proportional to the tonne-kilometre which, on the contrary, are low. This means that below a distance of 100 km, even 150 km, a full rake cannot effectively compete with a truck, barring exceptions. On the contrary, the longer the distance, the more cost effective rail transport becomes.

10.3 Transport by Waterways

Waterway transport can be profitable when the quarry sites on the one hand, and the utilisation centres on the other, are situated close to a waterway amenable to traffic of heavy-tonnage craft. This is notably the case of the Seine, Rhine and Saône in France.

We shall mention in particular the case of the Paris region which is catered to by waterways to the extent of about 10 million tonnes per annum. The means employed range from the 1000 t self-propelled craft to 6000 tonnes pushed convoys. The latter consist of two to eight barges moved by a 1700 hp pusher. Their length is 185 m, width 11.4 m and bay 3 to 3.5 m.

This modern craft enables waterway transport to compete with road transport for distances beyond 100 km. Of course, this means of transport must be backed up by sufficiently powerful loading and unloading installations to reduce the duration of operations to one day at the most.

Waterway transport can be validly utilised only on a properly maintained waterway and for distances of at least 100 km. Navigation on the Seine is permitted by night but some transport companies generally associated with quarrying companies prefer limiting the traffic to clear daylight hours. (Figs. XIV.35).

Fig. XIV.35. Storage of materials before loading the barges

11. UNLOADING

11.1 Unloading from Trucks

Unloading accounts for 85 to 90% of the movement of aggregates, starting from the centres of production to those of utilisation.

There is scarcely any general rule since the trucks will bring the aggregates either to areas of fill, or those of spreading with compaction or to the receiving zone of the batching plants for preparation of gravels stabilized with bitumen or cement.

The flexibility of the truck will provide an answer to all the problems posed and the only precaution to be taken is prevention of segregation.

11.2 Unloading from Trains

11.2.1 Rakes carry aggregates either to treatment plants that are directly linked to the railway, or to stations from where the materials are redispatched by road to the job site.

In both cases the equipment most often used for unloading is a belt conveyor mounted on the truck and specially designed for this purpose. The conveyor is fixed on the truck through an articulated frame, which enables change of its position with respect to that of the truck is terms of direction as much as inclination. The movements required for jockeying into the desired position are effected by hydraulic jacks connected to a hydraulic station (Fig. XIV.36).

Fig. XIV.36. Conveyor for hopper wagon unloading. CW3 equipment

This equipment, whose use is now well established, enables unloading rakes at an industrial rate of 300 t/h. Of course, it should be backed by a number of road transport vehicles adequate for maintaining this rate downstream.

11.2.2 When the user receives large quantities of aggregates at a fixed location, he might need to organise a virtual station for train reception. The most commonly adopted procedure is to make the wagons line up above a pit where they unload by gravity. A belt conveyor collects the aggregates from under the pit and transports them to the stockpiling installation, where a place is earmarked for each aggregate. Each pile is constituted as the material falls from the belt conveyor; this makes it imperative to stockpile only single sized aggregates if segregation is to be obviated.

Another method consists in installing, parallel to the incoming railway line, a belt conveyor above which the unloading traps of the wagons open successively. As in the preceding case, this belt transports the aggregates towards the stockpiling installation proper.

The fixed receiving station entails large investments but nonetheless constitutes the most economical means of unloading and stockpiling wherever huge quantities of aggregates are received (several hundred thousand tonnes per annum).

11.3 Unloading from Boats, Barges etc.

Trials have been conducted and schemes elaborated. Use of bucket wheels has been mooted but the almost general rule is to unload by grab bucket with manual supervision. In seaports the boats which bring the sands and fine gravels dredged or sucked in from the open sea are unloaded by making them into a slurry.

APPENDIX

The design of a conveyor takes into account the factors mentioned in the following tables.

1. Maximum Slope for Conveyors

Material	Bulk density (in t/m³)	Maximum slope	
		Angle (in deg.)	Slope (in %)
Crushed slate	1.2 to 1.5	18 to 20	32 to 36
Dry clay	1.2 to 1.6	25	47
Wet clay	1.4 to 1.8	18	32
Basalt	1.6 to 1.8	20	36
Bauxite	1.3 to 2.2	20	36
Soft limestone	1.2 to 1.4	18 to 23	32 to 42
Hard limestone	1.6 to 1.8	18 to 23	32 to 42
Crushed shells	1.6 to 1.8	20	36
Diorite	1.5	20	36
Dolomite	1.5 to 2.7	20	36
Rounded pebbles	1.5 to 2	13 to 15	23 to 27
Granite	1.5 to 1.8	20	36
All-in gravel	1.5 to 2	15 to 20	27 to 36
Screened gravel	1.5 to 2	15	27
Rounded gravel	1.5 to 2	13 to 15	23 to 27
Crushed gypsum	1.3 to 1.5	18 to 23	32 to 42
Calcined gypsum	1	15	27
Crushed limestone	1.4 to 1.6	18 to 20	32 to 36
Pulverised limestone	1.2 to 1.3	22	40
Quartz, quartzite	1.6 to 1.8	20	36
All-in rocks	Variable	18 to 22	32 to 40
Crushed rocks	Variable	16 to 20	29 to 36
Screened rocks	Variable	16 to 18	29 to 32
Dry sand	1.5 to 1.9	15	27
Wet sand	1.7 to 2.1	20	36
Stripping sand	1.5 to 2	16	29
Moulding sand	1.5 to 2	24	45
Schists	1.4 to 1.6	18	32

2. Minimum Width of Belts

Width of belts, in mm	Dimensions of blocks (in mm)		Graded big pieces
	Big pieces with 90% fines	Big pieces with 50% fines	
400	110	80	60
500	160	110	90
650	220	150	120
800	300	200	160
1 000	400	270	220
1 200	500	340	270
1 400	600	400	320
1 600	700	450	380
1 800	800	540	450
2 000	900	600	500

3. Possible Flow Rates

Theoretical flows by volume D_{vl} (in m^3/h) for:
- horizontal transport
- dynamic collapse angle of the material 15°
- speed of belt 1 m/s

Belt width, in mm	Profile of section								
	Plane	V-shaped 15°	V-shaped 20°	Trough-shaped with 3 equal rollers 20°	Trough-shaped with 3 equal rollers 25°	Trough-shaped with 3 equal rollers 30°	Trough-shaped with 3 equal rollers 35°	Trough-shaped with 3 equal rollers 45°
300	11							
350	17							
400	23	43	48	42	46	50	52	57
500	38	72	80	72	79	85	89	97
600	58	108	120	108	121	129	135	147
650	69	129	143	131	144	154	162	176
700	81			156	170	182	191	208
800	108			207	228	245	258	279
900	139			270	295	316	333	360
1000	173			337	369	397	418	452
1200	255			498	545	586	615	665
1400	351			686	755	810	857	920
1600	465			908	997	1072	1125	1217
1800	592			1160	1274	1370	1430	1554
2000	735			1445	1585	1703	1790	1933

4. Schematic Diagram of Conveyor and Conventional Signs

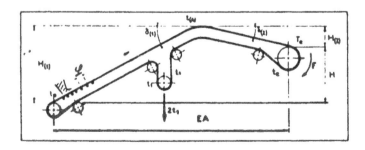

Fig. XIV.37

EA   Horizontal distance of transport, m

EA'   Total length of transport along profile, m

H   Total level differential (positive or negative), m

I   Width of belt, m (or mm)

v   Speed of belt, m/s

D_1 Maximum possible flow rate (theoretical), t/h

D   Actual flow rate, t/h

Q   Linear mass of mobile parts (without load), kg

b   Linear mass of belt, kg

c, c_c Linear masses of load, kg

C   Coefficient of length

f   Coefficient of restraint due to tradesmen

k   Coefficient of drive tension

P_v Power consumed with no load, kW

P_c Power consumed by horizontal movement of load, kW

P_d Power consumed by change in level of load (positive or negative), kW

P_a Additional power consumed by localised resistance, kW

P   Total power consumed, kW

P_m Installed power, kW

F   Total tangential force, N*

t_o Tension at point of loading if belt is hauled; at point of dropping if belt is self-driven

*N: symbol of force in SI units: the Newton 1 N = 0.102 kgp of the system MKps.

5. Calculation of Powers Required

• *Power consumed*

— Movement of empty belt:

$$P_v = \frac{9.8\,C}{1000}\ f \cdot Q \cdot v \cdot EA \quad \text{in kW} \qquad \text{... (1)}$$

— Horizontal movement of load:

$$P_c = \frac{2.73\,C}{1000}\ f \cdot D \cdot EA \quad \text{in kW} \qquad \text{... (2)}$$

— Vertical movement of load:

$$P_d = \frac{2.73}{1000}\ D \cdot H \quad \text{in kW} \qquad \text{... (3)}$$

If the conveyor is: — ascending, H is positive,
                       — descending, H is negative.

• Total power consumed

1st case: Horizontal conveyor

$$(H - 0) \; P = P_v + P_c \quad \text{in kW} \qquad \qquad \text{... (4)}$$

2nd case: Ascending conveyor (H positive)

$$P = P_v + P_c + P_d \quad \text{in kW} \qquad \qquad \text{... (4a)}$$

3rd case: Descending conveyor (H negative)

P_d being negative, there is, *a priori*, uncertainty in the value of P owing to the imprecision regarding the real value of f, of which it is only known that it is between f and f/2.

Accordingly the limiting values of P are calculated with the help of the following two formulae:

$$P = P_v + P_c + P_d \quad \text{in kW} \qquad \qquad \text{... (4a)}$$

$$P = \frac{P_v + P_c}{2} + P_d \qquad \qquad \text{... (4a')}$$

and the highest absolute value of P calculated as per (4a) and (4a') in kW is adopted.

Power at no load P_{vi} (in kW) for a speed of 1 m/s
(with coefficient of braking due to the non-localised frictions: f = 0.022)

Width, in mm	EA Horizontal distance of transport (in m)									
	10	20	30	40	50	60	70	80	90	100
400	0.3	0.4	0.5	0.6	0.6					
500	0.4	0.5	0.6	0.7	0.8	0.9	0.9	1	1.1	1.2
650	0.5	0.6	0.7	0.8	1	1.1	1.2	1.3	1.4	1.5
800	0.7	0.8	0.9	1.1	1.2	1.4	1.5	1.7	1.8	2
1 000	0.9	1.1	1.3	1.5	1.7	1.9	2.1	2.3	2.5	2.7
1 200	1.1	1.4	1.5	1.9	2.1	2.4	2.6	2.9	3.2	3.4
1 400	1.5	1.8	2.2	2.5	2.9	3.2	3.5	3.9	4.2	4.5
1 600	2	2.4	2.9	3.3	3.8	4.2	4.6	5.1	5.5	5.9
1 800	2.4	2.9	3.5	4	4.7	5.2	5.6	6.3	6.8	7.2
2 000	2.9	3.4	4.2	4.8	5.5	6.1	6.7	7.5	8.1	8.6

Width, in mm	EA Horizontal distance of transport (in m)									
	120	140	160	180	200	220	240	260	280	300
400										
500	1.4	1.6	1.7	1.9	2.1					
650	1.7	1.9	2.1	2.4	2.6	2.8	2.9	3.1	3.2	3.4
800	2.2	2.5	2.8	3.1	3.3	3.6	3.8	4	4.2	4.4
1 000	3.1	3.4	3.8	4.2	4.5	4.8	5.1	5.4	5.7	6
1 200	3.9	4.4	4.8	5.3	5.8	6.1	6.5	6.9	7.2	7.6
1 400	5.2	5.8	6.4	7.1	7.7	8.2	8.7	9.2	9.7	10.2
1 600	6.8	7.6	8.4	9.3	10.1	10.7	11.4	12	12.7	13.3
1 800	8.4	9.3	10.3	11.4	12.4	13.2	14	14.8	15.6	16.4
2 000	9.9	11.1	12.2	13.6	14.7	15.7	16.6	17.6	18.6	19.5

For a distance EA and a width of belt figuring in the Table, directly read the value P_{vl}.

For a distance EA and a width of belt not figuring in the Table, read the value corresponding to the immediately higher value of the distance or the width. To obtain P_v, multiply P_{vl} by the speed v (in m/s).

Horizontal transition power of load P_c (in kW)
(with coefficient of braking due to the non-localised frictions: $f = 0.022$)

Flow in t/h	EA Horizontal distance of transport (in m)									
	10	20	30	40	50	60	70	80	90	100
10	0.1	0.1	0.1	0.1	0.1	0.1	0.1	0.1	0.1	0.1
20	0.1	0.1	0.1	0.1	0.1	0.1	0.2	0.2	0.2	0.2
50	0.2	0.2	0.3	0.3	0.3	0.4	0.5	0.5	0.5	0.5
100	0.4	0.4	0.5	0.6	0.7	0.7	0.8	0.9	1	1.1
200	0.7	0.9	1	1.2	1.3	1.4	1.6	1.8	1.9	2.1
500	1.7	2.1	2.5	2.9	3.3	3.6	4.1	4.5	4.9	5.3
1 000	3.5	4.3	5	5.8	6.6	7.2	8.2	8.9	9.7	10.5
2 000	7	8.6	10	11.6	13.2	14.4	16.4	17.8	19.4	21
3 000	10.5	12.9	15	17.4	19.8	21.6	24.6	26.7	29.1	31.5

Flow in t/h	EA Horizontal distance of transport (in m)									
	120	140	160	180	200	220	240	260	280	300
10	0.1	0.1	0.2	0.2	0.2	0.2	0.2	0.2	0.2	0.2
20	0.2	0.3	0.3	0.3	0.4	0.4	0.4	0.4	0.5	0.5
50	0.6	0.7	0.7	0.8	0.9	1	1	1.1	1.1	1.2
100	1.2	1.4	1.5	1.6	1.8	1.9	2	2.2	2.3	2.4
200	2.4	2.7	3	3.3	3.5	3.8	4.1	4.3	4.5	4.7
500	6	6.7	7.4	8.2	8.9	9.5	10.1	10.7	11.3	11.8
1 000	12	13.5	14.9	16.3	17.7	19	20.3	21.5	22.6	23.6
2 000	24	27	29.8	32.6	35.4	38	40.6	43	45.2	47.2
3 000	36	40.5	44.7	48.9	53.1	57	60.9	64.5	67.8	70.8

Flow in t/h	H Change in level, positive or negative (in m)														
	1	2	5	10	15	20	25	30	40	50	60	80	100	200
10	0.1	0.1	0.2	0.3	0.4	0.6	0.7	0.8	1.1	1.4	1.6	2.1	2.7	5.5
20	0.1	0.1	0.3	0.6	0.8	1.1	1.4	1.6	2.2	2.7	3.3	4.4	5.5	10.9
30	0.1	0.2	0.4	0.8	1.2	1.6	2.1	2.4	3.3	4.1	4.9	6.6	8.2	16.4
50	0.2	0.3	0.7	1.4	2.1	2.7	3.4	4.1	5.5	6.8	8.2	10.9	13.7	27.3
80	0.2	0.4	1.1	2.2	3.3	4.4	5.5	6.6	8.7	10.9	13.1	17.5	21.8	43.7
100	0.3	0.6	1.4	2.7	4.1	5.5	6.8	8.2	10.9	13.7	16.4	21.8	27.3	54.6
125	0.3	0.7	1.7	3.4	5.1	6.8	8.6	10.3	13.7	17.1	20.5	27.3	34.2	68.3
160	0.4	0.9	2.2	4.4	6.6	8.7	10.9	13.1	17.5	21.8	26.2	35	43.7	87.4
200	0.6	1.1	2.7	5.5	8.2	10.9	13.7	16.4	21.8	27.3	32.8	43.7	54.6	110
250	0.7	1.4	3.4	6.8	10.3	13.7	17.1	20.5	27.3	34.2	41	54.6	68.3	137
300	0.8	1.6	4.1	8.2	12.3	16.4	20.5	24.6	32.8	41	49.2	65.6	81.9	164
400	1.1	2.2	5.5	10.9	16.4	21.8	27.3	32.8	43.7	54.6	65.5	87.4	109	218

Flow in t/h	H Change in level, positive or negative (in m)													
	1	2	5	10	15	20	25	30	40	50	60	80	100	200
500	1.4	2.7	6.8	13.7	20.5	27.3	34.2	41	54.6	68.3	81.9	109	137	273
600	1.6	3.3	8.2	16.4	24.6	32.8	41	49.1	65.6	81.9	98.3	131	164	328
800	2.2	4.4	10.9	21.8	32.8	43.7	54.6	65.6	87.4	109	131	175	218	437
1 000	2.7	5.5	13.7	27.3	41	54.6	68.3	81.9	109	137	164	218	273	546
1 250	3.4	6.8	17.1	34.2	51.2	68.3	85.6	103	137	171	205	273	342	683
1 600	4.4	8.7	21.8	43.7	65.5	87.3	109	131	175	218	262	350	437	874
2 000	5.5	10.9	27.3	54.6	81.9	109	137	164	218	273	328	437	546	1 092
2 500	6.8	13.7	34.2	68.3	103	137	171	205	273	342	410	546	683	1 365
3 000	8.2	16.4	41	81.9	123	164	205	246	328	410	492	656	819	1 638

• *Power for level change for load* P_d *(in kW)*

The foregoing Table shows that the powers required for change in level are quite high for large rates of flow and appreciable changes in level. In the case of intermediate values—most often the case—a double interpolation is needed for determining P_d with an acceptable precision.

In which case it is faster to do the calculation with formula (3):

$$P_d = \frac{2.73}{1000} \; D \cdot H$$

The order of the exact result value can then be checked in the Table and compared with those of the other terms: P_v, P_c (P_a) which are generally lower than itself.

REFERENCES*

AFNOR. 1982. Handling and Transport of Commodities.

AFNOR. 1986. Conveyor Belts.

AFNOR SOMIA. 1982. Encyclopaedia of Continuous Material Handling.

Arquié, G. 1970. Segregation. In: *Aggregates* (Chapter XIV). Compagnie française d'Edition.

Biaggi, P. 1958. *Rubber-lined Belt Conveyors*. Dunod.

Boudan, J.E. 1984 and 1985. Continuous weighing upon the belt. *Infovrac*, nos. 16. 17, 18 and 19; pp. 63–66, 47–51, 47–50, 49–51.

Boudan, J.E. 1986. Know how to correctly utilise belt-mounted weighers. *Infovrac*, no. 24, pp. 63–65.

Cartier, E. 1981. Industrial weighing. *Electronique Industrielle*, no. 10, pp. 76–81.

Documentation 'Belt conveyors! Method of Design. Kleber.

Dossier. 1978. Industrial batch weighing. *Industries et Techniques*, no. 362, pp. 107–125.

Haynes, D.O. 1965. *Materials-handling Equipment*. Compagnie française d'Edition.

Mauviel, P. 1984. Conveyor-mounted radiometric weigher. *Infovrac*, no. 14, pp. 42–44.

Reimber, M. and A. 1971. *Silos. Theory and Practice*. Eyrolles, Paris.

Ruban, M. 1987. *Quality Control in Road Construction. Textbook on Roads*. Chap. 2. Weighing. Presses de l'ENPC.

* All references in French—General Editor.

Siderurgical Standard. 1975. Specifications for construction of belt conveyors for iron-making works.

Tabary, M. 1970. *The Belt Conveyor*. Compagnie française d'Edition.

Torchet, B. 1985. Trials conducted on continuous weighing of granular materials conveyed on the belt, 21 pp. Report presented at the seminar on 'Inspector of bulk weighing installations', Paris.

Torchet, B. and D. Le Coz. 1982. Measuring the yield of dry aggregates: continuous measuring of wet yield by weight. *Bull. de liaison des Laboratoires des Ponts et Chaussées*, Special number XII, pp. 121–125.

Yernaux, M. 1961. *Stockpiling of Bulk Products in Quarries and Sand-pits. Work and Technics.* Editions Science et Industrie.

Yernaux, M. 1970. Stocking, reclamation and handling. In: *Aggregates* (Chapter XIII). Compagnie française d'Edition.

Yernaux, M. 1974. *Automation in Quarry and Sand-pit Installations*. Editions Science et Industrie.

XV

Engineering Problems in Aggregate Processing Plants

Andre Pantel and Serge Ravillard

1. INTRODUCTION

The term 'engineering' may be defined as follows: 'Overall study of an industrial project in all its aspects (technical, economic, financial, social), co-ordinating the special studies of several teams of specialists'.

In the case of an aggregate manufacturing plant, this means designing a set of crushing, grinding, screening, washing, dust-removing, handling, storing etc. equipment capable of manufacturing aggregates. It will not suffice for this purpose simply to co-ordinate the various suppliers of this equipment; it will also the necessary to interlink the operation of the equipment and in particular ensure homogeneity of the entire installation.

But above all, one should be fully knowledgeable about the end-uses of the aggregates produced.

Two highly different cases arise in this regard. The first, but not most frequent, is that of a quarry operator who wants to set up a new installation, either because he is opening up a new quarry (initial deposits exhausted) or because for various reasons he wishes to renovate an outdated installation from top to bottom. This case will rarely arise in a straightforward manner: very often, in fact, the quarryman will demand that a particular section of the installation, which he does not consider outdated (or the replacement of which appears to him *a priori* too expensive) be reused in the new plant. In such a case the authority in charge of the engineering works encounters some obligatory constraints and instead of designing an 'ideal' installation, is compelled to make compromises that sometimes are open to criticism.

A frequent variation of this case often arises in areas already well equipped with quarries; it is the situation of a quarryman who has an old installation, no longer suited to the consumption of the area served by him; the latter has changed, either increased or the aggregate sizes required are no longer the same or no longer have the same relative importance; most often, these various changes occur together. The quarryman has thus to adapt his production to the new requirements of his customers. He wishes, of course, to alter as little as possible the general framework of his installation and to take advantage of this arrangement for improving his productivity. In this case also the engineering office will encounter certain constraints.

An important risk must then be avoided, namely of a non-homogeneous installation, i.e., one having parts of highly different capacities. Certainly, quite often the primary processing section, for instance, does not have the same capacity as the secondary section. But if this difference becomes too great, inevitably some bottlenecks will arise. The temptation then is to introduce when advantageous for increasing production and eliminating a bottleneck, an additional device of relatively small capacity. The bottleneck will thereby be shifted, and in order to eliminate it again, another additional device introduced at another location of the installation. Gradually, an originally well-designed plant becomes a hodge-podge of bits and pieces whose purpose becomes muddled. Moreover, there is the risk of such being costly, as it is practically impossible to secure the necessary spare parts since the equipment is no longer standardised.

The second case is that of the opening of a new quarry meant to serve the construction site of a large work: new road dam, hydroelectric plant, nuclear station etc. This case arises particularly in developing countries and is the one wherein the role of the in-charge of the engineering works is the most comprehensive and clear because less constrained by the contingent requirements of the quarry operator (we are not taking into account here the cost and time limit constraints which are obviously always present). This is the more frequent case nowadays. It is the one we shall discuss in this chapter, without failing to recall the first case by showing that the advice given for the former is adaptable for the latter.

If the end-use of the aggregates is the first point to know, the second is obviously their source, i.e., the deposit. We shall return to this, merely emphasising here that it is absolutely imperative to fully know the deposit, a fundamental need that unfortunately is all too often neglected.

We shall not deal with an installation of loose materials, such as a deposit of massive rocks, nor with a quarry of hard rocks, such as relatively soft limestone; a polluted alluvial deposit will pose other problems unlike a bed exploited under water. The examples can be multiplied *ad infinitum* but that would serve no useful purpose. Rather, it appears to us essential to emphasise the *absolute necessity of knowing the deposit*. We shall go further: this knowledge should not be superficial; so many factors associated with the deposit are involved that it is imperative to *prospect it properly*. (see chapter III.)

2. EXAMPLES OF AN INSTALLATION FOR PROCESSING MASSIVE ROCKS

Rather than dealing in the abstract with a general case of an installation meant for a job site, we think it advisable to describe a concrete example.

2.1 Role of the Quarry Manager

Let us first look at the role of the quarry manager since it is he who is mostly involved upstream.

• *He should establish and make known his needs*

— *Quantitative*, that is:

    — Properly assess the actual densities, exceptional profiles, various losses;
    — Try to calculate the inflation of quantities;
    — Convert into tonnes all the types of cubic metres which do not have the same density.
    These tasks are not as easy as they look!

— *Qualitative*, namely

    — Mechanical strength of the finished products (abrasivity, resistance to impacts and to wear and polishing);
    — Their grading (including tolerances and separation sizes to be specified);
    — Their cleanliness;
    — Their shape;
    — Their angularity.
    Generally, the technical specifications themselves give this qualitative definition. But not always, and in any case not always judiciously. The manager should then make relevant choices. This is his job. In no case should he rely for that purpose on the engineering team per se.
    — *Momentary*, that is furnish the data which will enable determination of output of the station in finished products:
    — Working hours;
    — Probable efficiency (weather forecast, social habits, trouble-shooting service);
    — Attempt to assess production peaks.

• *He should indicate his resources*

Deposit selected:
— Thickness of exploitable deposit;
— Moisture content;
— Discontinuities of the massive rocks;
— Grading and cleanliness of gravels and sands, geological nature;
— Weathering;
— Susceptibility to weathering;
— Chemical analysis if need be.

Method of exploitation envisaged:
—Dry or underwater extraction;
— Size of largest blocks of massive rocks;
— Transport machinery for pit runs;

— Homogenisation if need be.

Discarding and grading out certain fractions:

This operation, so very important, concerns really the manager. It largely conditions the quality of the products delivered and their cost price.

• *He should furnish all other information permitting the manufacturer to adjust 'to measure' the future station:*

— Topography of the places;
> — Possible shifting of the station;
> — Possible extension of the station;
> — Storage, reclamation and evacuation of the finished products;
> — Water and electrical supply;
> — Disposal of waste waters.

Thus the manager's job in designing a plant is not easy. He has to submit an actual 'roster of specifications' to his supplier.

2.2 Specifications

The case under discussion pertains to an earth and rockfill dam in North Africa for which we give below the requirements, resources and philosophy of operation relating to aggregates for the job site.

2.2.1 REQUIREMENTS WORKED OUT AS FOLLOWS

1,160,000 t of unprocessed transition materials 0/D
+ 1,110,000 t of highly processed aggregates for concretes, filters and drains

i.e., 2,270,000 t to be prepared in about 40 months but at a uniform rate.

NB

— *Transition materials:*

Materials situated between the core of the structure and the upstream and downstream fills. The materials T3 and T8 mentioned in Table XV.1 refer to the position in the dam.

— *Filters:*

These preclude the risk of gradual erosion due to the fact that the shell of a dam is always traversed by riprap.

— *Drains:*

These play the role of an anticontamination layer which prevents the elements of the subsoil from rising into the downstream fill and polluting it.

2.2.2 RESOURCES CONFINED TO THREE DISTINCT DEPOSITS

Those specified by the contractor-in-charge. All require extraction with explosives.
   • **A limestone quarry** to be opened at a distance of 12 km from the job site. This limestone shall be of suitable density, strength and homogeneity.

   Apparent density: 2.73.

   Compressive strengths (Rc): 150 MPa (standard deviation σ 20 MPa).
   But the limestone contains marly inclusions, making prescreening to about 40 mm obligatory.
   The few samples subjected to chemical analysis indicate 8% silica.
   This material can be used for the manufacture of aggregates for concrete.
   • **A deposit of sandstone,** situated at less than 2 km from the future crushing station.
   This sandstone is not permissible in the manufacture of concretes but may be used for the manufacture of the upstream transition T3 and for filters and drains. Because of its proximity and the absence of earthy inclusions, and in spite of its obviously higher abrasivity, it appears more advantageous than the limestone. We notice, however some scattered and clearly low characteristics on absorption.

   Abrasivity LCPC: 1050 g/t
   Particle density γ: 2.26 to 2.53, average 2.42 Mg/t
   Moisture content (w): 0.29 to 0.88%, average 0.68%
   Seismic velocity: 3500 to 4200 m/s
   Rc on dry basis: 55 to 102 MPa, average 77 MPa (σ = 17 MPa)
   Rc saturated: 22 to 74 MPa, average 47 MPa

   In short, we take this material both for the transition T3 and for the filters + drains but with full knowledge that in the case of the latter it may have to be substituted with limestone.
   • **A clayey sandstone deposit** at a distance of 2 km from the future station.
   Its description covers in fact a heterogeneous sandstone consisting of real, well-shaped sandstone lenses embedded in a softer, clearly less abrasive and non-siliceous material.
   We presume that the clayey sandstone will yield considerable sand on crushing. Its characteristics, however, are still more scattered than those of first sandstone.
   Our first drillings show that the material is not very abrasive for a sandstone.

   Particle density γ: 2.26 to 2.66, average 2.54 Mg/t
   Moisture content (w): 0 to 4.2, average 1.93%
   Seismic velocity: 2200 to 3500 m/s
   Rc on dry basis: 8 to 81 MPa, average 42 MPa (σ = 26 MPa)
   Rc saturated: 3 to 85 MPa average 32 MPa

   Meant only for downstream transition T8, this clayey sandstone actually constitutes the most profitable and acceptable material for this purpose. No prescreening is required.
   None of the above-described three deposits poses a problem of thickness. Intuiting that the output of the installation will not be such as to necessitate a very

large primary crusher, we plan quarrying in such a way that the largest blocks make 800 mm.

Lastly, our advance research on natural sand for concretes proves infructuous.

2.3 Additional Data on Products to be Obtained

2.3.1 GRADING ENVELOPES

• The installation should be capable of complying without difficulty with the average curves of the grading envelopes in such a way that the variations with respect to these ideal gradings are the result of the actual adjustments (wear of the jaws) or of the feed, but not from the chronic incapacity of, for instance, sand-manufacturing equipment.

• In view of their simplicity, the *transition materials* T8 and T3 will be produced in a single aggregate size 0/80 without prescreening or systematic control on a mesh 80. But they should compulsorily pass through a gyratory crusher and contain the minimum coarse elements from blasting or from a primary jaw crusher.

Those for concretes, filters and drains are to be reconstituted from the elementary grades. We recommend the following separation sizes:

— for concretes: 2-5-16-32-63 mm
— for filters and drains: 2-16-63 mm

The fraction < 16 mm shall be screen washed.

• **Special case** of the 0/5 fraction

Let us recall the grading envelope recommended for concrete sand, as it appears perfectly judicious.

Passing	0.08 mm	:	0.3	
	0.16	:	5–10	
	0.315	:	18–30	Modulus of fineness: 2.90 ± 0.2
	0.63	:	32–50	
	1.25	:	50/70	
	2	:	67/84	
	2.5	:	75/90 .	
	5	:	92/100	

This sand being crushed, and consequently regular, we opt for a single separation size at 2 mm on a conventional screen. The 0/2 fraction will be passed through a cyclone and air dried.

The 0/2 entering into the composition of the drains will be dealt with in exactly the same manner.

2.3.2 PROGRAMME

An accurate table of requirements is given in Table XV.1.

2.3.3 OTHER INFORMATION

• At the price of not-too-heavy earthworks, it is possible to set up the installation on several levels and thereby take advantage of these variations for supply and stockpiling.

Table XV.1: Aggregate requirements (kilotonnes)

	0/2	2/5	5/16	16/32	32/63	About 63/80	Total	Nature Months of use
Ungraded aggregates								
Transition T8 (0/80)	(35%) 164	(10%) 47	(17%) 80	(14%) 66	(16%) 75	(8%) 38	470	Clayey sandstone Month 28 to 44 inclusive
Transition T3 (0/80)	(20%) 138	(8%) 55	(22%) 152	(15%) 103	(22%) 152	(13%) 90	690	Month 33 to 53 inclusive
Total ungraded aggregates	302	102	232	169	227	128	1160	
Graded aggregates								
Filter P9	(100%) 62						62	
Filter F10	(55%) 209	(25%) 95	(20%) 76				380	Hard sandstone 1600 kt — Month 48 to 53 inclusive
Drains D11 and D12		(30%) 140	(35%) 164	(20%) 94	(15%) 70		468	Month 27 to 53 inclusive
Various concretes	(39%) 78	(10%) 20	(13%) 26	(25%) 50	(13%) 26		200	Month 27 to 53 inclusive
Total graded aggregates	349	255	266	144	96		1110	Limestone inclusive: Month 13 to about 40
					Together		2270	Month 13 to 53 inclusive (41 months)

Note: Drains 11 and 12 have separate grading envelopes. However, as their average curves are very similar, we take into consideration a single production for all the drains.

The area that can be developed is sufficiently large to enable a ground installation, with prestock and stocks of finished materials.

• Stockpiling will be planned linearly, on tunnels for reclaiming noble aggregates from the secondary section in order to ensure, under proper conditions, mixing of the filtering and draining bed aggregates on the one hand, and prebatching of the aggregates for concrete making on the other.

• Construction authority shall confirm the obligation of rescreening and rinsing the aggregates for concrete except the 0/2, just before introducing them into the concrete mixing plant. As there is no concrete on a large scale, there is no need for cooling.

• Schedule and outputs. It has been possible to define fairly well the utilisation peaks indicated in the detailed programme. For this reason, we agree to provide systematically for double-shift crushing, up to a maximum of 325 hours of production per month.

— Day shift   : 10 h × 0.7 (efficiency):   7 h effective

— Night shift : 10 h × 0.6:   6 h effective

    ———————————————
    13 h/day × 25 day/month
    = 325 h/month

For the busiest months the rate of feed of the installation works out to:

$$\frac{125,000 \ t}{325 \ h \ max} = \text{about } 400 \ t/h \text{ minimum}$$

NB: We prefer a single station of 400 t/h processing alternately clayey sandstone, sandstone and limestone over two more modest stations specialised for processing different materials.

The effective output of materials for the transitions on the one hand, and of highly noble aggregates on the other, will be extremely modulated. For the busiest months, these outputs will be:

— 80 t/h for the manufacture of aggregates for concrete;

— 125 t/h for the manufacture of F9 and F10 filters and D11 and D12 drains;

— 140 t/h for the manufacture of F10 filters and D11 and D12 drains.

• *Miscellaneous:*

— The equipment shall be of the semi-fixed type on concrete bases.

— Feeding: normal feeding by dumpers of 40 t and if need be by trucks of 17 t.

  — Desirable stockpilings:

  — in silos: 0/2 for concrete, 300 t

  — in the open (uncovered stocks).

• Other aggregates for concrete making: 2/5, 5/16, 16/32, 32/63: 4000 t per aggregate size.

• Aggregates for filters and drains: 0/2, 2/16, 16/63: 4000 t per aggregate size.

• Transitions T8 and T3, maximum stocks taking into account the area.

NB: Buffer stocks are very important to enable adherence to the programme and to safeguard against drifts in production and consumption if any.

Water will be available in sufficient quantity, provided a part of it is recycled. The slurries will be stored in a thalweg, as a final deposit, by simply pumping the charged water.

Electricity will be supplied through two generating sets:
— 1 for primary crushing,
— 1 for secondary crushing.

3. PROBLEM POSITING

The above example was chosen to enable replies to any questions posed and to develop the main engineering problems specific at each stage of the investigation of a materials-manufacturing installation.

The complexity of the situation shows how important it is that the 'Specifications Roster' summarising the demand be clear and accurate.

As a matter of fact, we daily observe just how difficult it is to anticipate the performance of equipment working under well-defined conditions. We may as well confess the impossibility of knowing such when the conditions are unspecified.

In the absence of a specifications roster, we advise our users to at least reply to the questionnaire of the manufacturers (see Table XV.2). Such questionnaires are generally intentionally condensed to expedite completion.

4. STUDY OF MATERIALS-MANUFACTURING INSTALLATION

4.1 Preliminary Studies

After analysis of the client's roster of specifications giving details about the deposit reserves, the raw materials and the requirements, the engineering department of the manufacturer of crushing-screening equipment can initiate a study of the materials-processing instalation.

4.1.1 ANALYSIS OF RAW MATERIAL

In our example the materials to be processed are taken from three separate deposits. The study of the reserves carried out by the user shows that these deposits are quite sufficient for guaranteeing supply to the site:
— A limestone deposit consisting of marly inclusions selected for the production of aggregates for concrete.
— A sandstone deposit, abrasive, and not polluted selected for production of transition materials T3 and for filters and drains.
— A clayey sandstone deposit envisaged solely for the downstream transition T8.

While selecting the equipment and the processing plans, it is necessary to equally take into account the possibility of processing any one of these materials. This being so, it is advisable to study in particular the following problems:
— Stockpiling of various materials.

Table XV.2: Technical questionnaire

Client or job site : . . . .
Address : . . . .   Telephone : . . . .

1.   MATERIALS TO BE PROCESSED
      11— Extraction : in quarry   : . . . .
                         in borrow pit : . . . .
                         on dry basis : . . . . or in water : . . . .
      12— Nature : . . . .
      13— Density: Particle : . . . . Apparent : . . . .
      14— Hardness: Los Angeles: . . . . Deval : . . . . Micro-Deval : . . . .
      15— Abrasivity : Free silica : . . . ., %
                       or abrasivity factor . . . ., g/T
      16— Cleanliness : (% of soil, clay etc.). . . . .
      17— Impurities : (wood, rags, scrap, sulphates etc.) . . . .
      18— Moisture : . . . . , %
      19— Grading of pit-run material :
           Quarry: Larger blocks on three dimensions: . . . .
           Borrow pit: Curve with grading envelope
           (if possible, furnish laboratory analysis results or samples)

2.   OUTPUTS AND PRODUCTS TO BE OBTAINED:
      21— For a long-duration permanent installation:
           Actual daily production to be ensured : . . . ., T/day
           Daily work : . . . ., h/day. Weekly work : . . . . . ., days/week
      22— For temporary installation:
           Volume to be processed : . . . .
           Duration of work : Months : . . . . Days : . . . . Shifts : . . . . . .
                              Hours/shift : . . .
      23— Theoretical efficiency : . . . .
      24— Aggregate sizes : Number : . . . ., % respective : . . . .
                           Grading envelopes (to be attached)
           For riprap: Sizes or weight...
      25— Specify : Sand equivalent, flakiness coefficient etc. (if possible, furnish an extract
           of the specifications roster in which the characteristics are detailed)

3.   DETERMINATION OF INSTALLATION
      31— Fixed, semi-fixed, mobile, temporary, permanent
      32— Method of feeding : by trucks : . . . ., T
                             by loaders : bucket : . . . ., m^3
      33— Elimination of wastes : . . . . . . on which mesh : . . . . . .
      34— Open stockpile of precrushed materials.
           Volume that can be drawn : . . . . Number of extractors : . . . .
      35— Primary screening before or after open stockpiling : . . . .
      36— Possible pits : . . . . Max. depth : . . . .
      37— Method of stockpiling finished products:
           in the open : desired capacity for each product : . . . .
           in hoppers : desired capacity for each product : . . . .
      38— Mixing : . . . . . Total output, and product-wise : . . . .
      39— Washing : . . . . of which products : . . . .
      310— Scrubbing . . . . .

311— Dust removal. . . .

312— Defillerisation. . .

313— Existing equipment to be reused or modified? : . . . .
If yes, provide information and plans.

4. SITE

41— Temperature, max : . . . . Min : . . . .

42— Degree of humidity : . . . .

43— Maximum wind : . . . . . . , km/h

44— Altitude : . . . . , m

45— Terrain: Unless otherwise stated, the installation will be set up on a horizontal ground. If otherwise, furnish a topographic survey report.

For items 5 to 8, do not forget to clearly specify the limits of the supplies.

5. ELECTRICITY

51— Source of power : . . . .

52— Voltage : . . . . , Frequency : . . . .

53— Cabinet outside or within the electrical premises?

54— Lighting of site. . .

55— Wiring. . . .

56— Classes of engines : . . . .

57— Types of rheostats : . . . .

58— Assembly : supervision or complete. . . . .

59— Commissioning. . . . .

6. ASSEMBLY

61— Assembly: supervision or complete. . . .

62— Vulcanisation. . . .

63— Commissioning. . . .

7. DISPATCH

71— Packing, specify type of transport : . . . .

72— Dispatch from factories. . . . . . .

73— Other possibilities (FOB, C & F, CIF etc.) : . . . .

8. MISCELLANEOUS

81— Shelter buildings, roofs, hand transport of materials, hooding of screens, conveyors. . . . .

82— Travelling crane(s) or maintenance pulley block(s). . . . .

83— Finishing paint at site (in addition to our standard). . . . . .

84— Spare parts. . . . .

85— If motors are not to be supplied, specify types and shaft-end dimensions : . . . . . .

86— Armour-plating or special cladding in sheets.

9. OTHER INFORMATION

— Tendency to clog (clayey sandstone) and consequent effect on flow of materials.

— Hardness and abrasivity (sandstone) vis-à-vis selection of grinding equipment.

— Cleanliness of the pit run (limestone with marly inclusions) making a prescreening to 40 mm obligatory.

Other factors that have also to be taken into account:

— Tendency to cleavage, which favours production of flat materials, in order to determine the intake of secondary crushers.

— Rock strength, in order to define the power of equipment.

—'Pit-run' curves.

Comparing the information collected regarding the methods of quarrying and the nature of the beds processed in the case under study with data from similar cases, we conclude that the three deposits will produce 'pit-run' curves very close to those defined in Table XV.3.

Table XV.3: Pit-run

Deposit	Aggregates		0/2	2/5	5/16	16/32	32/63	63/100	100/800
Limestone	Concrete	%		15% elimination			5	10	70
Hard sandstone	T3 Filters and drains	%	5	5	5	4	7	7	67
Clayey sandstone	T8	%	19	6	11	7	10	9	38

We would like to emphasise at this stage the importance of a geological study, which in no case should be neglected. A proper knowledge of the deposit enables the user to do the following and concomitantly safeguards him against serious disappointments:

— Choose equipment and circuits most suitable to the materials, in accordance with the desired end-products.

— Define the conditions and modalities of exploitation (stripping, extraction, transport etc.).

— Ensure, thanks to sufficient reserves of the deposit, regular feeding of the installation during the entire period of exploitation.

4.1.2 DEFINITION OF NEEDS

The production rates of the various categories of materials were determined by the user in the example under study according to requirements during the busiest months, i.e.

— *Primary Section*

— Feeding = 400 t/h
— Case No. 1—'transition T3' manufacture 400 t/h

— Case No. 2—'transition T8' manufacture 400 t/h

— *Secondary Section*

— Case No. 3—manufacture of only concrete from the 13th to the 40th month = 80 t/h

— Case No. 4—manufacture of the filter 10 + drains 11 and 12 from the 23rd to the 32nd month = 140 t/h

— Case No. 5—manufacture of filters 9 and 10 + drains 11 and 12 from the 41st to the 52nd month = 125 t/h

The quantities and outputs are accurately determined by the user on the basis of a well-defined temporary market.

For a fixed installation or long-term exploitation, requirements cannot be so accurately assessed, however. In these cases the user has to carry out a very strict market study in order to define the qualities and quantities of aggregates to be obtained, which obviously cannot be done by the equipment manufacturer.

Furthermore, he has also to take into account the changes in the wishes of his customers over time, which may lead him to choose a flexible or progressive installation.

4.1.3 STUDY OF PRODUCTS TO BE OBTAINED

Most installations should comply with the well-defined specifications laid down by the construction authority such as grading, shape, angularity, cleanliness, strength etc.

The definitions of these qualities having already been dealt with in this book (Chapter IV), we shall study here only the grading aspect of the products to be obtained, which is demarcated by the following zones as defined by the construction authority (Table XV.4).

The conventional solution for including a product in a grading envelope consists in separating within the 0/D interval a certain number of aggregate sizes and mixing this product afterwards in compliance with the required percentages.

In the concerned case the sizes specified are:

— for concretes: 2, 5, 16, 32 and 63 mm,

— for filters and drains: 2, 16 and 63 mm.

The quantities, per category of materials, are calculated as a function of:

— the percentages defined by zones (A),

— the outputs laid down by the production programme (B),

— the various cases of production (simultaneous production of several products for instance) (C).

For values A, B and C: see Tables XV.5, XV.6, XV.7 and XV.8.

In order to obtain these various productions, it is necessary to envisage a mixing installation which uses, in addition to the traditional grinding-screening operations, huge storage and extractors under stockpiles.

This type of installation being generally very costly, it is advisable, insofar as possible, to obtain a product entering into a zone directly, without resorting to mixing.

Table XV.4: Grading envelopes

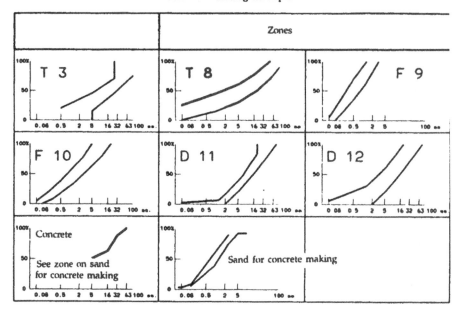

Table XV.5

(b) Aggregates S	0/5	5/16	16/32	32/63	(c) Totals JX	Note' A D
FABRICATION DE Production of 80 ton/hour'S aggregates for concrete making						
%	50	12.5	25	12.5	100	(A)
TH.	40	10	20	10	80	(B)

Table XV.6

Aggregates	0/80μ	80μ/2	2/5	Totals	Note
		Average curve of 0/5 sand			
%	2.5	77.5	20	100	(A)
TH.	1	31	8	40	(B)

Table XV.7

Aggregates		0/2	2/5	5/16	16/32	Totals	Note
			Production of 140 tons/h of filters F10, drains D11, D12				
F 10	%	60	18	22		100	(A)
	TH.	32	10	12		54	(B)
D 11 D 12	%	5	20	37	38	100	(A)
	TH.	4	17	32	33	86	(B)
Totals	%	26	19	31	24	100	(C)
	TH.	36	27	44	33	140	

Table XV.8

Aggregates		0/2	2/5	5/16	16/32	Totals	Note
			Production of 125 tons/h of filters F9, F10 and drains D11, D12				
F 9	%	100				100	(A)
	TH.	10				10	(B)
F 10	%	60	18	22		100	(A)
	TH.	20	6	7		33	(B)
D 11 D 12	%	5	20	37	38	100	(A)
	TH.	4	16	31	31	82	(B)
Totals	%	27	18	30	25	100	(C)
	TH.	34	22	38	31	125	

Table XV.9

		Transition materials T3 and T8 (cumulative curves)					
Aggregates		0/5	0/16	0/32	0/63	0/100	Note
T 3	%	15	36	50	65	100	(A)
	TH.	60	144	200	260	400	(B)
T 8	%	30	46	60	77	100	(A)
	TH.	120	184	240	308	400	(B)

In the case under consideration we are seeking a manufacturing process for transition materials T3 and T8, without prescreening or control, which will yield a final grading entering the required zones.

4.1.4 DETERMINATION OF PROCESSING EQUIPMENT

• *Selection of type of equipment*

Having analysed the raw material, accurately defined the needs and taken cognizance of the requirements regarding the quality of the products to be obtained, the engineer-in-charge of the design of the crushing installation should look into the choice of the processing equipment.

There are two methods at his disposal for this purpose:

— Either he proceeds by comparison, based on the experience of equipment functioning under similar conditions (analogical method);

— Or he carries out grinding tests on samples of materials.

In the latter case caution should be exercised for the following reasons:

— The sample taken from the quarry is not always representative;

— Tests are carried out on limited quantities, which does not permit detection of such phenomena as clogging, wear etc., which may manifest only after a certain period of use.

Interpretation of the results of these tests should be done carefully by an experienced technician who can use, if need be, the analogical method for supplementing or confirming his conclusions.

Given such information, it is easy to determine the type of equipment needed, referring as needed to the following Tables, which give the characteristics as well as the fields of application of the main crushers (Table XV.10) and grinders (Table XV.11).

For further information on the choice of crushers, see Chapter X.

In the example under development we have chosen the following, taking into account the hardness and abrasivity of the sandstone deposit:

— A double-toggle jaw crusher for processing the materials of the primary post,

Table XV.10 Characteristics and fields of application of various types of crushers

Type	Degree of hardness	Abrasivity	Degree of moisture content	Maximum reduction ratio	Main fields of application
With jaws	Semi-hard to very hard	Abrasive	Dry or slightly moist, non-sticky	8/1 to 10/1	Quarry materials, ores, blast furnace slags
Gyratory	Semi-hard to very hard	Abrasive	Dry or slightly moist, non-sticky	6/1 to 8/1	
With toothed rolls	Soft to semi-hard	Not very abrasive	Dry or moist, even very sticky	4/1 to 5/1	Limestones, iron ore, coke, clays and soils, various slags
With rigid hammers	Soft to semi-hard	Not very abrasive	Dry or moist, non-sticky	15/1 to 25/1	Limestones, gypsum, potassium salts, friable ores

Table XV.11 Characteristics and fields of application of various types of grinders

Type	Degree of hardness	Abrasivity	Degree of moisture content	Maximum reduction ratio	Main fields of application
Gyratory cones	Semi-hard to very hard	Abrasive or not	Dry or slightly moist, non-sticky	6/1 to 8/1	Quarry materials and various ores, carbides, abrasives
Plain rolls	Semi-hard to very hard	Abrasive or not	Dry or moist, even sticky	3/1 to 4/1	
Percussion grinder with horizontal axis	Friable to semi-hard	Slightly or not abrasive	Dry or slightly moist, non-sticky	20/1 to 30/1	Coal, limestone, plaster, clay, fertilisers, chemical products, wastes, all categories of agricultural products
Percussion grinder with vertical axis	Semi-hard to hard	Slightly or fairly abrasive	Dry or slightly moist, non-sticky	2/1 to 3/1	Production of quarry sand

— Cone grinders for processing the materials of the secondary and tertiary sections.

• Definition of size of size-reduction machinery

The types of machinery to be envisaged having been determined, the engineer should concentrate on the definition and size of the size-reduction machinery. For this purpose he should carry out a study of the aggregate sizes balance.

This study takes off from the grading of the 'pit run'. It enables determination at each stage of grinding of the quantities of materials to be preserved or to be processed for arriving ultimately at the production of the desired quantities of aggregates indicated in Tables XV.12 to XV.17.

A study of the granulometric balances permits, in addition to the definition of the size-reduction machinery, determination of the following equipment needs from the analysis of quantities and gradings of the products to be processed:

Table XV.12

Aggregates	0/2	2/5	5/16	16/32	32/63	63/100	+100	Totals
Manufacture T3 Rate: 400 t/h of 0/100 sandstone								
Pit run	5	5	5	4	7	7	67	100 %
	20	20	20	16	28	28	268	400 TH.
							−268	Feed VB 11.09
Jaw crusher VB 11.09 R: 200	2	2	2	4	10	13	67	100 %
	5	5	5	11	27	35	180	268 TH.
						−13	−180	Feed BCS: 4'1/4 STD
Cone grinder BCS 4'1/4 STD, extralarge R: 25	6	3	25	48	18			100 %
	12	6	48	93	34			193 TH.
Balance	9	8	19	30	22	12		100 %
	37	31	73	120	89	50		400 TH.

Table XV.13

Aggregates	0/2	2/5	5/16	16/32	32/63	63/100	+100	Totals
Manufacutre T8 Rate: 400 t/h of 0/63 clayey sandstone								
Pit run	19	6	11	7	10	9	38	100 %
	76	24	44	28	40	36	152	400 TH.
							−152	Feed VB 11.09
Jaw crusher VB 11.09 R: 120	4	2	4	7	17	36	30	100 %
	6	3	6	11	26	55	45	152 TH.
						−91	−45	Feed BCS: 4'1/4 STD
Cone grinder BCS 4'1/4, STD extralarge R: 25	6	3	25	48	18			100 %
	8	4	34	65	25			136 TH.
Balance	23	8	21	26	22			100 %
	90	31	84	104	91			400 TH.

Table XV.14

Aggregates	0/2	2/5	5/16	16/32	32/63	63/100	+100	Totals
Pit run	15% elimination				5	10	70	100 %
	or: 60 TH of wastes				20	40	280	400 TH.
							−280	Feed VB 11.09
Jaw crusher VB 11.09 R: 200	3	3	5	5	14	20	50	100 %
	8	8	14	14	40	56	140	280 TH.
						−96	−140	Feed BCS 4'1/4 STD
Cone grinder BCS 4'1/4 STD, extralarge R: 40	5	5	14	24	42	10		100 %
	12	12	33	57	99	23		236 TH.
Balance	6	6	14	21	46	7		100 %
	20	20	47	71	159	23		340 TH.

Table XV.15

Aggregates	0/2	2/5	5/16	16/32	32/63	+63	Totals
Manufacture of 'Concrete only' 80 t/h of 0/63 Limestone							
Requirements	40	10	12.5	25	12.5		100 %
	32	8	10	20	10		80 TH.
Tapping	6	6	14	21	46	7	100 %
	5	5	13	19	42	6	90 TH.
					−32	−6	Feed BCS 3' STD
Cone grinder BCS 3' STD, extralarge R: 12.5	12	12	52	24			100 %
	4.5	4.5	20	9			38 TH.
				−32	−8		Feed gyradisc 36"
Gyradisc cone grinder 36"	40	30	30				100%
	16	12	12				40 TH.
			−13.5	−3			Feed BB 16 × 35
2 Rod mills BB, 16 × 35 process	100						100%
	16.5						16.5 TH.
Loss: fillers	−10						
Balance	40	10	12.5	25	12.5		100 %
	32	8	10	20	10		80 TH.

Table XV.16

Manufacture F10, D11 and D12 — 140 t/h of 0/32 sandstone	0/2	2/5	5/16	16/32	32/63	+63	Totals
Requirements	26	19	31	24			100 %
	36	27	44	33			140 TH.
Tapping	9	8	19	30	22	12	100 %
	13	12	29	45	33	18	150 TH.
					−31	−18	Feed BCS 3' STD
Conical Grinder BCS 3' STD, extralarge R: 12.5	8	8	66	18			100 %
	4	4	32	9			49 TH.
				−22	−23		Feed gyradisc 36"
Gyradisc conical grinder 36"	26	24	50				100%
	12	11	22				45 TH.
				−17			Feed BB. 16 × 35
2 rod mills BB, 16 × 35 wet process	100						100%
	17						17 TH.
Loss: fillers	−10						
Balance	26	19	31	← 24 % →			100 %
	36	27	44	31	2		140 TH.

Table XV.17

Manufacture P9 and F10, D11 and D12 — 125 t/h of 0/32 sandstone	0/2	2/5	5/16	16/32	32/63	+63	Totals
Requirements	27	18	30	25			100 %
	34	22	38	31			125 TH.
Tapping	9	8	19	30	22	12	100 %
	12	11	26	40	30	16	135 TH.
					−30	−16	Feed BCS, 3' STD
Conical grinder BCS, 3' STD extralarge R: 12.5	8	8	66	18			100 %
	4	4	30	8			46 TH.
				−25	−17		Feed gyradisc 36"
Gyradisc conical grinder 36"	26	24	50				100%
	11	10	21				42 TH.
			−3	−14			Feed BB. 16 × 35
2 rod mills BB, 16 × 35 wet process	100						100%
	17						17 TH.
Loss: fillers	−10						
Balance	27	18	30	25			100 %
	34	22	38	31			125 TH.

— Screening equipment,
— Sand washing and processing equipment,
— Materials-handling equipment etc.
These we shall assess below while studying flow sheets.

4.2 Study of the Installation

Before taking up a detailed study of the various sections, it is necessary to determine the 'type' of installation from among the most common cases, as summarised below.

4.2.1 CHOICE OF TYPE OF INSTALLATION

• *Portable installations'* **to be provided in the following** cases

— Type of work: Construction of roads for instance.
    — Duration of work: about 6 to 12 months.
    — Movements: frequent.
    — Type of installation: equipment fitted on trailers.
    — Buffer stocks: non-existent or limited to 1 or 2 hoppers.
    — Stocks of finished products: in the open—capacity 150 to 200 m^3 per category.
    — Conveyors: semi-fixed (portable conveyors, contrary to a common notion, are more difficult to carry from one site to another).
    — Efficiency (ratio between the theoretical capacity of the installation and the actual production obtained): less than that of the fixed or semi-fixed installations because it is not possible to absorb by interposition of adequate buffer stocks, the feeding irregularities.
    — Civil engineering: of lesser importance.
    — Hoisting equipment: non-existent (site equipment is used) or limited to a crane hoist used only for replacing normal wearing components.

• *Semi-fixed installations*

To be envisaged on the assumption of a time-bound important site job, the working conditions of which do not differ very much from those given below:
    — Type of work: dam, airport etc.
    — Duration of work: about 2 to 6 years.
    — Movements: eventual, at the end of the work.
    — Type of installation: machinery installed on semi-fixed frameworks, laid on compacted ground, or a longitudinal frame by means of skids.
    — Buffer stocks: on the ground (it is advisable to provide for an effective capacity equal to about 1 day's production).
    — Stock of finished products: on the ground—capacity 150 to 200 m^3 per category, or in hopper—unit capacity of 20 to 60 m^3.
    — Conveyors: semi-fixed up to 30 m, or fixed for lengths above 30 m.

— Efficiency: presence of buffer stocks enables improving the service factor compared to portable installations which lack them.

— Civil engineering: moderately substantial.

— Hoisting equipment: provide for crane hoist or monorail system to ensure maintenance of major equipment.

• *Fixed installations*

To be provided in the case of a quarry with large reserves and located in an environment which assures the sale of finished products.

These installations are provided to ensure the smoothest possible functioning, to facilitate monitoring and access to equipment, to minimise duration of maintenance operations etc., and engender special conditions of use as summarised below.

— Type of use: permanent.

— Duration of use: 10 to 30 years or more.

— Type of installation: machinery laid on mass concrete or on large frames (e.g., metal superstructures above a battery of silos).

— Buffer stocks: largest possible capacity. Generally, provide for a lively stock corresponding to 2 days' production, which represents a total reserve (gross stock) of 8 days.

— Stocks of finished products: very substantial, consisting of large-capacity hoppers and/or stocks in the open with automatic reclamation.

— Conveyors: fixed.

— Efficiency: size of buffer stocks and maintenance facilities enable improving the service factor compared to semi-fixed installations.

— Civil engineering: substantial.

— Hoisting equipment: gantry cranes above the key equipment, monorail system above the others.

Finally, in the case of fixed installations we must also take into account:

— The growing nature of the market. For this purpose, the installation should enable, without too much modification, incorporation of additional grinding equipment.

Similarly, it is advisable to plan on a very wide scale the screening outfits

— With due regard for the environment which has an influence on:

• orientation of the installation (consideration of the prevailing winds for protecting near-by houses or stocks of finished products in the open);

• planning of the installation (buildings covering equipments, producing noise and dust, covering the stocks, dust removing equipments etc.).

— The means of access which permit:

• totally safe flow between the working face and the receiving hopper of the primary station;

• easy approach to the main equipment for facilitating replacement of heavy and bulky parts;

• rational dispatch of finished products which have to be carried under the most economical conditions possible (roads, railways, canals).

It is obvious that crushing installations are not systematically classified into one or the other of the above-defined categories.

It is sometimes necessary for various reasons to mix the types. This is the case in our example where, in agreement with the user, provision has been made for:

— A crushing section of the 'fixed' type, taking into account the size of the equipment and the capacity of the receiving hopper.

— Grinding and screening sections of the semi-fixed type to minimise the importance of civil engineering infrastructure.

Areas for storing and delivery of finished products of the 'fixed' type, necessitated by the size of the stocks.

Fig. XV.1. Primary crushing section

4.2.2 DETERMINATION OF FLOW SHEETS

Further to the detailed examination of the main factors studied in the above sections, it is possible to tackle an actual study of the various sections constituting the materials-manufacturing plant.

• *Primary crushing section* (see Fig. XV.1)

This section is of prime importance in a quarry. Its purpose is to regularly feed the final processing plant with materials which can be admitted by the secondary crushers (0/200 to 0/350 maximum).

The equipment constituting the primary crushing section is so chosen that the following observations, schematised in Table XV.18, are taken into account.

Receiving hopper

This is the first of the equipment to experience any irregularities of the quarry, due essentially to the rotation of transport machinery.

In order to permit passage from the discontinuous to the continuous process, it is advisable to provide for the largest possible hopper. Generally, a capacity equal to 1.5 to 2 times, sometimes more, the capacity of the largest trucks is chosen.

The shape of the hopper is determined on the basis of the nature of the pit run. If the pit run is moist and clayey, a hopper with highly inclined walls should be provided. If the run is dry and not much polluted, it is advisable to have a

Table XV.18

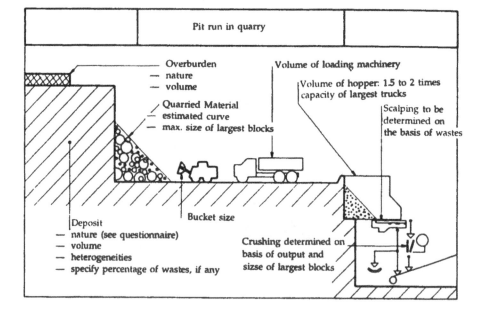

hopper in the form of a 'stone box' to preclude wear as well as deformation of the largest materials due to impact.

The hopper is made:
— either of concrete (fixed installations);
— or of metal (portable and semi-fixed installations).

Provision of a retaining wall for the ramp or an access bridge to a receiving hopper made of metal is becoming quite common;
— or of mixed design (or semi-metallic), consisting of concrete walls with metal floor and front face.

The last solution was adopted in our example for the following reasons:
— The 'concrete' was accepted by the user because it permitted him to restrict the quantity of supplies imported.
— To facilitate its manufacture, this concrete had to be of simple design, which is the case in a 'mixed' solution.
— A mixed solution permits large-capacity hoppers.

Extractor

Its function is to extract the materials from the hopper and to regularise the feed rate of the installation. The three most commonly used extractors are:

1. Distributor with racks

This equipment is simple but strong and requires very little maintenance. It is much cheaper than a feeder with a metal guard (ratio of 1 to 2). It is required in fixed and semi-fixed installations as well as in small or medium-size portable installations whenever the 'feeding' and 'scalping' functions have to be independent.

For the foregoing reasons, we chose a DAT 10.38 extractor for our example, possessing the following characteristics:
— Effective width 1000 compatible with the feed rate required and the size of the largest blocks;
— Length 3800, permitting use of a large-capacity hopper.

2. Apron feeder

This type of feeder is frequently used either in portable installations because it enables reduction in height of ramp of the receiving hopper or in industrial installations where its purchase price, as well as maintenance costs, are more readily amortised.

3. Vibrating apron feeder

This is used either in the 'feeder' version or, more often, the 'feeder-scalper' version.

This equipment, which is not very bulky as it simultaneously carries out the function of feeding and 'scalping', is particularly suitable for use on a portable unit.

—*Direct feeding of the crusher*: It is possible to envisage direct feeding of the crusher when there is no problem of scalping and wherever the outputs processed

(2000 t/h and more) necessitate gyratory crushers, which is rare in public works quarries.

For further details on feeders, see Chapter XI.

Scalper

The role of this equipment is to eliminate waste products (soils, clays etc.) or to relieve the crusher by short-circuiting small-size materials. The main types of scalpers are described below.

1. Giant grizzly with twin unbalances (EDB)

Very strong equipment, used either in quarries, in large-capacity primary crushing, or in installations for screening rocks for riprap.

2. Scalper with unbalances (CSB)

Also used in quarries under conditions less stringent than those of the preceding equipment and suitable in most cases. For this reason we selected the CSB for our installation.

The machinery chosen is a CSB 1328 with 2 decks as its dimensions (area = 1300 × 2800 mm) and accessories (100 mm and 40 mm) enable meeting the problem posed.

3. Bar grizzly

Very efficient with wet and sticky products. Hence it is mainly used in installations for processing rounded materials.

4. Vibrating apron type feeder (ATV) with scalping grid

This type does not afford a high degree of efficiency of screening. In the most difficult cases (desired separation size less than or equal to 40 mm), it is necessary to provide an additional screening equipment for screening the scalped products once again.

Crusher

The problems raised by the intake, or depth of feed spout, the flow of the materials to be crushed, the size of the largest blocks and the setting necessitated by the intake of the secondary crusher, led us to choose a double toggle jaw crusher of the type VB 11.09, width 1100 mm, intake 900 mm, maximum possible output of 380 t/h, with a setting of 200 mm for the required output.

The various possible combinations of the equipment constituting a primary crushing post are summarised in Table XV.19.

Taking into account the need to eliminate wastes in the case of production of materials for the manufacture of concretes or to reintegrate them in other cases (manufacture of transition materials, filters and drains), we chose arrangement no. 1 for our installation.

At the exit of the primary crushing section, we shall have materials which are readily handled with a belt conveyor.

Table XV.19

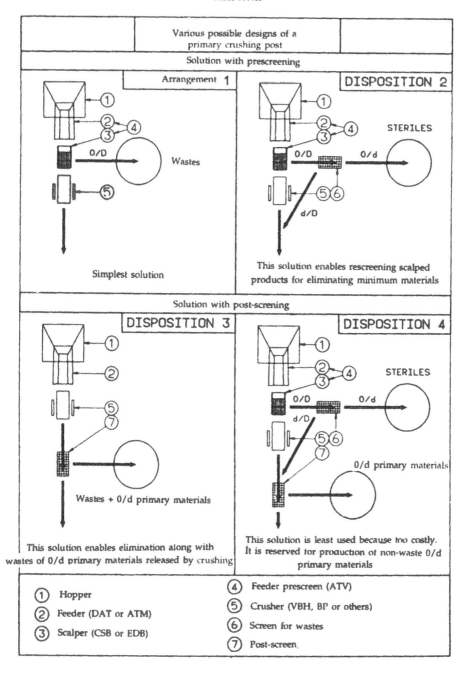

	Various possible designs of a primary crushing post	
	Solution with prescreening	

Arrangement 1 / **DISPOSITION 2**

O/D — Wastes

Simplest solution

STERILES
O/D   O/d
d/D

This solution enables rescreening scalped products for eliminating minimum materials

Solution with post-screening

DISPOSITION 3 / **DISPOSITION 4**

Wastes + 0/d primary materials

This solution enables elimination along with wastes of 0/d primary materials released by crushing

STERILES
0/D   0/d
d/D
0/d primary materials

This solution is least used because too costly. It is reserved for production of non-waste 0/d primary materials

① Hopper
② Feeder (DAT or ATM)
③ Scalper (CSB or EDB)
④ Feeder prescreen (ATV)
⑤ Crusher (VBH, BP or others)
⑥ Screen for wastes
⑦ Post-screen.

To avoid rapid deterioration of the conveyor belt, it is advisable to provide for a feeder between this equipment and the crusher.

• *Stockpiling of precrushed materials* (or prestock) (see Fig. XV.2)

The role of the stock of precrushed materials and its sizing have been defined in Chapter XIV, which may be referred to for additional information.

The setting up of this stock can be envisaged upstream or downstream of the secondary crusher. In the first case the product to be stockpiled may reach 250 to 350 mm in size, which necessitates the use of storage (belt conveyor) and destorage (extractors, culverts, large-width belt conveyors) equipment.

In the second case, corresponding to the solution adopted in our example, we have provided a secondary crusher of the BCS 4'1/4 ST type (see Fig. XV.3), situated upstream of the stock, permitting reduction of the product to be stored to about 100 mm. This type of crusher is generally preceded by a scalper (see Fig. XV.4) whose purpose is to relieve the crusher from the primary materials of sizes comparable to those of the materials which it produces and, if need be, by a buffer hopper which enables functioning under load.

The various possibilities of constitution of a prestock are represented in Table XV.20.

In the example under study the need to produce transition materials T3 and T8 into a single grade 0/80 without prescreening or control, led us to choose arrangement no. 4.

Moreover, the need to stock several types of materials calls for the use of a stacker.

Taking into account the selections made in the preceding paragraphs (size and arrangement of equipment), the flow sheet adopted for the equipment located upstream of the prestock will be the same in the three cases of production (see diagram and flows in Table XV.21).

By virtue of the prestock it is possible to feed the plant at a regular rate for preparing aggregates for the manufacture of concretes, filters and drains. This regularity is of prime importance because the quality and quantity of the materials to be produced depends on it.

• *Aggregate manufacturing plant*

Taking into account the desired products (see Section 4.1.3, Tables XV.5–XV.8), it was possible for us to define the required size-reduction machinery (see Section 4.1.4, Tables XV.12–XV.17).

Let us now concentrate on defining the other processing equipment as well as the flow sheet of the secondary section.

Let us recapitulate:
— we are required:
   • to manufacture materials for concrete (separation sizes 2, 5, 16, 32 and 63 mm) and materials for filters and drains (separation sizes 2, 16 and 63 mm);
   • to screen wash fractions less than or equal to 16 mm;
   • to cyclone and dewater the 0/2 fraction;

Fig. XV.2. Prestock   Fig. XV.3. Cone grinder, type BCS 4'1/4 ST

Fig. XV.4. Primary grinding-screening section (BCS 4'1/4 ST + CVB 1540 II P)

Table XV.20

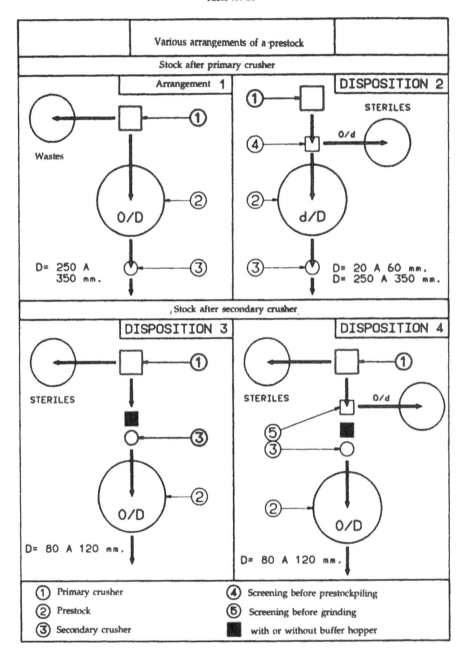

① Primary crusher	④	Screening before prestockpiling
② Prestock	⑤	Screening before grinding
③ Secondary crusher	■	with or without buffer hopper

Table XV.21

	Primary section Flow Sheet	

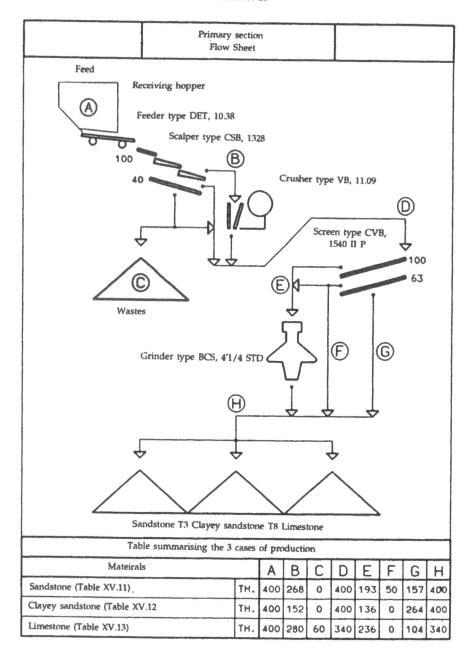

Feed

Receiving hopper

(A)

Feeder type DET, 10.38

Scalper type CSB, 1328

100

(B)

40

Crusher type VB, 11.09

(D)

Screen type CVB, 1540 II P

100

(C)

63

Wastes

(E)

Grinder type BCS, 4'1/4 STD

(F) (G)

(H)

Sandstone T3 Clayey sandstone T8 Limestone

Table summarising the 3 cases of production									
Mateirals		A	B	C	D	E	F	G	H
Sandstone (Table XV.11)	TH.	400	268	0	400	193	50	157	400
Clayey sandstone (Table XV.12	TH.	400	152	0	400	136	0	264	400
Limestone (Table XV.13)	TH.	400	280	60	340	236	0	104	340

• to separately stock materials for manufacturing concrete on the one hand, and materials for manufacturing filters and drains on the other (4000 t per aggregate size);

• to provide for an open stockpiling for all materials except concrete sands which we stock in silos (300 t).

— we selected the following grinders in our granulometric study:

• Symons cone grinder, type BCS 3' ST extra large. This equipment was chosen on the basis of:

— its large intake, permitting opening if need be of equipment of the primary section;

— its high reduction ratio, permitting production of maximum 0/16 mm materials in closed circuit;

— its capacity to grind hard and abrasive materials under the best conditions.

• 36" Gyradisc grinder capable of obtaining a large percentage of sand as well as chippings of very good shape obtained by attrition due to the special shape of its jaws, which slows down the flow of the materials.

• Two wet process rod mills, type 16 × 35, used for producing the quantities of fine materials required for meeting specifications laid down by the construction authority.

Extraction from under prestock

The aggregate manufacturing plant takes off from under the prestock of limestone materials for manufacturing aggregates for concretes and from under the prestock of materials of hard sandstone for the manufacture of aggregates for filters and drains.

This reclamation takes place through electromagnetic vibrating extractors which are particularly well suited for ensuring the regularity of feed.

The equipment is installed inside a metal culvert. This principle of a tunnel has been adopted because it is faster to implement and less expensive than the 'concrete' solution. Moreover, it is possible to retrieve the equipment at the end of the job. The materials are evacuated at the discharge end of the extractors, towards the screening section by means of a belt conveyor.

Screening section

The sizing and description of the various types of screens having been given elsewhere (see Chapter X), we shall confine ourselves to specifying the type, sizes and role of the screens selected for our installation.

Let it be noted, however, that the designing of a screening surface is fairly complicated and in spite of its appearance of mathematical rigidity, it should be interpreted with considerable caution; that is why we feel it wise to entrust this task to screen manufacturers who are well equipped to solve these problems.

Distribution screen

This is a vibrating screen with unbalances, type CVB 1845 (width: 1.80 m, length: 4.50 m) with 4 decks provided with square meshes for 63, 32, 16 and 5 mm separation sizes.

Classification screen

This is a vibrating screen with unbalances, type CVB 1845 provided with square meshes for 16, 5 and 2 mm separation sizes.

Washing of chippings

This solution, which consists of washing the quarry materials, was until recently reserved for the production of quality materials because elimination of waste products by prescreening (before the crusher) or by post-screening (after the crusher) was generally considered sufficient. This is no longer the case nowadays when most installations are provided with units for rinsing crushed chippings.

For further information, see Chapter XII (Washing of Aggregates and Processing of Sands).

Installation of screens

We can think of three types of screening installations: 'fixed' (or at a height), 'semi-fixed' (or on the ground) or 'combined'.

Installations at a height: These generally include one or more screens installed above a battery of bins. The screened products are directed by gravity into the various bins. The grinding section, if any, is fed by some of these bins. The ground products are recycled on the feeding conveyor of the screening section.

This arrangement has the following advantages:

— minimising the number of conveyors (storage by gravity);

— facilitating monitoring as the screening and grinding equipment are concentrated at one place;

— sheltering the products from inclement weather because this type of installation is generally covered and well protected.

The main disadvantage is a relatively low storage capacity, which is why it is often supplemented by belt conveyors fed by the spillover from the buffer bins whose function is to ensure storage in the open of excess materials, while concomitantly permitting direct loading of the trucks under the bins.

Ground installations: In these installations each screen is placed on a framework of the semi-fixed type. The different aggregate sizes are conveyed either to the stock or to grinders.

This type of installation is particularly envisaged when a large quantity of products is required over a limited period of time. For this reason we have adopted it in the installation taken as an example. Its main advantages are:

— minimum civil engineering work required;

— possible extension or modification for the purpose of another application;

— easily retrieved for another site job.

Combined installations: To enable the possible extension of the screening section, often considered the most common bottleneck in an installation, installation 'on the ground' is done, with the screened products directed by means of belt conveyors into cyclindrical silos of large capacity.

This type of installation is mainly used in the largest fixed quarries.

Fine grinding section

When we want
    — to process a rich sand into fine elements,
    — to correct a curve for complying with grading envelope,
    — to convert excess products into sand,
specific grinders have to be used, such as sand grinders of the gyradisc type and rod mills in the case of hard and abrasive materials or impact breakers or hammer mills in the inverse case.

This equipment is generally installed at the end of the grinding circuit. The materials to be fed do not exceed 30 mm and the output grain size can be about 2 mm.

NB: In the case of a rod mill, it is the time of passage of the products into the tube which determines the fineness of the products to be obtained (the more the feed rate is reduced, the finer the sand).

There are two cases of application of rod mills:
    — Use in the dry process enables obtaining very fine sands rich in fillers (0 to 80 μm).
    — Use in the wet process, envisaged mainly for the production of sands for concrete manufacturing which should contain the minimum of fillers. This is possible because the effect of water mobilises the materials inside the tube and thus increases output while concomitantly avoiding the production of fine particles due to excess grinding.

In our example, the materials being relatively hard and abrasive, we chose a 36" gyradisc and two 16 × 35 wet-process rod mills (diameter 1600, length 3500) so as to have a grinding capacity sufficient to produce the quantity of 80 μm/800 μm laid down for the manufacture of sand for concrete required on this job site (see Fig. XV.5).

Fig. XV.5. Rod mills 16 × 35 VH.

Sand-processing section

The sand-processing section calls for very special attention because, in the case of manufacture of concretes, bituminous mixes or other quality products, it is necessary to process the materials using different methods of separation or handling, i.e., instead of mechanical means, hydraulic (most common solution because the cheapest and least sensitive to product variations) or pneumatic (solution envisaged only for the processing of very dry products) are used.

We recall below the operations to be envisaged in the case of processing sands by hydraulic or pneumatic means.

Sand processing by hydraulic means

To obtain good quality sands, the following problems need to be solved:

— Achieving a higher separation size and washing the sand; these two functions are generally carried out on the sorting screen.

— Separating the water contained in the washed sand to permit simultaneous elimination of the very fine particles of clay (0 to 40 µm) contained in the fillers. This operation is generally carried out in a paddle decanter (DEA) which enables mixing, recovery and drainage of the sand while concomitantly permitting elimination of fillers, which are carried away by the water (see Fig. XV.6).

This operation can also be carried out in a single block sand-processing unit (see Fig. XV.7) consisting of a receiving tank at whose foot is installed a pump meant for sending the pulp (water + sand) into a hydrocyclone, whose role is to separate the water and sand and to eliminate excess fillers.

In order to stock the sands under proper conditions, a vibrating dewaterer is used whose function is to reduce the moisture content of the pulp to about 12 to 15%.

This last solution was adopted in our installation because we felt that the mixing of the sand inside the pump and the cyclone was an important factor in improving particle disintegration and thereby facilitating clay elimination.

This type of installation can be supplemented, if need be, by a screening panel whose function is to separate the sands into two categories (generally 0/630 µm and 630 µm/2 mm or more), which are separately stockpiled and then retrieved and mixed on request for obtaining the desired grading curve.

There are, of course, other sand-processing equipment, such as screw classifiers, hydroseparators, sandsort etc. For further details, see Chapter XII (Washing of Aggregates and Processing of Sands).

Sand processing by dry means

The main purpose of this processing is to correct the fines content of the ground sands.

This is the case, for instance, for a sand obtained through hammer mills. In principle, this sand contains too many fines; we should therefore consider defillerising it (i.e., removing the fines).

Fig. XV.6. Decanter with vanes, type DEA 520.

Fig. XV.7. Sand-processing unit

Table XV.22

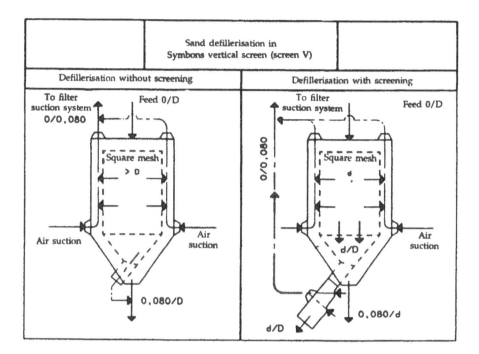

	Sand defillerisation in Symbons vertical screen (screen V)	
Defillerisation without screening		Defillerisation with screening

This operation is generally carried out in air selectors which enable, when they function properly, achieving an acceptable 80 μm separation size provided the processed product is very dry (permissible maximum moisture content = 0.5%).

Since the price of this type of installation is quite high, it has become more and more common to carry out this defillerisation operation inside a Symons vertical screen (screen V). The advantage of this equipment is that it can take on products which are slightly less dry (H = 1% maximum).

The various possibilities of using such equipment are summarised in Table XV.22.

NB: The simultaneous operation of screening and defillerisation can be envisaged only under certain conditions, i.e., to achieve intermediate separation sizes of the sand only in the upper portion of the zones (about 2 mm).

Taking into account the options adopted for solving the problems of preparation of finished products, the processing diagram of the secondary post was prepared as shown in Table XV.23. This part of the installation is visible in Fig. XV.8.

The circuits used are identical in the three cases of production whereas the outputs vary according to the requirements.

Table XV.23

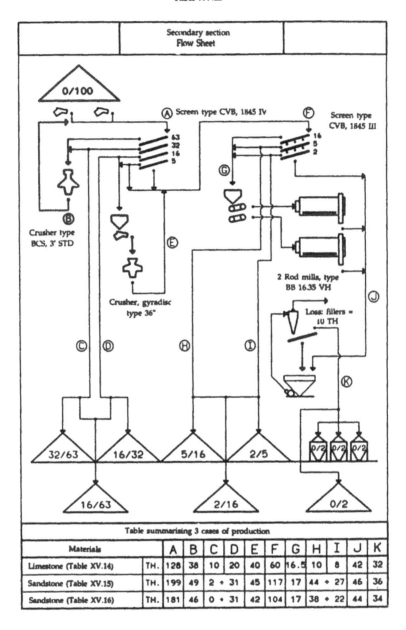

		A	B	C	D	E	F	G	H	I	J	K
Limestone (Table XV.14)	TH.	128	38	10	20	40	60	16.5	10	8	42	32
Sandstone (Table XV.15)	TH.	199	49	2 • 31		45	117	17	44 • 27		46	36
Sandstone (Table XV.16)	TH.	181	46	0 • 31		42	104	17	38 • 22		44	34

Table summarising 3 cases of production

Materials ... A B C D E F G H I J K

• **Stockpiling of finished products**

To enable the programme envisaged to be carried out, possible deviations in production or consumption notwithstanding, the user wanted large stocks of finished products to be available in his installation. For this reason we have provided:

Fig. XV.8. General view of aggregate manufacturing plant

— A storage for the concrete manufacturing needs, consisting of 4 stocks in the open of a total capacity of 16,000 tons of graded (calibrated) products.

— A storage for the filter and drain manufacturing needs, consisting of 3 stocks in the open with a total capacity of 12,000 tons of graded products (see Fig. XV.9).

— An assembly consisting of 3 silos for storing sands, representing a total effective capacity of 500 tons, supplemented by a safety stock in the open of about 2000 tons (see Fig. XV.10).

These stocks were set up in line to expedite rapid evacuation of the materials by means of belt conveyors installed inside tunnels.

Fig. XV.9. Linear stockpiling of filters and drains

Fig. XV.10. Silos for stockpiling sands

Fig. XV.11. Rinsing of fine gravels and feeding of concrete manufacturing plant

Fig. XV.12. Hopper for loading filters and drains

Table XV.24

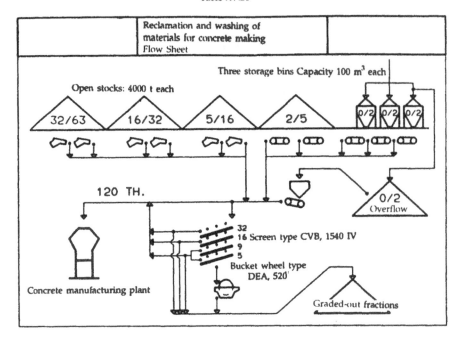

	Reclamation and washing of materials for concrete making Flow Sheet	

Three storage bins Capacity 100 m³ each

Open stocks: 4000 t each

32/63   16/32   5/16   2/5   0/2 0/2 0/2

120 TH.

0/2 Overflow

32
16 Screen type CVB, 1540 IV
9
5

Bucket wheel type DEA, 520'

Concrete manufacturing plant

Graded-out fractions

Table XV.25

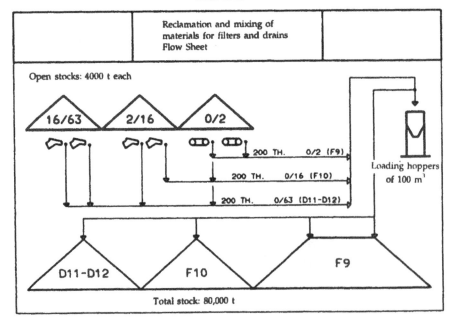

	Reclamation and mixing of materials for filters and drains Flow Sheet	

Open stocks: 4000 t each

16/63   2/16   0/2

200 TH.   0/2 (F9)

200 TH.   0/16 (F10)

200 TH.   0/63 (D11–D12)

Loading hoppers of 100 m³

D11–D12   F10   F9

Total stock: 80,000 t

Part of the materials was extracted from under the stock of products meant for the manufacture of concretes by means of belt feeders (for sands) or electromagnetic vibrating feeders (for fine gravels) at the unit rate of 120 tonnes per hour.

After extraction the materials were rinsed on a vibrating screen with unbalances type CVB 1540 (1.50 × 4.00) before being sent to the concrete manufacturing plant (see Fig. XV.11). This rinsing is compulsory in the case of manufacture of high-quality concretes as it is one of the necessary conditions for obtaining the highest possible mechanical strength.

The flow sheet of the materials of the stockpiling section, reclamation from under the pile, rinsing of the fine gravels for concrete manufacturing etc. were envisaged in our exemplar installation as per the flow sheet presented in Table XV.24.

The other part of the materials is to be extracted from under the stock of products meant for the manufacture of filters and drains by means of belt feeders to ensure mixing of the desired zones F9, F10, D11 and D12.

The mixed products are then discharged into a loading hopper.

The feed rate (200 t/h) as well as the capacity of the hopper (150 t) were so determined as to ensure continuous loading of the dumpers (see Fig. XV.12).

In order to guarantee a continuous supply to the job site in the event of a sudden breakdown of the installation, an open stock of 80,000 t was envisaged.

The flow sheet of the materials of the stockpiling section, mixing, loading of the dumpers and storage of the excess filters and drains is presented in Table XV.25.

For all details regarding the problems of stockpiling and loading, see Chapter XIV.

After studying the actual circuit of the materials, the additional devices required for optimising the working of an installation have to be defined.

4.3 Additional Devices

4.3.1 CLEAR WATER SUPPLY AND DISPOSAL OF CHARGED WATERS

For washing or rinsing the materials as well as providing clean water to the rod mills used in the wet process, about 350 m^3/h water are required.

In the case under consideration this poses no problem because, since it is an installation for preparing materials required in the construction of a dam, presumably sufficient water will be available.

But such is not always so, in which case part of the charged waters has to be retrieved after decantation and used in closed circuit. This solution nevertheless requires some additional water to compensate losses due to evaporation or retention of part of this water in the finished products.

Natural decantation, when large areas are available, makes it possible to solve this problem at the least cost. In this case the time of stay of the charged waters

Fig. XV.13. Settling tank

before being discharged into the river or recycled in the materials washing circuit is very important. So, to reduce the decantation time, flocculating agents are used.

The flocculants are injected either into natural settling tanks or into clarifiers, which enables recycling about 80% of the waters to be treated (see Fig. XV.13).

The residual slurries are generally stored in tanks. Considering that they can constitute a health hazard if not thoroughly dried out, it is sometimes necessary to improve their consistency by treatment in thickening equipment (belt filter or tassster).

4.3.2 AUTOMATION OF ELECTRICAL INSTALLATIONS

In most cases materials-manufacturing installations obtain their power supply through national distribution networks, voltage 380 V and frequency 50 Hz.

It is sometimes necessary to use other sources of supply such as generator sets. This principle was adopted in the installation under consideration, i.e., a generator set for the primary post (installed capacity 450 kW) and another for the secondary section (installed capacity 700 kW).

To avoid disappointments, it is necessary to indicate accurately the source of power, voltage and frequency envisaged for a particular installation, as well as the level of automation desired.

To facilitate the choice of degree of automation, let us recall the main stages which define the passage from manual control to computer control.

— Full manual control: The operator directly supervises the functioning of the equipment, for which his instruments consist of sight, touch, hearing and possibly his sense of smell. It is he who decides and he who implements his decisions.

— Semi-automatic control: The operator possesses certain instruments (ammeters, manometers, thermometers etc.) which enable him to assess more rapidly and more accurately the working conditions of the equipment.

— Centralisation of controls: All the data collected by the control equipment is assembled in a central post where the operator transmits his decisions according to the information received.

— Automatic control of equipment: The equipment is interlinked in such a way that each can only be started or stopped in a predetermined order. This partial automation does not do away with the presence of an operator, however.

— Conventional automation: The station master starts up the installation from a keyboard for each operation. From that moment onwards the entire operation becomes automatic.

— Integral automation: Full automation through computers. As on date, this solution is envisaged only in the largest production stations.

The above-listed schemes make possible an automation which, even if partial, will nonetheless always pay from the point of view of optimisation of production conditions, a fact reflected in cost price.

The advantages of automation are several:
— elimination of errors and wrong manoeuvres,
— reduction in labour costs,
— elimination of clogging risks,
— reduction in stoppage time,
— rapidity of adaptation to manufacturing changes,
— detection of anomalies,
— uniformity in quality of products,
— simplification of manoeuvres.

The profitability of automation vis-à-vis the system of manual control may vary from 10 to 20% depending on the solution adopted.

For other details regarding the automation of installations, see Chapter XVI.

4.3.3 MONITORING AND ACCESS TO EQUIPMENT

The environment of each equipment should be carefully studied. Further, the main maintenance operations must be carried out under proper conditions. For this purpose, it is useful to provide:
— Correct accesses to permit the monitoring and maintenance of machinery.
— Sufficient height under the support of the conveyors to permit cleaning.
— Sheets and their readily dismantable wearing plates.
— Monorails to facilitate the transport of certain accessories (screen grates, engines etc.).

— Travelling cranes for the maintenance of main equipment. However, since it is not always possible to provide each workshop with a gantry crane, in certain cases buildings could be so designed as to permit access of an overhead travelling crane.

4.3.4 POLLUTION TREATMENT

The problems of environmental protection are becoming more and more important these days; noise, dust and water pollution can no longer be ignored(see chapter II, XIII).

Selection of the site for an installation is a determining factor. Efforts should be made to see that it is insulated either by concealment in rolling terrain, or an interposing screen of trees between it and the neighbourhood, or the direction of the prevailing winds taken into account to deter the spread of dust. Obviously, prevention of dust production is preferable but when this is not feasible, its dispersion, either by settlement with a spraying system or collection by pneumatic process.

Water mist spraying is interesting because less costly and unlike other dedusting installations does not entail heavy consumption of power. On the other hand, since this system aids retention of fine particles on the chippings, it cannot be used at the final stage of production when cleanliness standards are applied. Therefore, it is at the primary and secondary crushing sections that this solution is generally employed.

In all other cases dedusting by pneumatic process which consists in collection of dust through suction at points of emission is carried out.

This dust is entrained by a current of air at a speed of about 20 m/s inside pipes which take it to a filtration installation. The most commonly used filtration process is that of filter bags. A very high efficiency is obtained with such bags. The ejected air contains less than 30 mg dust, which is less than the limit permitted by current legislation (max. 50 mg/m^3).

Other systems of dust filtration also exist (e.g., cyclones). For further details on dedusting, see Chapter XIII.

4.3.5 SAFETY

Crushing installations can be the source of numerous accidents if suitable precautions are not taken.

In addition to the general rules and regulations regarding the incorporation of safety measures while designing machinery and installations, we give below certain essential precautions to be taken into consideration while designing an installation.

— Provide, above the primary crushers, either a gentry crane fitted with a special hook, or an arm of an hydraulic crane fitted with a hook, to nullify any worker's temptation to climb down into the crusher to remove blocks that might be obstructing the equipment.

— Protect all rotating parts by means of protective housings.

— Comply with rules governing the use of belt conveyors (emergency stops along the conveyors, protection preventing access to winding drums etc.) (see chapter XIX).

— Cover the bins with decks or grids to prevent anyone from falling inside them.

— Avoid accumulation of dust which may conceal some possible traps.

— Strictly enforce the safety rules relating to electrical installations.

4.3.6 SETTING UP THE INSTALLATION

Setting up an installation depends in most cases on the nature of the site.

When the site shows wide differences in level, efforts are generally made to utilise them to the maximum for increasing the capacity of stocks while concomitantly reducing the length of the conveyors.

Trying to utilise gravity alone to facilitate the flow of materials under preparation is futile as such a procedure does not permit backward movements in the case of a closed circuit operation. It is precisely for this reason that equipment of the same section is very often placed at the same level. This solution also permits easy monitoring and maintenance.

The secret of a profitable installation is often linked with the proper design of the connecting components, which should not be considered as auxiliary equipment.

We have taken into account the above observations for the best utilisation of the profile of the ground on which our exemplar installation has been set up.

— The feed ramp of the primary section is at level 1012 m. This level was chosen to facilitate the arrival of the pit-run products from the quarry.

— The level of the primary section (1000 m) was determined on the basis of the height of the crushing post (12 m).

— The level of the secondary section (995 m) was chosen to optimise the volume of the prestock.

— For the same reason stocks of the finished products were set up at level 900 m whereas the feeding sections of the concrete batching and mixing plant on the one hand, and the loading sections of the filters and drains on the other, were set up at level 980 m to minimise the length of the connecting conveyors.

The overall schematic plan (Table XV.26) summarises the above selection. It also enables, since the equipment is labelled, ready understanding of the circuits defined in the preceding sections.

5. CONCLUSION

All the systems developed in this chapter show that in the study of a project, a selection has to be made at every stage and an unwanted snag thereby avoided.

We must also take into account other cases, such as installations for the treatment of alluvial materials for which a large number of parameters need to be considered (see Table XV.27), which have a determining influence on the choice of circuits, their sizing, the cost and the setting up of processing plants.

Table XV.26

Ali Chouari dam, Morocco. Overall schematic plan

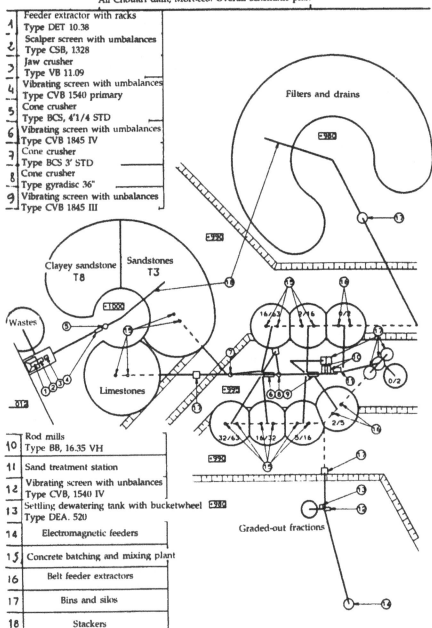

1	Feeder extractor with racks	Type DET 10.38
2	Scalper screen with umbalances	Type CSB, 1328
3	Jaw crusher	Type VB 11.09
4	Vibrating screen with umbalances	Type CVB 1540 primary
5	Cone crusher	Type BCS, 4'1/4 STD
6	Vibrating screen with umbalances	Type CVB 1845 IV
7	Cone crusher	Type BCS 3' STD
8	Cone crusher	Type gyradisc 36"
9	Vibrating screen with unbalances	Type CVB 1845 III

Filters and drains

Clayey sandstone
T8

Sandstones
T3

Wastes

Limestones

10	Rod mills	Type BB, 16.35 VH
11	Sand treatment station	
12	Vibrating screen with unbalances	Type CVB, 1540 IV
13	Settling dewatering tank with bucketwheel	Type DEA. 520
14	Electromagnetic feeders	
15	Concrete batching and mixing plant	
16	Belt feeder extractors	
17	Bins and silos	
18	Stackers	

Graded-out fractions

Table XV.27: Pit run in gravel pit

	Pit run in gravel pit	

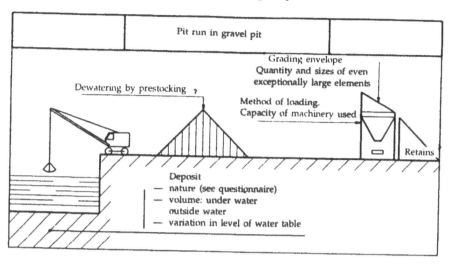

Dewatering by prestocking ?

Grading envelope
Quantity and sizes of even
exceptionally large elements

Method of loading.
Capacity of machinery used

Retains

Deposit
— nature (see questionnaire)
— volume: under water
outside water
— variation in level of water table

Thus for a successful study of an installation for preparation of materials, it is advisable to rely upon a specialised manufacturer capable of efficiently guiding the user in the selection of machinery, design of the circuits and installations and in the commissioning, supervision and maintenance of these installations.

The most important among the above have been modernised with computer-aided drawings and computer-aided designs for solving the most complicated problems under the best conditions of time limit and quality.

XVI

Automation of an Aggregate Processing Plant. Definition and Methods

J.M. Arnaud and A. Maldonado

INTRODUCTION

Industrial processing of aggregates calls for recourse to means which make possible uninterrupted manufacturing in the transformation of a raw material into marketable products.

Investment in machinery in the aggregates industry is generally very high. Hence it is imperative that the entrepreneur make the best use of the resources at his disposal.

What steps should be taken then for timely catering to market needs, which fluctuate throughout the years, without concomitantly being forced to stockpile finished products at the quarry? How to ensure a steady and consistent quality of production to win customer confidence when the process of manufacturing is inevitably disrupted by a number of factors?

These questions are not unanswerable. But the solutions depend on the choices initially made by the entrepreneur at the time of opening the deposit and the procedures adopted for exploitation and processing. We shall delineate below the factors involved and discuss the problem of automation of production. Accordingly, the dynamic process of transformation of raw material into aggregates will be dealt with in the following sections:

— automation: definition and concepts,
— pertinence and maintenance of an automated process,
— stages in the induction of an automation scheme.

1. AUTOMATION: DEFINITION AND CONCEPTS

By definition, automation of a quarry involves implementation of various measures to ensure constancy of functioning of the aggregate processing plant at a particular 'level' chosen by the operator. Minor deviations from the selected 'functioning goal' in quantity or quality of production are quickly corrected through automatic modifications in the setting of the equipment without manual intervention.

To start with, the process of quarry automation requires elaboration of certain concepts and application of a methodology, which we shall discuss in the following order:

— identification of the problem,
— automation tools,
— conduction of the automated process.

1.1 Identification of the Problem

The aggressiveness of the exploited deposit vis-à-vis the processing machinery, and its varied responses to mechanical impingement, tend to destabilise the working of an aggregate processing plant. Depending on the wear of its grinding elements (they may have to be changed after less than a month in use), the size and petrographic nature of its feed, the output of a crusher can vary from 100 tonnes per hour to 50 tonnes per hour.

The quality and quantity-wise management of the outputs of various processing machines is often the main problem encountered by the person in charge of manufacture. Accordingly, after identifying the problems to be solved and the objectives to be fulfilled, it is imperative to draw up a set of specifications for the automation expert.

Such a document will have several components:

— Delineation of objectives: the enterprise should explain its targets of production, its information system and take into consideration disturbances in working or aberrations in manufacture identified earlier.

— List of the plants and machinery which will be placed under automatic control: what in particular are the possible limitations incidental to their use (ranges of control, electrical power to be used, matching of equipment etc.)?

— Automations desired: this presupposes prior knowledge on the part of the entrepreneur about the measures to be taken in the various processing stages for optimisation of their production. This task often falls to the automation expert to expedite. The enterprise going in for automation should likewise specify its requirements in terms of 'man-machine' dialogue (screen display, filing, remote control etc.).

— Aids required for implementing automation: manuals pertaining to automation mechanisms, programs, guaranties etc.

— Personnel training for commissioning and maintenance of tools.

The automation expert will then propose the tools to be installed.

1.2 Automation Tools

The tools of automation are of different types.

— Technological, relating to the actuators (hydraulic jack, motor, etc.) meant for carrying out an on-line command pertaining to the process, for example, modifying a trap opening, a speed, a state of being (start, stop) or an adjustment.

— Electrical, in the case of certain sensors, which after making a measurement, deliver an electrical voltage proportional to the value measured.

— Electronic, when the apparatus captures, transmits and exploits the data made up of variations in electrical quantities. In this case, it is the actuating instruments of an automated system. Programmable rubric automats and industrial microcomputers come under this category. Programmable automats are eminently suited to the so-called 'sequential' control (successive start/stop of different machines, with due regard to safety aspects) and accomplishing local control loops. As for the microcomputers, they are the first choice for overall plant control and for monitoring, which enables the operator in the control room to have a user-friendly interface with the manufacturing process.

— Computer type, with respect to software, which covers all the operations involving the use of computers and numerical data processing.

Once incorporated into a control loop (Fig. XVI.1), all the tools of automation constitute a single system of command. Accordingly, this system ensures the following functions:

— conversion of real time physical phenomena into a digital or analog representation;

— use of these data for determining the changes to be effected in the manufacturing procedure (process);

— modification of the condition of the organs of command (actuators);

— identification (not always the case) on the control panel installed in a control room and manned by an operator of the production variables or the characteristics relevant to the command system (supervision).

Figure XVI.1 depicts a control loop. It links the quantity measured and the command through the intermediate of a PID controller*. This type of control is employed, for example, to maintain a particular filling level in a bin.

The principle of control is to compare the value as entered in the instructions with the measured value furnished by the sensor. The difference thus obtained is utilised by the control mechanism (integrated into a programmable automat or a

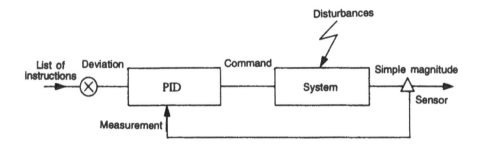

Fig. XVI.1. Simple control loop

*PID: Integral and derivative proportional controller.

microcomputer) for passing a command to the process so as to bring it closer to the point of functioning as entered in the instructions. This command may be proportional to the deviation, its integral and/or its derivative, in the case of a PID controller.

Sometimes a more complex (more 'intelligent') control loop works through a model of the system to be controlled. This enables presetting the command (Fig. XVI.2) and precludes unstable operation.

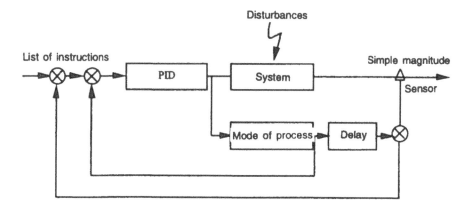

Fig. XVI.2. Intelligent control loop (Smith's regulator)

1.3 Operation of Automated Process

Automatic control of a quarry consists, in the first place, in 'controlling' the communication between the tools of automation after installing the system. The technical solutions employed during the last few years for satisfying such communication needs are well developed. Accumulated experience has led to the development of systems appropriate to the local intermachine communications, which constitute local industrial networks.

The choice of a network suited to the technological level of the processing plant or the communication needs of the enterprise is crucial. One should avoid multiplicity of cableworks over time and the obligation of having to manage systems which are technically heterogeneous and require more access procedures for maintenance operations.

The topology of a network constitutes its architecture. For a quarry it is often in the form of a loop or a bus (Fig. XVI.3).

The transmission support of the network is a coaxial cable, optical fibre etc., as the case may be. The proper functioning of the network presupposes a co-ordination in time of the exchanges (methods of access) has been effected. Generally, recourse is had to a deterministic method which gives the right to speak for a particular duration to each station (sensors, controllers etc.).

After installing local control loops and ensuring their interconnection, the user should fix the value for the plant's settings. Then the points for the plant's opera-

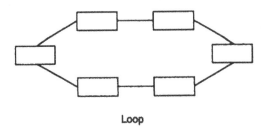

Loop

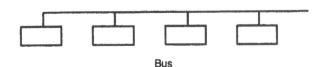

Bus

Fig. XVI.3. Quarry networks

tion are either calculated or determined. How to utilise all the resources of the plant for achieving a production of the desired quality at economical costs and in optimum time? It is worthwhile to take into account the condition for a stable working of the plant and the link available between the equipment used in the quarry. This work can be done only with a simulator, which will determine 'off line' the optimum setting of the equipment depending on the optimisation criteria dictated by the operator. The simulator will come up with the appropriate results if it is fed with an extensive database.

The interdependent relation between quarry equipment is exemplified in Fig. XVI.4 with a diagrammatical representation of one processing unit (loop type processing). This unit comprises two crushers, C_1 and C_2, delivering against a target C_f. R_1 and R_2 are the settings of the crushers. G_f is the product desired depending on production G_1 and G_2. It is then found that:

$$G_f = f(G_1, G_2)$$

$$G_2 = f(R_2, A_2)$$

$$G_1 = f(R_1, A_1)$$

$$A_2 = f(G_1, R_f)$$

Only R_f (reduction in the screening fabric) changes slowly. The simulator will then calculate the R_1 and R_2 which will enable achieving the optimum production desired.

These relations of interdependence, called 'models', are the representation of a machine's behaviour in the form of mathematical equations. It then becomes possible to program equations in a microcomputer, whereby a tool becomes available for the simulation of a plant.

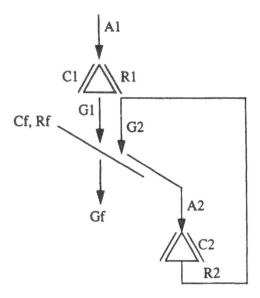

Fig. XVI.4. Loop type functioning

2. APPROPRIATENESS OF AN AUTOMATED PROCESS AND ITS SUSTAINED OPERATION

The foregoing section shows that use of a microcomputer, in conjunction with the model of a plant's machinery, enables provisional calculations and determination of the plant's operating points.

For the optimisation to be pertinent, the simulator has to work on an extensive database. Calculation of the particle size of any crushed material is done, for example, by programming while consulting a file. The models adopted by automation experts 'linearise' the grading curve of crusher output between two points (two sieves). The particle size of a product resulting from the mixture of several lots of materials is equal to the linear combination of the particle sizes of each lot.

It appears necessary to make special mention of two particular tools: the VDG 40 and the software LUCIE. These off-line tools are indispensable to those who wish to control and optimise quarry production runs.

2.1 Base Tools Necessary for Control of Quarry Production Runs

Special mention must be made of the VDG 40 which enables obtaining within a few minutes the grading curve of a 2/40 mm rounded or crushed product. These curves, obtained from a series of measurements, can eventually feed the database needed for the LUCIE software.

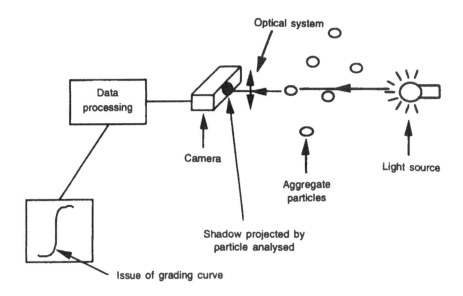

Fig. XVI.5. Diagram of the principle of VDG 40.

The grading curve is obtained from an analysis of the image of the sample's particles passing between a CCD camera and a light source (Fig. XVI.5). Recourse to this intelligent sensor is indispensable to enterprises which wish to control their production runs more expeditiously.

LUCIE is a software developed in the WINDOWS environment. It works on IBM PC or compatible having at least 4 MB RAM and VGA screen. It comprises:

— a user-friendly graphic interface enabling schematic representation of the plant, using icons corresponding to different types of machines;

— a database characterising the working of each of the machines. This database can incorporate the grading curves given by VDG 40;

— a simulation algorithm enabling generation of the granulometric balance sheet for every point of the plant;

— an optimisation algorithm enabling automatic calculation of the optimum settings of the crushers, with due regard to the production targets set by the operator.

Figure XVI.6 depicts the screen image of the simulation of a quarry with LUCIE software.

2.2 Automated Process

To ensure the sustained operation of a process as automated, it is especially expedient to constantly monitor the working of the sensors and the quality of the measurements made. The measurements made are liable to undergo different disturbances depending on the type of sensors, as shown in Table XVI.1.

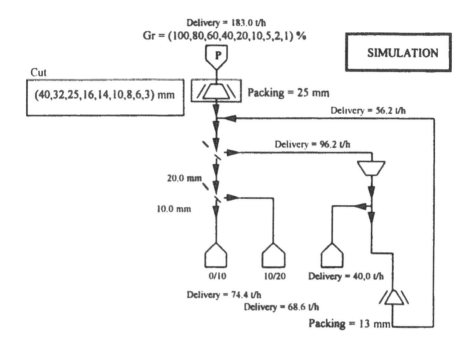

Fig. XVI.6. Simulation of a quarry using LUCIE software

Table XVI.1: Capabilities of different sensors

	Characteristics	Sensor delivering analog signals	Sensor delivering digital signals
1.	*Disturbances*		
	Feed voltage	Influence on precision	Risk of non-availability
	Ageing of components	Influence on precision	No influence
	Electromagnetic parasite	Defects difficult to detect	Easy detection
	Insulation defects	High sensitivity	Moderate sensitivity
2.	*Metrology*		
	Precision	Moderate	Can be excellent
3.	*Transmission*		
	Multiplexing	Difficult	Very easy
	Upon single support	Difficult	Easy
	Data output	Makes all information available	Easy, but for sampling of data
4.	*Others*		
	Data storage	Not possible	Very easy
	Evaluation of sensor installation	Simple tests	Tests simple, but sometimes prolonged
	Maintenance	Moderate	Autocontrol arrangements possible

It must always be kept in mind that the sensors have to function in an aggressive industrial environment. Vibrations, dust and the presence of water severely tax the machines and sensors. A regular surveillance of them is mandatory.

Furthermore, personnel must be trained. Developing the competence of quarry personnel is logically consistent with upgrading the quality of the work and the competitiveness of the enterprise.

3. STAGES IN INTRODUCTION OF AN AUTOMATION SCHEME

It is but prudent that the project for automation of control of a processing plant proceed in stages. The personnel should be given sufficient time to learn to rely upon the newly inducted equipment. Indeed, the personnel should gradually master the tools of automation so that their responses to signals become the habituated reflexes of professionals.

3.1 Stages of Automation

Quarry automation should thus be introduced through successive modules.

Generally, automation of a plant commences with the primary crusher. Feeding of the primary crusher should be uniform. The feeder, which introduces at the crusher mouth boulders dropped by the dumpers into the receiving hopper, should operate opportunely. Risks arising from an arching effect should be taken into account and all risks of blockage during feeding precluded.

The next stage of automation applies to the finished product. Expediting loading and dispatch is indeed worthwhile since intermediate storage of finished products costs dearly. Minimalising these costs is commonsensical.

The third stage consists of inducting the LUCIE simulator and establishing the optimal point of operation of the plant. Part of the advantage from incorporating this stage will be lost if some of the regulating controls have been disregarded, for example, crusher feeding.

3.2 Example of Application

By way of example, we list hereunder devices (cascade type intelligent control loops) actually exploited in the automation of the secondary section of a crushing plant (Fig. XVI.7).

The objectives were as follows:
— make the crushers operate at constant power,
— avoid saturation-delivery on the screen to preclude recycling the materials needlessly,
— monitor the output of the tertiary hopper.
Several problems had to be taken into account by the automation expert:
— the feeding rate of the feeders is rather difficult to master;
— conveyor belts cause delays between the crusher and the screen;

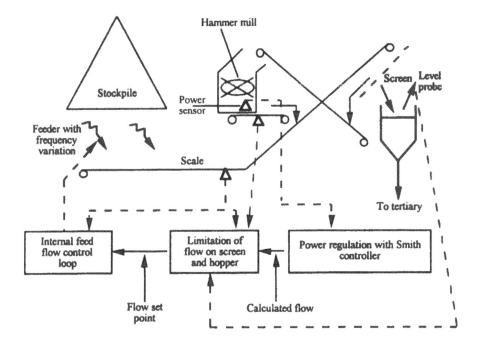

Fig. XVI.7. Example of automation of a secondary crusher station

— variation in frequency of the feeders changes not only the power consumed by the crusher, but also the delivery rate to the screen.

The solutions applied were as follows:

— Control of the power supplied to the crusher was effected by considering a mathematical model of the process to be controlled. This course was resorted to preclude the appearance of oscillations which would have occurred with a conventional control (presence of delay caused by the conveyor belts).

— A local loop incorporating a weigher positioned as close as possible to the feeders helped to resolve the problem of non-linearity in the feeding rate of the feeders and adjustment of frequency f command to these feeders.

— The ultimate feeding rate value resulting from overall control is shown in Fig. XVI.7.

The instrumentation used comprised the following sensors:

— Geosica weighers,

— Chauvin-Arnoux sensor of the power consumed,

— Siemens level sensor,

— TXS 47-40 telemechanical controller,

— IBM compatible PC.

Accordingly, the system of total control comprised:

— An internal loop utilising the data from the weigher at the prestock exit. It enables a feed equal to a given value (feed instruction), despite using vibratory feeders.

— This value is calculated through a second control which takes into account the crusher power, rate of delivery on the screen and the level in the bin.

— This control contains a prediction module for taking into account delays arising from the length of the conveyor belts.

Induction of this control device enabled:

— minimising the cost of some investments,

— increasing production rates of the plant by 20%,

— control of the particle size of the materials resulting from a steady operation,

— determination of the crusher's setting, as need be, off line with the simulator's help.

3.3 Return on Investment. Practical Case

We shall demonstrate that the automation of a quarry enables achieving increases in production of the order of 20%. Given this, it then becomes possible to make estimates of the return on investment provided one knows the fixed and running costs.

On the basis of a rise in investment of the order of 350,000 F through the aforementioned stages (sec. 3.1) and for a quarry having a production of 400,000 t per annum (cost price of aggregates 30 F/tonne exclusive of taxes, sale price: 35 F/tonne), the pay-back period can be notably shortened, as shown by the example presented in Table XVI.2.

Table XVI.2: Return on investment

Production	Before automation	After automation
	400,000 t/annum	480,000 t/annum
	in 10³ Francs	in 10³ Francs
Fixed costs	6000	6000
Running costs	6000	7200
Total costs	12,000	13,200
Gross turnover	14,000	16,800
Gross profit	2000 (5 F/tonne)	3,600 (7.5 F/tonne)

The annual rise in the gross profit (3,600,000 – 2,000,000 = 1,600,000) is greater than the investment made at the end of the three stages (3 × 350,000 F).

4. CONCLUSION

Hence, incorporation of an automatic quarry control is the culmination of a procedure comprising several levels:

Level 1: installation of actuators and sensors after identification of the problems.

Level 2: installation of control loops and sequential controllers.

Level 3: determination of the optimum settings of the machinery and circuits of the plant. Installation of intelligent control loops.

Level 4: working out preventive actions to ensure optimal control of the entire plant.

This calls for an extensive collaboration between the undertaking and the automation expert who should place at the service of the entrepreneur his expertise in modelling manufacturing procedures and processing the data which the sensors installed at the site will make available.

The results of a successful automation will manifest themselves quite fast. They will be in the form of:

— improvement in the quality of the finished products in all the manufacturing aspects of the materials;

— improvement in production: automation once and for all adjusts the settings and compensates the deviations in production;

— reduction in maintenance costs by enabling a more uniform loading of the machinery in terms of work;

— reduction in operational costs (fixed and running costs);

— improvement in working conditions through location of the control away from the processing plant.

It should also be emphasised that the process of automation sometimes goes beyond the purely technical, as it calls for a rigorous analysis of the manufacturing facility and management of the human resources at the very start of the industrial plan.

It relies on personnel training, requiring them to appreciate the need for introducing new techniques and motivating them towards this objective.

Therefore a scheme for automation is in keeping with a quality-oriented strategy of industrial activity.

REFERENCES*

+Arnaud, J.M., A. Maldonado and J.A. Ortiz. 1995. Integrated and hierarchized system for automated control of aggregate production. Unbound Aggregates in Roads. UNBAR 4 Nottingham (Great Britain) 17–18 July 1995, pp. 25–29.

Bagrel, C., J.F. Lafay and A. Maldonado. 1993. Should small aggregate production units in quarries be automated? *Bull. des Laboratoires des Ponts et Chaussées*, 186 (ref. 3693): 39–44.

Blot, G. 1987. Automation of particle size analysis. Presentation of a new sensor of shapes. The videogranulometer. *Revue Générale des Routes et des Aérodromes*, 639: 39–43.

+Delalande, G., A. Maldonado and J.Y. Monteau. 1991. Simulation and optimisation of an aggregate plant. IMAG World Congress, Dublin (Ireland) 22–26 July 1991.

French Standard Specification. 1994. Granulometric analysis. Flattening and elongation. Videogranulometer test, pp. 18–56.

Maldonado, A. 1991. Automation of quarries and its effects on the quality of aggregates. *Bull. des Laboratoires des Ponts et Chaussées*, 174: 39–45.

Maldonado, A. 1993. Total quality and processing of aggregates. *Bull. des Laboratoires des Ponts et Chaussées*, 188 (ref. 3790).

---

*All entries in French except those marked with a cross (+)—General Editor.

XVII

Mobile Plants for Production of Aggregates

Michel Deniau and Claude Archimbaud

1. INTRODUCTION

Crushing and screening plants other than those located at quarry sites under long-term exploitation were not known in France until a few decades ago except for 'movable machines hauled along to the working faces and manufacturing gravels on location', modern and mechanical versions of road-maintenance, or even simple and low-capacity machines shifted from one site to another in the alluvial valleys, and sometimes even from one pit to another, depending on need.

The growth and concentration over time and space of large construction works perforce gave rise to processing facilities of a new genre, inspired to be sure by existing techniques, but the features of which fulfilled two essential requirements:

— temporary nature (a few months to two or three years),
— high production rate.

Such facilities have come to be termed 'temporary installations', often shortened to 'mobile plants' or 'mobile units' (Fig. XVII.1).

The distinction sometimes made between temporary, semi-mobile or mobile installations is not always strictly valid since 'mobility' is mainly inversely proportional to the size of the installation, that is, in reality its performance in terms of quality and quantity.

The design of the components per se is in the process of evolution: some elements are mounted on wheels that suffice for towing them along (as in the case of aggregate coating plants), others are readily dismantlable.

These 'temporary installations' have acquired over several years a considerable importance, so much so that one cannot envisage a large job site complex without them. They have notably gained an enviable status in the exploitation of massive rocks, in quarries or excavations, and not just in alluvial sites.

Fig. XVII.1. Example of temporary installation or mobile unit

2. PRINCIPLES OF DESIGN OF TEMPORARY INSTALLATIONS

2.1 New Requirements and Functions

Apart from the development of large job sites, other reasons explain the increasing favour these installations have found:

— Increasing distance of these large job sites from the traditional sources of supply, which are often close to the high-consumption areas (industrial and urban sectors).

— Sharp increase in transport costs coincident with 'oil shocks' that dictated changes in a long-established economic order, even though the initial causes have almost disappeared by virtue of the mechanism of 'countershocks'.

— Taking into account 'environmental' problems, that is, the well-being, comfort, preservation of national wealth, aesthetics—all of which have drawn attention to the harmful effects of the permanent opening up of quarries, mining operations and hauling of materials. Such a consideration has also led to a rapid evolution in legislation (revision of the Mining Code, law on the Protection of Nature, revamping the law on notified installations etc.). This aspect of the question is treated in detail in Chapters II and XII.

— Advances made in knowledge of rocks and their properties, quarry working, controls, processing machinery, which have extended the idea of exploitable materials.

In simple terms, it can be said that a temporary installation helps to bring the place of production and the place of utilisation closer, and to match production and utilisation better, both quantity and quality-wise, as well as in real time. Sometimes the very structures of roads and specifications can be tailored to the deposits and the installations. But this kind of design has given rise to several errors and should be examined carefully.

2.2 Investor and Professionals vis-à-vis a Major Project

When the investor and his contractor-in-charge get hold of a construction project, one of the first questions they ask themselves is: 'From where do we procure the materials?'

Let us suppose that it is a 'conventional' highway section of some 30 kilometres. This section calls for more or less processed aggregates in quantities which can be estimated around:

— 500,000 m^3 or 1,000,000 tonnes for improved subgrade,

— 150,000 m^3 or 'special' materials (draining embankments adjoining highway structures, bearing surfaces, spoilbanks, draining materials etc.),

— 600,000 tonnes of aggregates for roads,

— 100,000 tonnes of aggregates for concretes, say roughly 1,000,000 m^3 or 2,000,000 tonnes in total.

These needs should be met over a period of about two years, which works out to a total production upwards of 80,000 tonnes per month over this relatively short period.

If the work should proceed simultaneously in at least two or more segments of highway, it can be seen that production rates for the purpose would far exceed the local capacity.

Local producers, anxious not to be left out in this momentary surge of activity, quickly realise nonetheless that neither huge investments in machinery, nor an unsustainable tapping of reserves would be justified, nor desertion of their traditional and regular clientele.

Investors and contractors-in-charge themselves are also well aware that their intrusion should not lead to an irreversible upset of the economic balances of the sector traversed, nor disturb any other domain, say environment.

This search for a compromise between an altogether 'imported' activity and depletion of local resources forms part of the contractor-in-charge concerns.

The predicated course of action consists of examining the following factors:

— Existing potential: deposits, processing plants, available quantities, qualities, installed production capacities, possibilities of extension, improvement and augmentation of capacity.

— Exploitable natural resources in the neighbourhood of the planned infrastructure: geology, environment, regulations.

— Accessory local products: by-products, processed binders, admixtures, correctives etc.

Once completed, this examination enables a decision regarding the 'best' tech-noeconomic policy for the complex: road-bed + improved subgrade + roads.

Sometimes it is concluded that the existing quarries, depending on the in-dividual case:

— have inadequate reserves,

— are too far from the places of utilisation,

— have an installation whose production is too low to meet the needs of the job site (over the above those of the usual clientele) and do not always produce the desired aggregate sizes.

It is in such cases that one has to resort to one or more temporary installations for making aggregates derived from:

— either wastes from the job site;

— or new quarries to be opened up, be reopened or extended for the needs of the job site, preferably close by;

— or even existing quarries by adding to the production machinery already in place (Fig. XVII.2).

Fig. XVII.2. Another example of a well-equipped mobile plant

2.3. Expectations from a Mobile Plant

In the light of the foregoing, these plants should therefore:

— Be mobile, that is, able to move from place to place and capable of easy and quick assembling and dismantling.

— Have a high production capacity to cope with surges of demand, and especially large primary crushers to ensure continuous feeding of the installation and capable of coping with oversize fragmentation at the extraction stage.

— Be quick to start up so as to minimise installation idle time.

— Be built of standard and rugged components since extended breakdowns can force halting of job-site operation, causing delays down the construction line.

— Be flexible, i.e., adaptable:

• To the terrain, sometimes to excavations or to quarries where one has neither the facility nor the time to adapt the geometry;

• To the deposit, i.e., incorporate facilities for corrections in manufacture, settings, recyclings etc. since the raw materials are necessarily less well known, and much more non-uniform and unpredictable than in a quarry long worked wherein the deposit and the working faces are documented and monitored.

— Incorporate from the beginning of the processing line a precrushing (or scalping) device for keeping out unclean materials. In some very special cases there are even installations for washing with water. But generally it is preferable to ensure conformity to specifications without resorting to washing, since either water may be in short supply, or settling of slurry might pose problems of environment and availability of space.

— Achieve the best product mix since sale of the by-products in generally more than unpredictable (and what is more, would be criticised, and rightly so, by the local producers).

It may by added that mobile installations are subject to a dispensation which imposes special restrictions: for example, sometimes quarries are licensed only for a specific project and for a limited period. The plants themselves should normally have the provision for a public enquiry, which in practice is prevented by the time-lags between the decision and the execution; hence recourse to the faster procedure of revocable temporary authorisation.

2.4 Preliminary Operations

It is desirable to minimise the proportion of improvisation which goes with this type of plant.

Certain preliminary investigations are available for this purpose, all of which at once serve to specify the choices in respect of the aggregates, organisation of the job site and detailed project engineering by professional agencies.

— Prospecting the deposits by drillings, core extractions, perforations with hammer drill, loggings, geophysical investigations (sonic or electric), rock tests etc.

— Cutting of experimental trenches with, in the case of massive rocks, blasting trials for roughly deciding the blast-hole patterns, nature of explosives and their quantity, and fragmentation; in the case of alluvial deposits, trials of pumping and lowering of water table or underwater extraction.

— Crushing trials on actual size blocks in an existing installation (permanent or temporary) to obtain some idea of the crushability of the materials, preferably followed by a practical application test which will permit, for instance, assessing the change in the material upon compaction and fix the specifications of the aggregates accordingly.

It is worthwhile to proceed side by side with the above with certain preparatory actions, such as:

— Acquisition of land needed for the installation, provisional stockpiles, reclamation of quarries and appurtenant lands etc.

— Application for licensing of the installation, if need be, in case the latter can be sufficiently defined at the planning stage.

It is obvious, and a general rule in such investigations, that the shorter the period imposed for execution of construction works, the more comprehensive the preliminary studies and actions should be.

It is also worthwhile paying attention to the volume of the overburden, the possibilities of its storage and reuse, the quarrying plan depending on the geology of the site, reuse of the spoils and the post-quarrying reclamation plan.

2.5 Main Problems Encountered

The problems most often encountered in respect of mobile plants can be divided into three broad categories:

• *Adaptation of plant to deposit*

Faced with competition, undertakings eager to get business often tend to underestimate the difficulties of the job site. The most frequent bottlenecks relate to the following:

— Steady supply from the deposit (need for sorting, changes in working face, inadequacies of fragmentation, unclean zones etc.).

— Proportion of wastes from prescreening (leading to short-feeding of the crushers and fall in efficiency).

— Installation oversimplified, necessitating addition of another crusher, indeed even one stage of crushing, in order to meet specifications.

— Fast wear of components arising from highly abrasive materials.

— Excess of fines due to the practical impossibility of washing or dedusting (more rarely, there may also be a lack of fines due to the fact that the installation is not provided with a hood), or a tendency of the rock to fillerise.

— Intermediate storage of raw materials and finished products, not anticipated at the project report stage but inevitable in actuality: raw materials are rarely fed as they arrive nor aggregates immediately dispatched upon production.

• *Quality control measures*

— High production rates, often 'shift' working, quality variations in the raw materials etc. should predicate a quality organisation plan just as rigorous, if not more so, than that obtaining in permanent installations. Such is often not the case and hence slip-ups, even rejections of delivery are known.

In practice, any temporary installation of some size should have a site laboratory paced to the horarium of the production proper.

Indeed it should be borne in mind that the quality of production is in a state of unstable equilibrium. All unforeseen deviations (clogging, wear, dysfunction, quality of raw materials etc.) degrade the quality of the product.

• *Health and safety*

The following are subscribed under this heading:

— Electrical safety.

— Mechanical safety: emergency stops, guard for hoppers and openings of crushers.

— Traffic safety: laying out lines for supply, removal weighing etc.

— Dusts: difficulty in obtaining water spray bars, apart from watering having the adverse effect of blinding the screens.

— Washing: disposal and decantation of waste waters.

— Protection: safety of third party.

— Noise (absence of hooding), exaggerated at night because not muffled by daytime din, which cannot be helped in the case of high output rates.

However, it ought to be recognised that:

— Industrial accidents are few, given modern advances in know-how.

— Quarries or excavations are generally located away from residential zones and hence their nuisance to the locals minimalised.

— The temporary nature of the installations engenders forbearance in those locals who might be discomfited.

Nevertheless, further thought needs to be given to the domain of health and safety since stipulations are becoming more stringent, and rightly so.

3. IMPORTANCE OF STUDY OF DEPOSIT

Contractors-in-charge are adamant that the quality of aggregates meet specific criteria; this can only be achieved by appropriate mechanical processing of the raw materials. This supposes installation of more or less sophisticated processing lines, incorporating several stages of reduction and size classification interlinked amongst themselves by the combination of handling, distribution and loading operations. Thus the production of aggregates, whether from crushed stone or an alluvial source, calls for very large investments.

But then, regardless of the quality or size of the machinery installed, difficulties do crop up in actual production, when the preliminary investigations enumerated in sections 2.4 and 2.5 have been skipped; such investigations are even more imperative in temporary installations.

An *a posteriori* examination of a few cases of actual exploitations revealed that the majority of difficulties which crop up during production stem from inadequate knowledge of the deposit. This knowledge is particularly indispensable, since the time needed for controlling the exploitation is short vis-à-vis the high production rates.

Therefore, detailed prospecting of a deposit should furnish all the data needed for its exploitation later. Problems of estimating the quantity and quality of the materials are not the only ones; it is very important to collect all the data requisite for planning the most rational utilisation of the entire materials extracted. A good study of the deposit should enable identification of the best method of exploitation and facilitate the choice of machinery for the crushing-screening installation.

That is why a knowledge of certain parameters is imperative to good management of exploitation of a deposit.

It would be pertinent to recall that these studies cater to three categories of users:

— *Contractor-in-charge:* he should know which part of the deposit can be excluded from a quantitative as well as qualitative point of view.

— *Entrepreneur,* producer of materials: he should find in the study in particular information which will first of all enable him to work out costs, identify the machinery needed, and the manner in which he will organise the work, and then envisage ways and means of dealing during actual production, which some variations in the deposit, either discernible or predictable.

— *Machinery supplier:* he should be able to deduce from the preliminary studies the nature of the machinery required and how to integrate the different equipment for resolving any problems that might crop up.

Given the foregoing we deemed it necessary to identify the main aspects of a deposit which should imperatively be known for proper designing of an installation and ensuring smooth management of an exploitation.

We shall not touch problems relevant to the geographic situation of a site (position, communications etc.) but confine ourselves to the deposit proper. We shall thus have to consider its geometrical parameters (overburden, reserve), conditions of hydrology, intrinsic characteristics of the materials, fragmentation of the rock face, grading of the materials present (alluvial) and the various types of pollution (see also chapter III).

3.1 Geometrical Parameters and Their Implications for Exploitation

It is obvious that a good knowledge of the total volume of a deposit is indispensable—both of the overburden as much as the usable materials.

3.1.1. GEOMETRY OF OVERBURDEN

Quite often stripping of the overburden, a highly onerous job, is insufficiently done; not only should the top soil be removed, but the weathered portion of the rock, which often presents a challenge.

Serious consideration should be given to the economic implications of the overburden wastes and the handling and dumping problems they entail.

How to judge the overburden volume?

An exploitable geological horizon does not necessarily signify considerable depth; hence the overburden thickness is a very important parameter. Any miscalculation can spell catastrophic economic consequences for the quarry manager:

— Higher volume of wastes to be stripped so as to avoid a large part of the blasted materials being removed in the prescreening before primary crushing.

— Inadequate height of working face and hence a larger number of blasts to be carried out for feeding the installation properly.

• *In deposits of massive rocks* delimitation of the overburden in the case of magmatic rocks is more complicated than with limestones, since the degree of weathering varies progressively from the weathered clay in place up to the unaltered rock (Fig. XVII.3). It is essential to know this progression well since the thickness of the overburden to be removed may be considerable before mining of the unaltered rock.

Fig. XVII.3. Mixture of clay and unaltered and weathered blocks in an overburden of magmatic deposit

• *In alluvial deposits* the overburden presents disorderly configurations, as a result of which the overburden to be removed at the time of extraction in order to secure a satisfactory platform of extraction, may be larger than anticipated (Fig XVII.4).

3.1.2 NATURE OF OVERBURDEN

A thorough knowledge of the nature of the overburden is paramount before organising extraction and management.

• *In limestone deposits* in which the overburden happens to bear minimal pollution, one may proceed with preparation of the deposit in two stages: first, a rough stripping to remove the earth, followed by clearing the platform. This method can make prescreening materials usable, which under other conditions of preparation would at best be suitable only as filling materials.

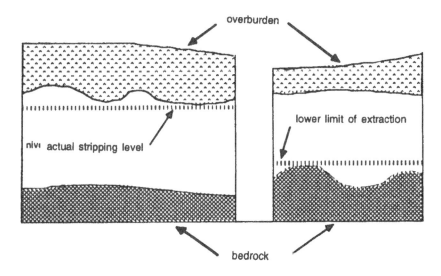

Fig. XVII.4. Alluvial deposit, irregularities of overburden and bedrock

• *In magmatic deposits* a thorough knowledge of the distribution of weathering enables proceeding in three stages: stripping the highly weathered zone, mining the less weathered zone for low-quality aggregates and finally the truly unaltered rock (Fig. XVII.5).

• *In alluvial deposits* the overburden is seldom reusable except when composed of relatively clean sand. Generally it comprises clays or highly argillaceous silts, which are stored for reclamation of the site.

3.1.3 MINABLE MATERIAL

The processing lines are designed for converting the 'normal' and known variability of the deposits into a homogeneity acceptable for aggregates. But abrupt and unexpected variations represent a factor of abnormal variability and cause very large, even serious, upsets in the functioning of the processing lines and the quality of the finished product. Therefore it is very important to make out these heterogeneities to enable directing extension of the working faces accordingly.

• *In limestone deposits* the benches are often narrow (Fig. XVII.6) and the changes in geological facies rapid. The problem is thus knowing the homogeneity and the continuity of the facies, as even minor faults may have caused displacement between the beds and changes in dip of the strata. Such data straightaway impinges upon a delimitation of the exploitable zone, choice of the point for opening up the extraction and siting of the installations.

• *In magmatic deposits* the heterogeneities are generally distributed in a manner more difficult to predict: veins of rocks that differ in nature from the main rock, faults, diaclases, crushed zones etc. (Fig. XVII.7).

Fig. XVII.5. Bowl-shaped weathering in an overburden of magmatic rock

Fig. XVII.6. Thin beds in a limestone deposit

Fig. XVII.7. Heterogeneities on the face of a magmatic deposit

• *In the case of alluvial deposits* estimation of the total volume should take into account the geometry of the bed-rock and its variations on the one hand, and the presence of heterogeneities which can make horizons or surfaces inexploitable, and involve neutralisation of one portion of the deposit on the other. These are mainly lenses, pockets of clay and consolidated horizons or huge blocks.

It is to be noted that for all types of deposits, putting up installations depends on the topography of the terrain, which can be turned to advantage for minimising and simplifying the materials-handling operation.

3.2 Hydrological Parameters

These are of paramount importance in the case of alluvial deposits, but it also happens that the presence of a water table in massive rocks leads to considerable water flows. It is imperative to know the variations in the water table level as these directly influence:

• progress in mining (problem of flooding or problem of waterlogging on the quarry head);

• mode of overburden removal according to whether it lies outside or in the water zone;

• mode of extraction of gravels;

• cleanliness of the materials;

• choice of installations when washing the materials is necessary.

3.3 Parameters Defining Qualitative Characteristics of the Material

It is well to recall a proven fact; while one can produce a mediocre or poor material from a good deposit, it is impossible to produce a good aggregate from a bad deposit. This parameter becomes important to the extent that knowledge of the raw materials quality conditions the whole equilibrium of the installation. It is then possible to avoid the following classical mistakes:

— Inadequacy of primary crushers consequent to poor assessment of the state of fragmentation, the hardness and hence the size of the blocks after mining.

— Excessive wear in some machinery arising from an erroneous consideration of the rock's abrasiveness and crushing strength.

These few examples illustrate just how necessary proper data on the deposits is.

3.3.1 ROCK MASSES

Knowledge of four important aspects is indispensable: the petrographic nature of the materials, their mechanical properties, the state of fragmentation of the materials *in situ* and the various types of pollution.

• *Petrographic nature and mechanical characteristics of the materials*

Microscopic examination of rocks and alluvia makes it possible to ascertain:

— *The structure of the rock*

Through this analysis one can accurately determine the nature, dimensions, shape and mode of the mineral assemblage of the rock. The mechanical properties depend by and large on this structure. For example, crushers are highly sensitive to the cohesion between grains; thus a hard and well-bonded rock is difficult to crush while an assemblage of very hard but poorly bonded minerals is readily broken. As a general rule, it can be said that the fineness of the grains, nature and quality of the interstitial cement, and distribution and arrangement of the minerals definitely influence the resistance of a rock to crushing and the resultant grain size.

— *Its mineralogical composition*

Based on a microscopic analysis of the minerals constituting a rock, it is possible to make a first approximation of its characteristics. Thus the presence in abundance of weathered minerals will entrain a notable fall in mechanical strength.

In the case of limestones a chemical analysis as a complement to the mineralogical one may well prove useful; silica, dolomite and sulphur contents can be detrimental to the functioning of some machinery and pose problems for end uses of the materials.

The strength and abrasiveness of the material serve as parameters in selection of the equipment, which ensures that the particle size of the output accords with the desired range.

It may be noted that the concept of the term 'strength' as used here is none too readily grasped. Tests such as compressive strength, Micro-Deval and Los Angeles (see Chapter IV) do illuminate the choice of equipment to some extent but there is no simple rule for applying such tests results to the machinery under consideration.

Likewise, the concept of abrasiveness based on silica content is only approximate; measurement with an LPC abrasimeter is preferable (see Chapter IV).

One may perhaps attribute this state of affairs to the fact that most of the current tests for aggregates do not fully reflect the behaviour or rock fragments in a size-reduction machine. In the absence of a perfect characterisation, which anyway does not exist, it is well to have the maximum possible information.

The geological and petrological data collected on the rock to be processed often enables selection of machinery in the light of experience with similar materials. Experimental crushing trials are highly useful.

• *Microfissuration and tendency to cleavage*

Some rocks may have oriented structures which create lines of weakness and well-defined rupture planes upon subjection to compression during crushing. Consequently, in primary crushing for example, flakes may be produced in preponderance, which might entrain disturbances in conveying the materials and feeding the secondary crushers whose openings might prove undersized for the purpose.

• *State of in-situ fragmentation of materials*

Knowledge of this governs the type of drilling, mode of mining, type and sizing of the primary crusher and scalping mesh prior to the primary crusher. For instance:

— If a primary jaw crusher is too big, it will first constitute an enormous investment, then work 'with extremity of the jaws' with a localised wear in this zone.

— If the crusher is too small, the well-known risk of the blocks jamming at the inlet is likely, with the concomitant problems of recovery by the hook, rupture in the mass and attendant breaks in production. If the oversized blocks are picked out before discharging into the primary crusher, they must be fragmented by blasting, which boosts mining costs.

• *Various pollutions*

The problem is knowing whether undesirable materials (clays, friable elements) are present in the pit run and if so, should they be removed and how. It is that much more difficult to resolve this point in the case of an abandoned working face or a virgin deposit, as it is generally easy for a quarry under exploitation.

There are two types of pollution: that connected with the origin of the massif and that which is 'accidental'.

— *Pollution traceable to the genesis of the rock* essentially comprises:

• In the case of limestones: intercalation of clayey or marly beds or lenses of a type different from the facies of the rock body. The two types of heterogeneity which have to do with the conditions of sedimentation happen to coexist in the same deposit.

• In the case of magmatic rocks: the existence of veins which repeatedly cut the main facies and which can more often be considered as pollutions because they are often more weathered than the enclosing rock.

— *'Accidental' pollution:* The presence of highly fractured or even crushed zones can greatly hinder an exploitation (Fig. XVII.8). If not detected in preliminary examination, it may happen that the working face falls into such a zone parallel to its longest stretch. Running through or skirting it is unavoidable. Since such zones most generally comprise highly weathered materials unfit for use, several days of profitable production may be lost.

Fig. XVII.8. Gouge zone polluting quarry working face

Faults seldom occur alone but rather form in association with diaclases, a hierarchical network of major fractures and secondary fissurations. This fractured body is the preferred zone for the penetration of surface weathering, forming bigger or smaller pockets of weathering which have to be removed at extraction. At the level of processing machinery the major disturbance will arise from the clayey nature of the materials, especially when they are wet. The blocks will be virtually coated, resulting in blinding of screens, made all the more serious if prior removal was not envisaged nor adequately provided for at the design stage of the installation.

Faults are not the only phenomena which can engender the appearance of weathering. Mention should also be made of *pockets of clayey dissolution*. These are relatively frequent phenomena is limestone massifs which exert a direct influence on the success of blasting schemes and the state of the material after mining, which in turn governs the manner of loading and the prescreening mesh for eliminating the clayey fraction depending on atmospheric conditions.

Elimination of pollutions

In the case of aggregates coming from rock masses, one or more eliminations before or after crushing should be provided in order to obtain a satisfactory cleanliness, tested with the sand equivalent. In a mobile crushing unit it is almost imperative to provide a prescreening in advance of the primary crusher. In some cases removal after primary crushing is necessary to improve the cleanliness of the material.

Accordingly, three schemes are possible:

— *Elimination in advance of primary crusher:* A 'bypass' (Fig. XVII.9) enables extraction of muddy wastes (0/60 for example) to extract the aggregate size (60/100 for example), which relieves pressure on the crusher and avoids pollution of the blocks inside it.

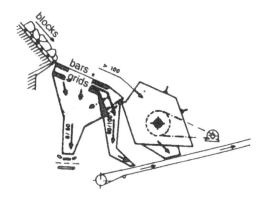

Fig. XVII.9. 'Bypass'

— *Elimination after primary crusher:* This arrangement removes the 0/d products after primary crushing. It can be advantageous (utilisation of wastes) but means a loss in production, a negative point in the case of a primary impact crusher.

— *Elimination on advance of and after primary crusher:* This solution, though it may combine the advantages of both types of elimination, is nonetheless very costly.

The cleanliness of crushed sands thus depends on two factors: the nature of the deposit and the processing methods. It is known that the various crushers available tend to concentrate on the finest fractions and the most friable and most weathered elements; hence the cleanliness of sands increases with the number of crushing stages and the number of eliminations of the fractions. But when the materials come from a portion of the deposit in which the rocks are weathered, it is inevitable that even under good atmospheric conditions an appreciable fall in the values of sand equivalent of the crushed products will result, and this even when one raises the cut-off point for elimination after the primary crushing.

3.3.2 ALLUVIAL DEPOSITS

• *Grading*

In respect of alluvial materials, the starting grading is a preponderant factor:

— When it is too sandy, one faces the problems of desanding, excessive sand and often also of the maximum mesh size to be chosen for obtaining materials which can be effectively crushed.

— When the starting material is too coarse, the opposite problems of very intense crushing and production of sand are encountered.

Thus it is imperative to know the grading curve and especially the size of the coarsest element, or more explicitly, the percentage of elements exceeding a given mesh unless the downstream crushers are oversized.

The pattern of change in grading in the deposit, both vertical as well as lateral, should be precisely known; wide variations have a direct influence on extraction and processing installations, for instance:

— When superposition of two horizons of widely different grading occurs (fine sand upon gravels) (Fig. XVII.10), one might have to extract them one after the other or, on the contrary, mix them depending on the needs.

— In the case of a deposit of meanders, the percentage of sands can change from 30% at one end of the deposit to 70% at the other. (see also Fig VIII.8). Ignorance of such a change could be catastrophic for a continuous supply whereas

Fig. XVII.10. Grain-size heterogeneities in an alluvial deposit

prior awareness would enable, for example, marking out two simultaneous extraction points.

Therefore, a pit run to be processed is often a zone enveloped by possible grading curves. The crushers are generally sized for the most hollow curve (the coarsest) and the screens for the most convex curve (the finest).

There can be gaps in the particle size distribution of alluvia, which are mainly of two types:
— Gap due to excess of sand (between 0.5 and 1 mm);
— Gap due to lack of fine elements (finer than 1 or 2 mm).

Elimination of gaps in the particle size distribution accordingly supposes installations for correction of grading of the sands. These installations imply in practice a treatment of the material in two steps: firstly, a cut of the 'raw sand' fraction into two or more intermediate fractions, then a blending of the different fractions obtained in specified proportions to yield the grading as per the specification zone. This equipment is complex and seldom used in mobile plants.

• *Cleanliness*

Directly related to grading, cleanliness has a direct influence on the installation: is washing called for and, if so, at what stage of the processing?

These questions, which have to be faced by the quarry manager, make a study of the pollution one of the essential objectives of deposit investigation. What are the pollutions incidental to an alluvial site? They are of two origins (Fig. XVII.11):
— If they are contemporaneous with the deposit, they pervade the entire material or occur in the form of pockets or interstratified lenses;
— If they are post-sedimentary, they correspond to phenomena which have taken place subsequently in the deposit.

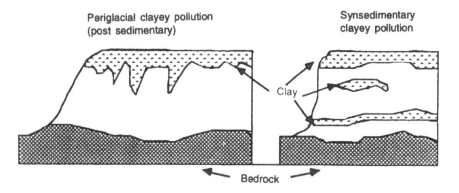

Fig. XVII.11. Alluvial deposit: Two different types of clayey "pollution"

Once these pollutions are marked out in the deposit, it becomes possible to explore their elimination through the processing circuit:
— Where the impurities are loose, a simple screening under water spray will suffice to bring them into suspension and make them pass along with the sands, later to be separated during classification operations.

— But more often the materials, gravels (i.e., 2/20) or sands, arriving at the processing plant are coated with clays or marls that should be treated with water and totally removed to obtain finished products of good quality. The solution then consists in sending the pit run through a scrubber which dislodges the clay adhering to the surface of the materials while wetting them.

To accomplish a good washing, access to water is imperative. When it is plentiful, the layouts of the circuits and the volumes used matter little so long as the slurry can be discharged from the plant. Contrarily, when water is scarce or discharge of waste water is not possible except after treatment, the water circuits should be studied in great detail to achieve rational utilisation with maximum efficiency.

On the subject of washing, mention should be made of a crucial point in the decision to open up an alluvial deposit: discharging of clayey waters into rivers is subject to water pollution regulation. Therefore it will often be necessary to provide settling ponds, which require sizeable land areas and cannot be improvised.

• *Mineralogical, petrographic and mechanical characteristics of the materials*

The problems are the same as those discussed in the case of rock masses but in addition one must ascertain the percentages of soft or weathered elements in the different aggregate sizes of the raw materials, as well as the exact nature of the hardness of the predominant elements.

3.4 Conclusions

Once the characteristics of the materials and the deposit are known, that is, the exploitability is judged to be good, it is possible to define the type of treatment likely to give the desired aggregates at the most attractive cost. This approach raises varied and complex problems calling for wide experience. An examination of the choice of installations takes for granted knowledge of the operations ensuring their feeding with the given materials; stripping and ultimate reuse of the stripped material; blasting or extraction carried out with adequate machinery and following a specified direction of advance; and handling as dictated by the topography and the structure of the deposit.

These 'defined' materials which feed the installation are such that they demand a minimum number of 'negative' operations (elimination of pollutions, washings) for conversion into aggregates.

4. TREATMENT OF MATERIALS IN MOBILE PLANTS

The designer of a mobile unit should be able to take into account all the data collected during the initial studies, then install at all levels of the processing line the machinery most adapted to the known or unknown variations in the deposit and the materials to be processed.

The machines meant for mobile plants in no way differ from those used in permanent quarries. But of all the existing ranges, some are more commonly used than others and it is precisely these that we propose to discuss briefly, since machinery per se has been treated in detail in Chapters VII to X.

Let us now examine the different functions, confining ourselves only to what is peculiar to the mobile units or the machinery widely employed in this type of installation.

4.1 Extraction and Transport of Raw Material

This initial phase comprises the operations which should be carried out upstream of the size reductions and classification into the desired elements.

4.1.1 STRIPPING OF OVERBURDEN

Careful survey of the deposit brings out weathered portions called 'steriles' which should be removed. This is effected by means of a bulldozer or power shovel and occasionally calls for some mining. A thorough stripping is absolutely essential since the possibilities of removing polluted materials are less than in a permanent quarry.

4.1.2 LOADING OF MATERIALS

After extraction, the materials reclaimed by the loading machines (loader or power shovel) are sometimes stacked to form buffer stockpiles which fulfil two essential functions: blending and regulation of feed to the primary crusher. It is desirable

Fig. XVII.12. Loading of mined materials in loader

to promote the idea of primary stockpiles, which are as yet seldom resorted to in mobile units (Fig. XVII.12).

4.2 Feeding the Installation

Before crushing, the pit run is discharged into a hopper from dumpers and the primary crusher fed through metallic apron type or vibratory feeders.

4.3 Prescreening—Scrubbing

This operation helps effect the initial operations of elimination of pollutions ahead of the installation. A scalping prior to primary crushing is compulsory in all mobile installations, barring very special cases. If a rocky deposit is little polluted, a prescreening, that is, one elimination before the primary crusher suffices. On the contrary, if the deposit is highly polluted, another elimination is carried out after primary crushing (see Sec. 3.3).

In the case of alluvial materials, when the raw material is highly polluted with a 'sticky' clay, wet treatment is necessary upstream of the installation, i.e., 'scrubbing' (Fig. XVII.13).

Fig. XVII.13. Scrubber

4.4 Primary Processing

The objective is to produce a material of consistent quality which should serve to feed the processing line for making aggregates. Therefore the primary processing (Fig. XVII.14) aims at reducing the size of the extracted raw materials to a particle size, generally 0/200 or 0/300 mm, which can be easily transported by belt conveyors. In a permanent installation this material is stored in the form of a primary

Fig. XVII.14. Mobile primary processing unit

stockpile called 'precrushed' which is generally reclaimed in a tunnel. A few mobile installations resort to this method of intermediate storage but a large number still work as the material arrives, with all the risks this entails.

Use is generally made of:

— Double toggle jaw crusher;

— Hammer crusher whose use remains confined to the primary crushing of non-abrasive materials (limestone type) since the cost incidental to replacement of the hammers soon becomes prohibitive.

4.5 Secondary Crushing

Products from the primary processing are generally too oversized for direct usage and hence have to be reduced in a second stage, termed secondary crushing. This more intensive reduction of the materials enables obtaining finer aggregate sizes. Several types of machines can be employed.

— Gyratory or cone crushers which give a product of consistent grading provided the crusher operates steadily at full load. This explains the increased interest in 'buffer hoppers' in temporary installations, which are arranged upstream of these machines to regulate their feed.

— Hammer crushers.

4.6 Tertiary Grinding

It may happen that the desired particle size is still not obtained even after the secondary crushers.

At this stage the question is thus one of producing materials which are richer in sand and finer elements (fines). To meet these requirements recourse is had to crushing machinery termed grinding mills.

— Cone or gyratory grinding mills with their 'fine grinding' devices,
— Rod mills (Fig. XVII.15),
— Roller mills (Fig. XVII.16).

Fig. XVII.15. Rod mill

4.7 Screening

Once the materials are reduced to the desired size by the crushers and grinding mills, screens are employed for separating them into 'fractions' so as to check the segregation of the 0/D and to utilise to the utmost all the particle size fractions. Since the products manufactured are more susceptible to bad weather and facilities for sheltering fewer than in permanent installations, screening of materials to less than 4-mm mesh should be avoided.

Fig. XVII.16. Roller mill

4.8 Machinery for Treating Sands

In some cases the materials for processing call for wet treatment. These are notably alluvial materials whose screening is carried out under water spray; also, the fractions passing the final stage of screening should be dewatered by a machine or treated at a special station. The machinery involved includes:

— *Paddle wheels,* which present the problem of achieving a size fraction around 100 μm, hence impoverishing the sand of its fine fraction.

— *Centrifugal classifiers,* employed for achieving size fractions down to 40 μm.

4.9 Materials Handling and Storage

The products have to be transported from one point of the installation to another, for example from crushers to screens. Furthermore, the manufactured aggregates need to be stored.

4.10 Conclusion

Mobile plants often combine on a single platform one or more processing machines. The functions most often combined are:

Fig. XVII.17. Crushing-screening unit

— Feeding and primary processing (primary units);
— Crushing, grinding and screening (secondary and tertiary units) (Fig. XVII.17).

Most of the machines incorporate all the back-up needed for their operation (motors, electric panel, aprons for feeding, sheets for flank etc.).

5. CONCLUSION

To manufacture aggregates which are homogeneous and conform to a specified quality is a technically complex task, whatever their end use may be.

In this field, as in many others, nature imposes her own law.

Once they leave their place of manufacture, these aggregates find their resting place in composite structures, which from the moment they are commissioned are exposed to multiple stresses: air, water, ice, chemical agents, attrition, erosion, compression, tension, fatigue—all of which contribute to the deterioration and destruction of the said structures.

One can never emphasise too much the quality requirement, which is occasionally overlooked when the techniques of embankments, roads or cements are perceived as 'rustic' compared to the so-called 'sophisticates'. But, beware! A few per cent of excess or less fines, an inadequate 'cleanliness', an unsuitable shape or angularity can spell the premature ruin of a road or highway structure.

The difficulty in the case of mobile plants is compounded due to the scarcity of time, if any, available for tuning the installation to the deposit, the unpredictabilities of the deposit, the quick pace of slippages etc.

By way of conclusion, here are a few axioms which may prove useful on occasion.

— Preliminary studies and tests are costly only when avoided.

— Deposits are uniform only on a geological map; otherwise they are discontinuous, heterogeneous and anisotropic media.

— Machinery is new only in the manufacturer's catalogue.

— An installation is not in stable equilibrium while in operation, but rather when out of order.

— Gravels are less hard than parts of steel but greater in number.

— Crushers and screens appreciate neither indigestion nor sudden pangs of hunger.

— The force of things draws not towards quality, but towards defects.

— Quality control is not a shackle, but an auxiliary motor.

— A good dish cannot be made with bad ingredients.

REFERENCES*

Archimbaud, C. 1978. Establishment, layout and exploitation of quarries. IRF Conf. for Middle East and North Africa, Cairo.

Archimbaud, C. and Y. Martin-Guillou. 1977. Influence of the characteristics of deposits on the design of installations. *Bull. Liaison Lab. P. et Ch.*, special issue IV, pp. 201–210.

Archimbaud, C., J.P. Joubert, A. Maldonado, A. Prax, L. Primel and J. Roy. 1973. Production of aggregates. *Travaux*, special issue, *The Road*, no. 464, pp. 46–62.

Gaud, P. 1980. Temporary installations for production of aggregates. *Aggregates*. Assoc. Amicale des Ingénieurs Anciens Elèves de l'ENPC, 1st ed., pp. 399–409.

Panet, M., L. Primel and C. Archimbaud. 1972. Deposits of materials and temporary installations. *Bull. Liaison Lab. P. et Ch.*, July–Aug. 1972, pp. 69–91 and *RGRA*, no. 477, pp. 57–74.

Rodenfuser, J. and J.P. Maquaire. 1970. Mobile and temporary installations. *Aggregates*. Part XVII: Mechanical Equipment, Quarries and Materials. Compagnie française d'Edition.

---

*All references in French—General Editor.

XVIII

Quality Control of Aggregates

Claude Tourenq, Alain Maldonado and Jean-Marie Gross

1. ORGANISATION OF QUALITY

Organisation of quality is a constant objective of the entire French industry to make it more efficient. It should imbue confidence, both within and without a firm that the results desired have been achieved unfailingly and the stated or potential needs of the users thereby satisfied.

Producers of aggregates are particularly bound by this policy of industrial quality, which has been officially sanctioned by publication in the Official Bulletin of the French Republic* of the new "Part 23" annexed to the General Technical Specifications for State public purchases.

"Part 23" deals with the supply of aggregates used in the construction and maintenance of roads and makes it obligatory upon every aggregate supplier to implement a quality organisation plan (QOP) which should be presented along with tenders. This enjoins upon the producer to institute certain measures, guarantee their actuality, ensure their regular implementation, and lastly promulgate them. This obligation increases the supplier's accountability to his client.

It appears that the obligation on the part of the producer to choose a QOP prompts him to think anew about his quarrying organisation, which in turn makes him better appreciate and optimise the processes of manufacturing the products. Once this choice is made, he becomes obliged to effect it in legiance with the purchaser.

It is important to note that:

— The obligations arising from the adoption of a QOP are contractually binding only in the case of public sector clients (State, civil bodies, notified public institutions).

— The public purchaser can only insist on an A, B, or C type plan.

— The adoption of the most elaborate scheme will take effect upon a special remuneration to the producer which in no way shall be construed as relieving the purchaser from his duty to inspect the same.

---

*This Bulletin includes detailed laws and official announcements.

The aggregate producer should choose from amongst the three types of QOP defined below.

1.1 Type 'A' QOP

This comprises the following minimum general requirements.
— Designation by name of the quarry-in-charge.
— When the deposit has not been proposed by the purchaser, specifics of the deposit in which the locations of extraction are marked shall be submitted.
— Processing and stockpiling of machinery shall also be specified, notably presenting:
 • Crushing-screening scheme and relevant equipment,
 • Specimen of delivery ticket mentioning identification and quantities of aggregates by aggregate size.

Therefore this plan takes into account the machinery in place but involves the purchaser for regulation and conduction of the manufacturing process (Table XVIII.1).

Table XVIII.1: Allocation of tasks between supplier and purchaser

Quality Organisation Plan			Operations
A	B	C	
Obligations of			Description of deposit and installation
Duties of		Supplier	Regulation of installation Functioning checks Log of internal checks
	Purchaser		Inspection of internal checks Tests for conformity Log of external checks
			Monitoring regulation of installation Inspection of external checks

1.2 Type 'B' QOP

In addition to the general/duties described under the type 'A' QOP scheme, QOP scheme 'B' includes the following which pertain to the production chain:
— Designation by name of the in-charge of internal checking.
— Carrying out the crushing machinery settings for defining its parameters of variation.
— Maintaining a daily log of internal checking carried out and assuring access to it at all times to the user.
— Standing instructions relating to selection of the materials at the working face or at extraction, checking of setting and proper functioning of the equipment.

This plan is based on knowledge of equipment functioning and their slow drift from the initial setting, with a view to devising a suitable corrective strategy.

1.3 Type 'C' QOP

In addition to the measures described under type 'A' and type 'B' QOP schemes, type 'C' includes checking operations outside the production chain, which are as follows:

— Designation by name of the in-charge of external checking.

— Day-to-day inspection of the log of internal checking.

— Mention of the type of tests to be carried out and their frequency with a view to checking conformity to specifications.

— Posting results in the log of external checking, with mention of the steps taken by the supplier to rectify any deficiency in conformity.

— Summation of how the results were utilised upon completion of contract.

2. ACHIEVING QUALITY

2.1 Control of Production

The quality of aggregates has a great influence on the good performance of the materials used in civil engineering and highway structures. This makes it obligatory upon the purchaser and the user to exercise due care regarding conformity to specifications. Laying down a quality control system should necessarily take into account the data specific to the aggregates and the manner of their production:

— For transport distances of the order of a few tens of kilometres, transport cost of the aggregates outstrips production cost. Hence there is an obvious economic need for implementing control operations at the place of manufacture where any decision for rejecting or outgrading will have less onerous consequences.

— Considering the present set of tests, the cost of a statistical quality control soon becomes disproportionate to the production cost of the aggregates.

— The quality of the aggregates is never totally unrelated to the natural variations inherent in the deposit, irrespective of the production methods employed

Consequently, if acceptance control was done by checking a fewer number of test results, it becomes imperative to double that number as part of control on the means of production so as to reduce the risks for the users (purchaser and contractor). In the absence of this, acceptance control could mean underwriting a lottery-like affair for the contractor or making it impossible for the purchaser to appreciate in the normal course the real quality of the product delivered (or both).

Practice also shows that it is necessary to orient acceptance control to match the requirements of the degree of sophistication in the means employed at the construction site.

Control of production rests essentially upon a thorough knowledge of the deposit and the processing installation. Indeed such information helps to elaborate and implement a strategy of quarrying and manufacture which yields a product matching the demand, by minimising drifts in quality over a period.

2.1.1 KNOWLEDGE OF THE DEPOSIT

The absolutely essential knowledge of a deposit comprises petrographic identifications (description of the different kinds of rocks encountered) and investigation of its structure (spatial arrangement of the beds, masses, discontinuities and weathering). At the end of such studies, the producer ought to possess all the data requisite for an assessment of the extent and orientation of the deposit, the characteristics of the minable rocks and the contours of weathering liable to be encountered, even if not detected in the investigation.

These elements are indispensable to the quarry manager for opening or developing a deposit, i.e., determination of depth and direction of quarrying, taking into account the data relating to:

— optimal exploitation of the deposit,
— minimisation of effects of polluted zones on mine workings,
— safety of quarrying personnel,
— uniformity in size of the blocks to feed the processing machinery.

When the quarry is under active exploitation, a map incorporating qualitative observations zone-wise can be useful for tailoring the quality of the finished product to the market requirements. Fig. XVIII.1 depicts an example of a geotechnical map whose updating frequency will be governed entirely by the rate of development of the working faces and the heterogeneity of the deposit.

Selection of the zone to be quarried is the result of a compromise between the picture of how a quarry working face would progress in the medium term (the upper levels should be quarried first, then the lower, and a medium cost of transports maintained at extraction) and the expedient wherein it is necessary to manipulate the different quality grades at the working faces, depending on the properties demanded of the aggregates. It is also necessary to take into account the constraints arising, on the one hand, from the capabilities of the machinery employed (crawler-mounted shovels, loaders, dumpers etc.), all of which do not have the same capacities nor the same mobility and, on the other, the time needed for quarry preparation (drilling, charging and blasting).

2.1.2 INSTALLATION KNOW-HOW AND ITS REGULATION

Investigation of the installation consists in identifying the function and capabilities of each of the machines employed in the processing line. The role and the object of the links and circuits which can be used have to be identified beforehand.

The adjustments possible in the equipment for qualitative and quantitative control of output should be known. This investigation has as its end product a project report setting out a scheme of the plant indicating the equipment, the machines, the circuits, their adjustments and the options of feed, as well as the various materials produced.

This description will be complemented by a databank on the output of crushers for different modes of feed and settings. The grading of the crushed products (in terms of aggregate sizes, flakiness index) depends on the operating conditions of the crushers and grinding mills.

Armed with this data, the production-in-charge will be in a position to select a production circuit and the settings of the equipment which will give the desired output under optimal conditions. In other words, he will be able to produce the aggregates at the desired output rate, minimise the production of materials falling outside the desired range, lower the cost of intermediate transport etc.

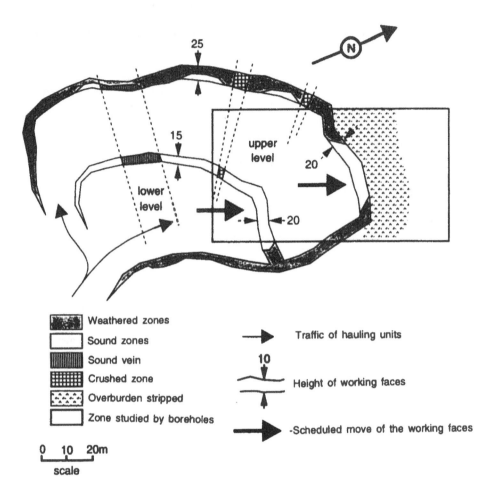

Fig. XVIII.1. Plan delineating sound and weathered zones in a quarry

This median functioning point of working can be usefully determined by means of a microprocessor having a suitable program (see Chapter XVI, Automation).

2.1.3 ON-LINE CHANGES IN THE WORKING OF A DEPOSIT

When a deposit is worked on more than one face and in case a gradual slip in quality is noticed after analyses, mix proportions of the raw material arriving at the primary crusher can be temporarily changed by altering the turn-around pattern of the dumpers.

If the change in quality is caused by a well-marked-out zone of weathering, it will probably be necessary to 'clean the working face'. The polluted portion of the material quarried will be consigned to the wastes to preclude contact with quality materials. However, it should never be forgotten that change in the mining plan is neither easy nor swift, which is why a rather comprehensive investigation of the deposit is essential in order to predict quality.

2.1.4 INTERVENTIONS IN THE INSTALLATION FOR MAINTAINING QUALITY

Once the settings are done, that is, once the system has been set on stream, one should:

— Check the quality of the materials to validate the choices made;

— Introduce a procedure for maintaining quality over a period, and the rate of output if the quality is satisfactory.

The validation procedure consists in carrying out a run of tests with reference to the specifications of the aggregates to be delivered, taking care to do the sampling for this purpose when the installation is working in a stable regime. This run of tests can be followed by a change in the setting as necessary to optimise the manufacturing operation.

Figure XVIII.2 and Table XVIII.2 show how the setting of two crushers can change the output of an entire installation.

Maintaining quality over a period consists of detecting slips in the manufacture and altering the adjustments of the equipment to effect absolutely essential corrections.

Slips in output quality are often caused by wear in the grinding components of the crushers and the screen fabrics. In all cases the throughputs of materials in the installation stand changed as also the feeding conditions of the machinery, and

Table XVIII.2: Simplified scheme of simulated quarry. A1, A2: adjustment in mm of crushers C1 and C2. 0, 4..., 10 14 mm: output in t/h

	A1	A2	0 4	4 6	6 10	10 14 mm	Feed
1)	Fine aggregate feed		Sandy feed				
	19	16	57.4	27.8	23.9	16	349 t/h
		10	60.7	25	21	18.5	
		6.5	66	25	21.3	13	
	8	16	60.6	22.1	21.7	10	
		10	62.4	20.4	20.2	11.2	
		6.5	65	21	20.3	8.3	
	19	16	66.9	31.2	27		349 t/h
		6.5	66.9	25.5	21.7		Recycling of 10/14 accumulation
	8	16	66.5	24.2	23.6		
		10	69.9	22.3	22		silo 1
		6.5	70.7	22	21.6		
2)	Coarse aggregate feed						
	8	16	78.7	33.5	16.9		Granulation and recycling of accumulation above C1, and surplus feed C2
		6.5	82.7	24.9	24.2		
	19	16	75	31.7	26.6		

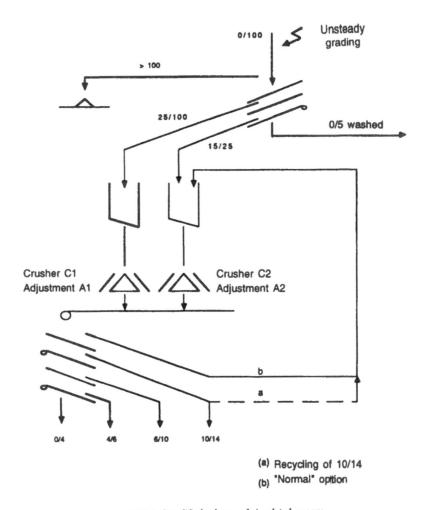

Fig. XVIII.2. Simplified scheme of simulated quarry

the plant foreman is faced with an altered performance regime that obliges him to implement new adjustments.

Therefore careful watch must be kept over the sensitive points of the installation after duly identifying them. Some examples of sensitive points, which differ in terms of how early or how late they may necessitate intervention are shown in Fig. XVIII.3.

The cut-off nominal size of the primary crusher output is to be adapted to the conditions of the primary raw material so as to maintain balanced circulating loads in the grinding circuit. The quality of the tertiary 0/D depends on the combined functioning of the crushers in place.

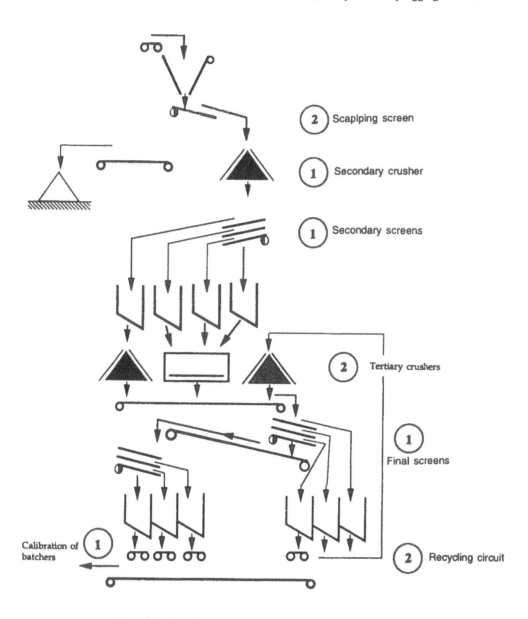

Fig. XVIII.3. Examples of sensitive points. ①—Slow drift; ②—Fast drift

The wear in screen fabrics is not necessarily fast but the service wing has to keep an eye on the distribution of the feed to the screen, which can be disturbed by an off-centring of the conveyor belt or a crushing-in of the chute.

Lastly, various kinds of breakdowns can happen unexpectedly and warrant shutting down some departments of manufacture or some equipment. When the main 'axis' of manufacture in the quarry can still be kept going, the manager of the manufacturing operation is confronted by an altered regime of work and is obliged to set up a new one, validate the new adjustments and maintain the quality of the output.

2.2 Check for Conformity (Fig. XVIII.4)

A conformity check on the aggregates produced is carried out either before their dispatch or after their delivery at the storage area in the job site. This check consists of drawing sampling increments and analysing them to ascertain with sufficient accuracy whether the aggregates from a well-defined batch conform to the specifications.

When analyses are carried out on the aggregates delivered to the storage area of the job site, the group in charge of checking is confronted with huge batches whose properties have to be determined fast. The determination of the sampling points is often difficult; not only does one has to fix the number of samples to be drawn but one should also be able to draw them from any point in the stock. It should then be possible to isolate the batch not conforming to the specifications from the sound ones already in place.

We leave it to the reader to imagine the problems involved in removing 1000 tonnes of accidentally polluted sand already dumped over 5000 tonnes of clean sand! We have already emphasised that it is always preferable to conduct the check for conformity at the locations of production. Ensuring that stockpiling is done at the right place in the storage areas will then suffice.

Non-conformity can also be detected rapidly and the number of tests reduced when the acceptance check utilises to some extent the results from the check at manufacture.

For example, it is not necessary to conduct Los Angeles tests for each of the batches at the acceptance stage if the nature of the deposit is thoroughly known and a simple visual examination or a small number of tests reveal that the zones of extraction have not changed. The test results from the previous batch characterise the new batch.

Likewise, for grading, flakiness index, Micro-Deval of coarse aggregate ..., a reduction in number of tests is possible if monitoring the quarrying accompanies monitoring the manufacture.

In the quarry the laboratory can orient its conformity check towards slips at the sensitive points of the installation and, if need be to the weightages accorded to certain specifications, with due regard to the resources at its disposal. Let us take, for example, aggregates for surface dressings which are specified by grading, strength and cleanliness. If the investigation of the quarry reveals cleanliness to be the only property liable to vary in the rainy season, the laboratory will check this property in particular, increasing the number of tests. The time that would normally have been spent on tests of other properties can be advantageously used otherwise, checking stockpiling for example.

The discretion to commit manpower and equipment only to what merits checking is indeed one of the preoccupations of the engineers charged with inspection in quarries. Corrective actions by them become that much easier if the processing line incorporates:

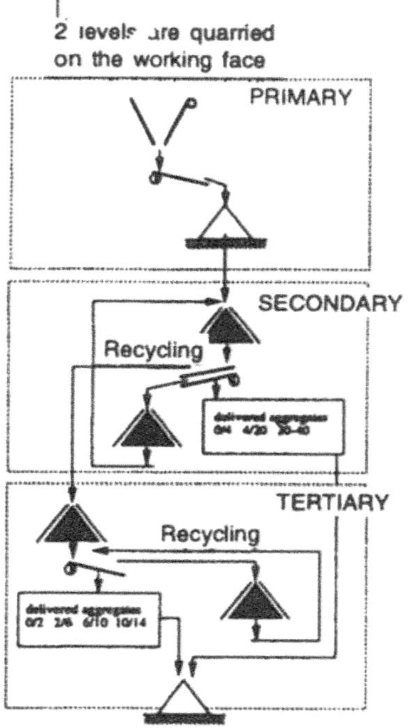

CORRECTIVE ACTION FROM THE LABORATORY

Monitoring of	Objective in view	Tests conducted in normal working regime
Quality of zones being quarried (as per geotechnical map of working faces)	Satisfy specifications on mechanical properties / Improve cleanliness of aggregates	Los Angeles and Micro-Deval tests per quarried zone
Diameter of materials removed relative to their water content and pollution	Prevent clogging of screen / Prevent pollution of sands	
Grain size of draw-off / Recycling function / Screening accessories / Feeding of screens	Consistency in cut-off nominal sizes and efficiency of screens / Balance of circulating loads	Granulometric analysis per product each week (for a medium rate of output) / Checking of hourly or daily throughputs
Recycling function / Screening accessories / Feeding of screens / Setting of crushers	Consistency in grading of products	Each week one granulometric analysis per product / 2 S.E. on sands / 1 shape test on 'd/D sensible' / Checking of hourly or daily throughputs
	Optimum hourly capacity	
	Homogeneity of aggregates / Segregation prevention	
Conditions of stockpiling	Maintenance of cleanliness through good stockpiling conditions	

2 levels are quarried on the working face

PRIMARY   SECONDARY   TERTIARY   Recycling

Fig. XVIII.4. Example of corrective action by a control laboratory at quarry.

— A certain number of sensors which halt manufacturing when the product does not conform to the desired properties;

— Automatic sampling devices.

This shows that, so long as it is reliable, on-line control can constitute a very effective aid to the check for conformity and thereby relieve the latter to some extent.

2.2.1 DEFINITION OF THE SCHEME OF TESTS

The number of test portions to be analysed depends on variability in production, the manner in which the sample is drawn and the accuracy desired in assessing the property under measurement.

It is possible to reduce the number of tests requisite for assessing the quality of one batch of aggregates, as will be shown in Sec. 2.3.1. For this, it is proper to multiply the sampling increments in this batch. If the number of sampling increments is multiplied by n, the number of tests required can be divided approximately by $\sqrt{n}$ (Table XVIII.3).

Table XVIII.3: Number of tests to conduct (probability 0.9) for an accuracy of ± x% when the standard deviation of the population to be analysed varies from 1 to 6% (Each test portion drawn from a mixture of 10 sampling increments)

x% \ σ	Normal production			Aggregates that have undergone segregation or abnormal production		
	1%	2%	3%	4%	5%	6%
0.5	13	40				·
1	3	13	26	40	60	100
1.5	3	6	13	21	31	40
2		3	8	13	19	26
3			3	7	10	13

In this case the technician will devote a greater part of his time to the samples rather than conduction of tests, which explains the interest in installing sampling devices, namely, to reduce his interventions.

Lastly, when the quarry output is high, assessment of the quality of supply should be quick and corrective action even quicker to preclude production meanwhile of tonnes of material not meeting the specifications. As a corollary, and given the fact that preparation and execution of tests take time, their number should be limited.

The bases upon which the number of tests to be conducted should be calculated for road-making materials are presented below as a guide:

— one granulometric analysis per 500 t of coarse aggregates,
— two granulometric analyses per 500 t of fine aggregate,
— two sand equivalents per 500 t,
— one shape test for every 5000 t.

The frequency of mechanical tests will be fixed with due regard to the peculiarities of the deposit, especially its homogeneity, with a maximum of one test per 2000 tonnes. These rates, which make no reference to the size of the batch, are not above criticism. As a matter of fact, the efficiency of statistical quality controls depends not upon the sampling rate, but upon the absolute number checked in a single manufacturing batch.

When the standard deviation is not known and Student's law for samples numbering less than 20 (of the order of 16) entrains a loss in efficiency, the supplier's risk is increased, as is the risk to the client because of the sample size. Hence it is necessary to confine oneself to the reduced standard law by choosing a sample size equal to or greater than 20.

Furthermore, what the size of a batch should be and how many tests should be conducted could be determined from an economic investigation. This investigation would take into account the probability of production of defective materials, the costs incidental to the inspection and rejection of supplies, the risk for a client in accepting a defective supply, and the risk for the supplier in finding a supply answering the specifications being rejected.

In any case, if the test results are to be compared the sampling method has to be laid down and adhered to.

2.3 Sampling: General Principles

2.3.1 DEFINITIONS

Practice has not yet succeeded in standardising the terms used; so, to ensure proper understanding of this and subsequent sections we define the meaning of the terms as used herein.

Batch. A batch is an individualisable unit from production or a unit offered for acceptance, which should present a certain homogeneity. For this whole batch to be homogeneous, it is imperative that no significant factor in production has varied during its manufacture. Thus only variations of minor influence on the material produced are acceptable. Attention should be drawn to the size of the batch: the larger the size, the higher the risk of variation in each factor. Contrarily, if the batch size is too small, the cost of control will rise since, as remarked in the preceding section, the number of sampling increments is independent of batch size.

Similarly, if the batch is too big, the consequences in the event of non-conformity can be onerous.

Sampling plan. This refers to the whole set of operations involved in obtaining the sample which enables assessment of the quality of the batch (Fig. XVIII.5).

Errors that can creep into a sampling plan are mainly of three types:

• **Random errors** which arise from splitting-up are inherent to the samplings and reduction and unavoidable. Random error is illustrated in Fig. XVIII.6 wherein the result from granulometric analysis of the test portion differs according to where the block lands.

• **Systematic errors** which recur if the sampling is repeated: they are imputable to the sampling method or the apparatus. An illustration of systematic error is given in Fig. XVIII.7. The sampling device does not sample part 2 of the jet, which is made up of fines.

• **Integration error**, that is, the error one commits while making a bulk sample combining several sampling increments without knowledge of the distribution law

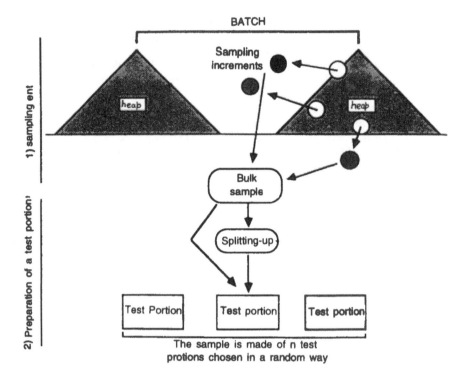

Fig. XVIII.5. Scheme of sampling plan

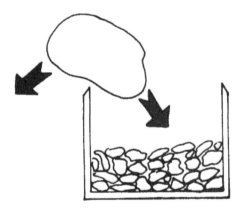

Fig. XVIII.6. Random sampling error

of the property to be measured in the batch under analysis. The integration error for the property is illustrated in Fig. XVIII.8.

The sampling is said to be "true" when the random errors are minimal.

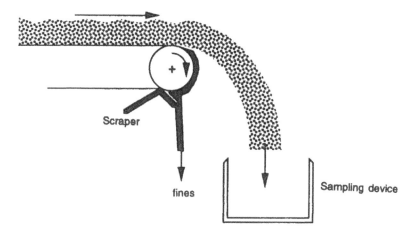

Fig. XVIII.7. Systematic error

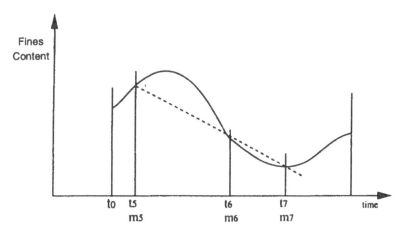

Fig. XVIII.8. Integration error

The sampling is said to be "correct" when systematic errors are excluded. In any case it is advisable to detect the systematic errors and remove the causes behind them.

The sampling is said to be "accurate" or "representative" if it is true and correct.

Therefore it can be said that the paths leading from the stack to the test portion are strewn with pitfalls in the form of errors. The main precautions to be taken for eliminating or minimising them are discussed below. We shall deal successively with the making of sampling increments and test portions.

2.3.2 UNIT SAMPLE

Collecting a representative sample can be done manually or by means of automatic devices designed for this purpose.

- *Unit sample mass*

The minimal mass of sampling increments has been so fixed as to exclude error owing to grading. The coarser grains in fact can thus be found here in sufficient number so as not to alter their percentage in the sampling increment.

Table XVIII.4 Gives the mass of the sampling increment for different aggregate sizes, which can be slightly exceeded with no adverse effect but has no justification for gross exceedance since reliability will not be practically improved on that score and the bulk sample will be abnormally voluminous.

Table XVIII.4

d/D	0/2	2/4	4/6	6/10	10/20	20/40
Mass sampling increment in kg	0.200	0.300	0.500	0.750	1.500	3.000

- *How to prepare a sampling increment*

If sampling is done from a stack, either a probe or a shovel with high sides can be used. The diameter of the probe and the distance between the sides and the height of the shovel sides should be 3 times larger than the size of the largest element to be sampled.

The lips of the device in contact with the material should be thin. The device should penetrate the stack uniformly with minimal jerks, this with the idea that all the materials in the batch will probably be sampled thereby. A spillover must not be allowed.

Lastly, the risks of systematic errors will be minimised if the locations of sampling can be selected without reservation. A stack can be rightly sampled only if all its parts can be reached. For deciding the locations of sampling, the stack should be notionally subdivided into comparable volumes and sampling done near the centre of each.

The weight of each sampling increment should not be less than the weight specified in the preceding section. Therefore, in some cases the sampling device should be plunged several times into the same volume to obtain the desired quantity but taking care to so proceed that the successive drawals are combined.

Often, channelling from a quarrying bench turns out to be absolutely essential since sampling should be done from the mass of material and not the debris from the surface or that resulting from excavation.

The rules governing sampling from stacks apply to sampling from wagons or trucks. In these cases the safety rules imposed by the difficulty in gaining access to the loads complicate the sampling rules.

Manual sampling can also be done on a conveyor stopped momentarily. With an appropriate device, conforming to the shape of the conveyor belt, the flux of material is cut over its entire width along two parallel planes. The materials situated between these two planes are carefully sampled using a small shovel and a small brush.

Very special care should be exercised in the collection of materials to avoid systematic errors. For the same reasons, lifting the sampling device before completing the sampling should always be avoided to preclude inadvertent sampling of fugitive elements. For this reason it is preferable to do sampling over the rollers.

Let it be noted that it is advisable to cut the flux of materials along the smallest section to minimise random errors (Fig. XVIII.9).

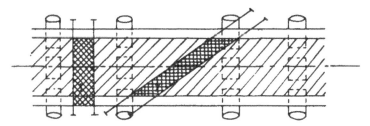

Fig. XVIII.9. Sampling increments in a moving stream of material

1—Small section: random errors minimised;
2—Large section: random errors significant.

This type of sampling has a few advantages:

— The sampling increment can be collected with particular care and each of the portions of the stream of material is accessible (it is theoretically possible to stop the belt movement at any moment).

— The sampling increment is free from systematic bias and its size can be predetermined, whatever the conveyor belt rate.

The distance between the planes cutting the conveyor belt is given by the following formulae:

$$e = p \times \frac{V}{D} \times 3.6$$

where D = belt conveyor rate in t/h; V = belt conveyor speed in m/s; p = mass of sample in kg; e = distance in metres between the parallel planes.

To minimise the mixing error, sampling from a stream of static material should be done when the rate of the belt conveyor is uniform and when it is uniformly loaded.

Here again, safety rules should be scrupulously observed to prevent any accidents which could take place at the time of manual intervention on a belt conveyor.

• *Automatic sampling*

Automatic sampling devices are installed at the discharge end of a belt conveyor. They are made up of a scoop with very thin lips, which move as needed, affixed to a trolley mounted on rollers and guided by rails (Fig. XVIII.10).

The scoop moves at a constant speed since, for a sampling to be representative of a stream of materials, the probability of sampling the 'tubes of stream' constituting this flux should be the same for each of the tubes. Therefore the scoop should enter, traverse the entire jet of materials and exit full.

Lastly, the avoid systematic errors, it is necessary that the scoop opening be at least three to four times larger than the largest element to be sampled and the speed of scoop motion not too high.

Fig. XVIII.10. Automatic sampling device

By definition, it may be said that:

— The sampling is transverse when the speed vector of the sampling device is perpendicular to the plane of movement of the particles constituting the flux of material.

— The sampling is longitudinal when the speed vector of the sampling device is parallel to the plane of flow of the particles.

— The sampling is alternate when the device traverses the flux of material successively at equal speed but in opposite directions.

— The sampling is oriented when the device traverses the flux of material always in the same direction.

The quantity of material sampled by means of an automatic cross-wise sampler is given by the following formula:

$$P = \frac{D \times l \times 1000}{3600 \times V}$$

where P = mass sampled in kg; l = width of scoop in cm; V = speed of scoop movement in cm/s; D = rate of belt conveyor in t/h.

When the quantities of materials sampled are appreciable, a division or splitting should be provided for, which may be done automatically with the help of devices belonging to the divider from which the scoop picks up directly.

Figure XVIII.11 gives the trend of the volume sampled by a device installed at the discharge end of a conveyor. The sampler moves perpendicular to the flux of the materials.

The volume sampled in forward motion is shown by a solid line and that sampled in reverse motion by a broken line. The two volumes are identical.

Monitoring the state of wear of the scoop lips becomes obligatory if the quality of the sampling increments is to be maintained.

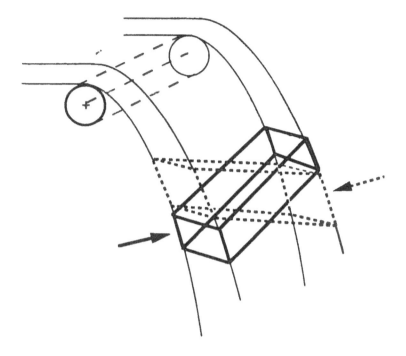

Fig. XVIII.11. Cutting of flux of material with a sampling device

2.3.3 PRACTICE

It should be squarely recognised that in actuality few quarries, if any, are equipped with automatic sampling device.

Stopping a belt conveyor creates a break in a continuous dynamic process, which could give rise to heterogeneities. The only rational approach consists of sampling from the transport vehicles, after loading, or from the stocks.

In both cases the ideal would be access to any point in the contents. But it must be admitted that the zones situated below the surface are most often inaccessible. Sampling increments are always taken from the surface and hence this surface should be sampled over a widely cut area.

Here, for inspection of the batches, it is essential to lay down a procedure and stick to it throughout the inspection period, the objective being not to alter the variance proper to the results and thereby be in a position to compare them *inter se*.

It is necessary to mix the sampling increments from one batch to obtain its mean composition in order to lay down a formula or to inspect a conformity.

It is advisable to examine the variation between the batches entering the homogenisation device of the manufacturing process; these variations reflect the heterogeneity of the supply as a whole.

Advances in automation procedures would contribute a great deal to mitigating these drawbacks by integrating the control processes into the manufacturing line.

REFERENCES*

Dauval, R. 1978. Some reflections on the sampling of lots from minerals in bulk and deposits. *Annales des Mines*.

Griveaux, B. 1974. Role of faults in quarrying of massive rocks. *BLPC*, no. 57, January–February.

Griveaux, B. and A. Maldonado. 1972. Typical scheme of quarry investigation. *BLPC*, no. 57, January-February.

Gy, P. 1971. Sampling of minerals. *Revue de l'Industrie Minerale*, September.

Gy, P. 1972. Sampling of split-up minerals. *Annales des Mines*, November.

Kohling, R. 1972. Evolving trend in the field of drawing and preparation of solid mineral raw material samples, Wattenscheid. *Aufbereitungs Technik*, no. 11.

Maldonado, A. 1972. Inspection of aggregates in quarries. *BLPC*, no. 60, July.

Panet, M., L. Primel and C. Archimbaud. 1972. Deposits of materials and temporary installations. *BLPC*, no. 60, July.

---

*All references given in French—General Editor

XIX

Safety*: General Rules, Example of Conveyor Belts

Alexandre Iatropoulos

1. GENERAL

Quarrying entails various types of activities which may lead to accidents during work.

National statistics place mining industries, in terms of number and seriousness of accidents, in the 'high risk' category just below building construction works and public works.

Thus professional and administrations have been motivated for the past several years to undertake efforts to improve this situation. Without doubt some laudable results have been achieved but much still needs rectification.

1.1 Safety Rules

Extractive industries should conform to the rules and regulations relating to mines and quarries laid down and periodically updated in accordance with the advances made, by the Ministry of Industries.

Regional Sickness Insurance Funds (Social Security) have the authority to ask for the enforcement of the Labour Code and the general rules or recommendations published thereby.

1.1.1 STATUTORY CONTROLS

The Ministry of Industry has approved certain specialised bodies charged with assisting the technical directors in charge of the works of extractive industries. The obligation of these bodies to intervene is in force everywhere in France.

The federation of professional trade unions of producers led to the creation of an association whose objective is the prevention of accidents in quarries.

---

*see Chapter VII for stripping, stoping, loading and transport.

The Regional Sickness Insurance Funds can ask for the intervention of the approved bodies to inspect certain special equipment:

— electrical installations,
— lifting equipment,
— reservoir of compressed air,
— radioactive equipment etc.

The task of the Industry Regional Director is to see that safety rules are followed in mines and quarries.

1.1.2 ADMINISTRATIVE DOCUMENTS

Compliance with the safety rules in force in a quarry calls for the preparation of various documents, some of which have to be forwarded to the staff for information and for training them in the application of these rules:

— Operating instructions
— Files pertaining to:
• specifications (for tyre-mounted machines),
• traffic on trackways,
• utilisation instructions,
• maintenance instructions.
— Instructions for site machines (shovels, bulldozers, special equipment).
— Driving permits (validated every year) issued to drivers, subject to a medical examination.
— Machine maintenance diaries (containing details about the periodic checks relevant to safety points).

1.1.3 OUTSIDE ENTERPRISES

Their interventions should be for the purpose of:

1. Declaration to the Regional Director of Industry (nature of the works, place, duration).

2. Preparation of a report at the initiative of the quarry operator containing the various instructions and special provisions, if any.

1.2 Nature of the Risks

The use of explosives, machines, belt conveyors, processing equipment and storage devices entails risks of a different nature.

The environment of the persons exposed can also exert an influence, for instance in works carried out on machinery mounted on barges on rivers or lakes, or works carried out in the vicinity of places with large level differences.

The risks are therefore multiple and the situations in which they occur highly variable.

1.3 Consequences of Accidents

It is proper and fitting to think first and foremost about the consequences to mankind. The seriousness of an accident, quite often extreme, should make us consider the importance of the preventive measures advocated.

As an example in 1989, in a single region alone, during a single semester (population of about 800 persons), there were:
— 3 deaths (machine accidents),
— 3 wounded, 2 seriously (conveyor belt accidents).

1.3.1 LEGAL RESPONSIBILITY

Serious or fatal accidents lead frequently to a thorough investigation in order to attach responsibilities.

The sentences delivered during the judgements are sometimes very severe and affect in particular the employers. However, the officials of the firm are also very often subject to punitive sanctions.

Equipment suppliers may be questioned when the conformity of the machines or equipment to specifications is not absolute and when, for this reason, risks are unduly increased.

1.3.2 ECONOMIC CONSEQUENCES

The financial costs of accidents taking place on the job are fully borne by the firm which has undertaken the project. The annual 'subscription' to be paid is determined according to the extent and seriousness of the accidents.

Since these subscriptions have to cover:
— care of the injured,
— stoppage of work,
— accident allowances, and
— social security,

the financial outlay is very heavy (of the order or 8 to 10% of the maximum salaries).

• *Prevention constitutes a management element of the firm*

Moreover, indirect costs, have also to be taken into consideration; these are sometimes much higher than the actual cost of the accident:
— broken equipment,
— disruption of production,
— substitute staff etc.

1.4 Prevention of Accidents during Work

A prevention of accidents policy should entail actions of different nature but it requires first and foremost:
— determination of the employer,
— involvement of the officials.

1.4.1 TRAINING

Necessary attention should be paid to this important point. Employees should be informed about the risk present in their environment and about the appropriate methods to prevent them.

The essence of safety depends on the behaviour of the concerned in the face of situations fraught with risks.

1.4.2 INFORMATION AND AWARENESS

The professional body in charge of assisting the firm can make a significant contribution in these areas by providing information on the accidents which occur in the profession and by making known the systems or actions which help to prevent them.

The method of intervention should be carefully studied in order to avoid whimsical and risky improvisations.

1.4.3 IMPROVEMENT OF TECHNICAL EQUIPMENT

This is a very vast field and warrants far more discussion than can be given here.

Let us not forget that technical improvements for the regularisation of function have a direct impact on risks since disruptions of any nature are the root of most accidents.

1.5 Identification of Risks

Risks can be classified into three categories:

1. Those related to *production actions* during which personnel behaviour is paramount:
   — running machines,
   — supervision of equipment,
   — setting of machinery,
   — disruption in functioning.
2. Those related to *maintenance action* during which competence and safety alertness are of utmost importance:
   — equipment maintenance,
   — repairing.
3. Those related to *movement of personnel*:
   — use of ladders,
   — use of staircase,
   — use of platforms,
   — use of approaches.

The care taken to achieve or improve the means of access and intervention is a decisive factor.

1.6 Technical Safety Provisions

It would take too long to review all the equipment described in the previous chapters and to explain all the technical details which can improve the safety of installations.

Moreover, we have just seen in Sec. 1.4.3 that most accidents are due to the machinery and that some take place during operations on conveyor belts. By way of example, we shall content ourselves with dealing with accidents taking place on conveyors since those due to transport machinery are unfortunately too general to land themselves to indications specific to the aggregate industry.

Of course, the manufacturers of these conveyor belts have tried to incorporate the safety aspect in the design of their machines, at least in their normal functioning. Unfortunately, accidents take place most often when the personnel try to rectify a dysfunction.

A detailed analysis of such situations has led to certain modifications of detail intended to preclude the origin of intervention. These modifications may appear to contradict some of the indications contained in Chapter XX. In our view, they result from the experience acquired from numerous exploitations wherein the working conditions, influenced by the capricious changes in the product transported and climatic conditions, become very different from those the designer had logically envisaged.

Safety devices stand to gain even if the personnel operate under conditions contrary to safety instructions.

2. SAFETY AND BELT CONVEYORS

A familiar tool of quarry operators and in public works sites, belt conveyors carry out the function for which they have been designed under very different conditions.

Their regular and relatively silent functioning, besides their apparently simple techniques, prompt the operating personnel to adapt them to a wide variety of uses.

Statistics on 'accidents at work' have shown that belt conveyors are the source of too many accidents, sometimes fatal or very serious. Why? All these accidents are associated with interventions during belt working for the purpose of resolving a dysfunction. In fact, we can distinguish two types of intervention:

1. *Repairs*, which require dismantling and which should be done only be following an instruction procedure, an essential condition for ensuring safety. Only a total cut-off of the source of driving energy can guarantee total stoppage of the machine.

2. *Common interventions*, which take place mostly when the machines are running because stoppage of a conveyor often leads to stoppage of a complex assembly or because running is necessary for carrying out the intervention (setting, cleaning, lubricating etc.).

2.1 Dangerous Points

Re-entering angles: They consist of the angle formed by the belt and the pulley (or idler) when both directions of movement converge towards the same point.

2.2 Protective Equipment

Re-entering angles can be neutralised. It is thus possible to render a conveyor harmless, but for that the protective equipment should be carefully selected. Protective casing, boxes or grids are often deceptive since it is generally necessary to remove them to carry out interventions while the machine is running. Exposure to risks then becomes great.

It is advisable to fill the space of the re-entering angles by means of fixed devices which permit carrying out the interventions during running of the machines, the risks then being neutralised.

The following diagrams show the manufacturing principles of these devices.

2.2.1 PROTECTION OF REAR PULLEY (Fig. XIX.1 and XIX.2)

The safety guard-iron should ensure three functions:
— Conceal the re-entering point;
— Scrape the return side to block the passage of any large fugitive elements;
— Scrape fine products or frost (ice) accumulated on the pulley.

The guard-iron should follow the rear pulley during its movements. It should not be articulated.

Fig. XIX.1. Protection with guard-irons which also ensure scraping of the pulley and elimination of products likely to disrupt functioning

2.2.2 PROTECTION OF HEAD PULLEY (Fig. XIX.3 and XIX.4)

The device should fill the space of the re-entering point. To give it the trough shape of the belt, a wooden block, cement mortar, plaster or any other filling material may be used.

2.2.3 PROTECTION OF SUPPORTING ROLLERS (Fig. XIX.5 and XIX.6)

Only those rollers should be protected in which the stress of the belt is such that it is not possible to lift it.

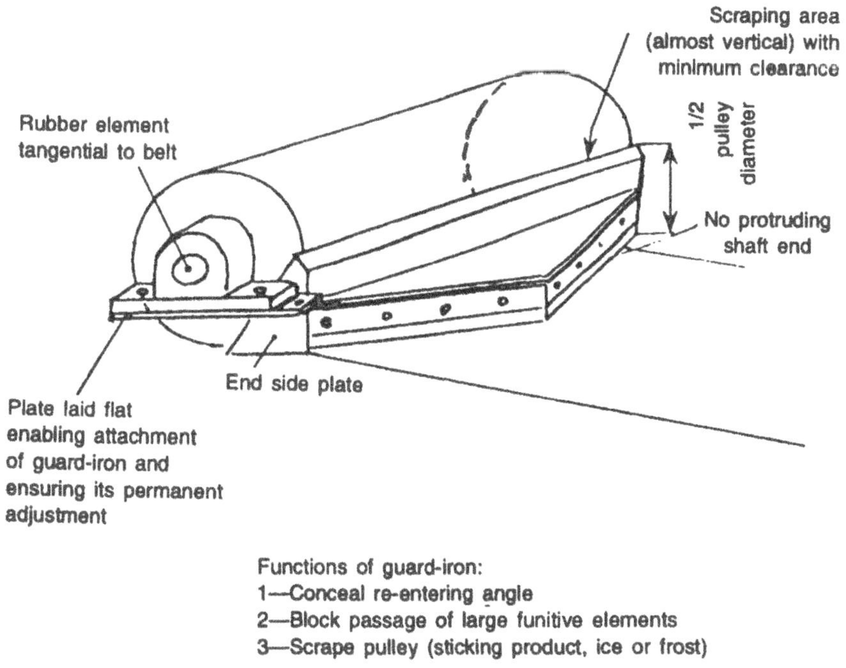

Rubber element
tangential to belt

Scraping area
(almost vertical) with
minimum clearance

1/2 pulley diameter

No protruding
shaft end

End side plate

Plate laid flat
enabling attachment
of guard-iron and
ensuring its permanent
adjustment

Functions of guard-iron:
1—Conceal re-entering angle
2—Block passage of large funitive elements
3—Scrape pulley (sticking product, ice or frost)

Fig. XIX.2.

Protection can be provided by placing flatirons which support the rubber elements against the impact of large elements.

2.2.4 PROTECTION OF LOADING POINTS (Fig. XIX.7)

Protection can be ensured by means of flatirons, as indicated above, but also by means of grids which should preferably not be removed.

Fig. XIX.8 and XIX.9 show the modification of a loading point initially equipped in a traditional manner.

2.2.5 PROTECTION OF AUXILIARY IDLERS (Fig. XIX.10 and XIX.11)

Protection is ensured by means of fixed angle irons which are provided in such a way that they constitute an obstacle in the re-entering point.

When the idler is sufficiently driven by the belt, the bar serves as a scraper, which prevents the development of mud-cake leading to drift.

Fig. XIX.3. Re-entering angle occupied by space of an adjusted crossbar. Self-adjusting top wooden portion

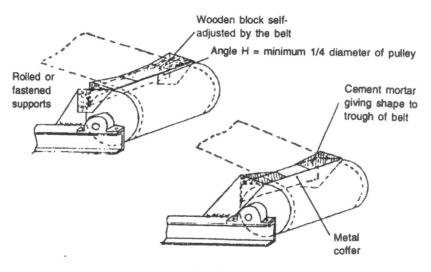

Wooden block self-adjusted by the belt

Angle H = minimum 1/4 diameter of pulley

Rolled or fastened supports

Cement mortar giving shape to trough of belt

Metal coffer

Fig. XIX.4.

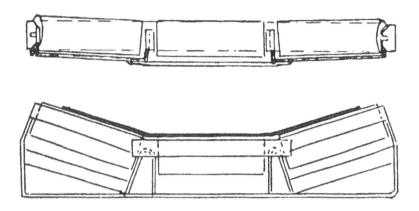

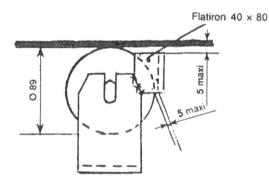

Note: This system of protection is suitable for products in which the grain size does not exceed 25 to 30 mm.

Fig. XIX.5.

2.3 Influence of Design of Conveyors on Safety

We have spoken of dysfunctions; they are of various types;
— Centring the belt (intervention on the return idlers, intervention on the rear pulley, intervention on the loading point).
    — Cleaning material deposited under the rollers of the loading point.
    — Cleaning material deposited under the conveyors.
    — Cleaning deposits on the framework coffers.
    — *Lubrication*: This is not a dysfunction but is often carried out while the belt is running, which is advisable for the ball bearings.
    — *Slipping of driving pulley*: This is a major source of accidents.

Fig. XIX.6. Protection of a loading point by means of flatirons placed on each roller support: 'undismantable system'

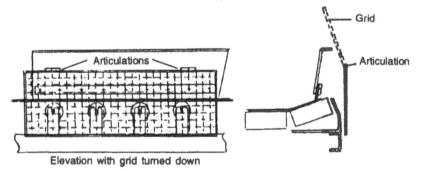

Elevation with grid turned down

Note: This system may be used when the grading of the products does not make it possible to opt for fixed protections (greater than 25 or 30 mm). It is advisable that the grids be articulated in such a way that removal is not possible.

Fig. XIX.7.

The foregoing enumeration is not exhaustive but reveals the advisability of improving conveyors in order to make them machines 'in stable working condition', requiring the minimum intervention.

Fig. XIX.8. Loading point equipped in traditional manner

Any advance in this direction would be very much appreciated since whenever the computer and the printer list out all the faults of an installation, it is to be feared that the conveyors may prove to be the great disrupters of an overall good working order.

2.4 Suggestions for Improving Conveyors and Safety

The following suggestions presently form the subject of an attempt to lay down specifications in collaboration with conveyor-belt manufacturers, which can be used by quarry operators during negotiations with manufacturers.

2.4.1 DRIVE MECHANISM

- Choose a generous pulley diameter with flat sides free from metal bracing.
- Introduce a stress idler with large-size external bearings.
- Aim at doing away with belt transmissions.
- Use a method of gradual starting (for large distances between centres) which limits the stresses on the belt to an acceptable value.

Fig. XIX.9. The same point modified:

— the sheet for holding back fugitive products has been dispensed with;
— the supporting rollers are individually protected by flatirons;
— the rear pulley is fitted with a guard-iron. All the elements are fixed and safety is ensured by means of guards or grids. Overall working is thereby improved.

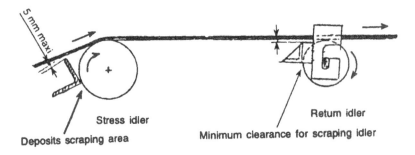

Fig. XIX.10.

Fig. XIX.11. Lower idler protected by a very simple device which ensures its scraping

2.4.2 CONVEYOR BELT CASE

The conveyor belt case should also be closed in the rear to make the re-entering angles of the pulleys inaccessible. Protection against rain improves mechanical efficiency and thereby prevents slippages.

2.4.3 REAR PULLEY

Used a pulley with sides flat and free from bracing and provided with external bearings.

2.4.4 TENSIONING DEVICES

It is imperative that these be reduced to their simplest form. Automatic tensions in the rear are often ineffective and contribute to decentring of the belt.

In the case of conveyors of great length, only head tension can be effective.

2.4.5 LOADING POINT

This is the point which calls for the utmost care in order to ensure correct centring of the materials on the belt. It conditions the self-centring of the carrier side through the 'centring angles' of the carrier idlers.

Efforts should be made to obtain maximum dust-tightness in order to exclude fugitive elements of the product.

It is very important to avoid 'trapping' materials that find their way under the loading point. There should be no metal sheet under these idlers (the safety guard-iron ensures cleaning of the belt and provides protection to the rear pulley). This observation is even more valid in the case of an extractor or feed regulator.

Damping idlers preclude belt slippage when the feeding jet is transverse.

Provision of a 45° troughing idler aids in centring the load.

Judiciously positioned guide-idlers can keep the heel of the belt in position and prevent its displacement. The loading area of box-type conveyors would be enhanced if just two rolled section beams are installed.

2.4.6 RETURN IDLERS

The spacing of these should be so chosen that the belt drives them vigorously be resting on them; this makes scraping them by means of a protection bar possible, which contributes to centring of the return side.

Conventional spacings do not allow for work under these conditions. Avoid positioning them at a right angle to an upper station or a crossbar. This might cause some blocking there by large fugitive elements.

2.4.7 CENTRING THE RETURN BELT

It is absolutely imperative that conveyors be provided with an automatic setting device. Inverse 'V' idlers constitute an efficient means in so far as the belt is highly stressed on the idlers.

Some suitable protection is required here.

2.4.8 BELT REVERSAL

This technque is very advantageous and simple to implement. It permits:

1) doing away with the bulk of the cleaning operation,

2) protection the return belt against rain,

3) extending the life of the idlers as they remain clean and regularising the work of the return belt.

It is to be noted that the front propeller of a belt reversal constitutes an automatic stretcher, which exerts a marked influence on the drive.

2.4.9 EMERGENCY STOP CABLES

These are required as per the rules. As their action is delayed by the inertia of the belt in motion, they are hardly of help to an accident victim. On the contrary, they are very convenient for limiting cramming or damage as soon as the need for cessation of functioning is felt.

The action on a cable can entail total stoppage of the installation, except for those machines which are likely to create start-up problems, such as grinders, crushers, mixers. It is a question of judgement.

All accessible areas of the conveyor should be provided with emergency stop cables.

Return by compensation bars should be preferred to pulleys.

2.4.10 ACCESS

The supervision and maintenance of belt conveyors require means of access to the various components. A side gangway up to 800 mm in width may suffice. Turning back the head is advisable to provide access to the two extremities of the driving pulley.

Conveyors of 1 metre width onwards are provided with two side gangways. The width of the gangways should be 0.60 m and the railing should be standardised (height 1 metre, smooth at plinth mid-height of 150 mm).

Mobile conveyors may be provided with railings that facilitate access to the head by walking on the belt when it is in a stop position (instruction procedure). However, we consider that up to 5 metres of height measured above the head pulley, mobile conveyors can be reached by a ladder or similar means.

Ground conveyors should be provided with a gangway for crossing every 100 metres.

2.4.11 STARTING ALARMS

Conveyors which are hidden to the operator who starts them (tunnels, ground conveyors etc.) should be provided with a timed starting alarm. The signal may be a sound one or possibly a light signal.

3. CONCLUSION

The overall management of an aggregate manufacturing plant requires that the safety aspect be taken into account in its entirely.

The incorporation of this concern not only in the design, but also in the operation of the equipment should enable achieving a situation much more favourable than presently exists.

Certainly, human behaviour is almost always in question in the genesis of accidents. However, detailed consideration of the reasons for such behaviour would enable creation of devices likely to reduce or remove their cause or the risks resulting therefrom.

XX

Aggregates in Newly Developing Countries

Michel Yernaux and Gerard Liautaud

'When building activity is on, there is all-round progress.' We might extend the spirit of this proverb to suit the last chapter of this book by remarking that when there is a great demand for aggregates in a newly developing country, this is a highly auspicious sign attesting to the existence of a programme of constructions in progress.

Now, which are these newly developing countries? Does this adjective signify that nothing whatsoever exists in them and that everything needs to be done from scratch? In other words, are they barren countries?

Far from it, indeed, in some cases they are countries whose civilisation is very ancient and very rich, but within which civic infrastructures have not been totally realised.

In the context of our interest in aggregates, newly developing countries are those in which a substantial programme of renovation or creation of civic structures, highways, railways, dams, factories, residential units etc. is under execution.

This renewal is such that the builders will meet with conditions of work totally different from those found in France.

In this last chapter we shall attempt a survey of the special problems which will confront aggregate manufacturers and their machinery suppliers and assess how working patterns applicable to a greater or lesser extent in France can be adapted to these special conditions.

1. SITUATIONS IN NEWLY DEVELOPING COUNTRIES

1.1 Absence of a General Policy

The preceding chapters commended a policy on aggregates for France which, by and large, is already in force. Obviously, it is out of the question to extend these recommendations to newly developing countries which form the subject of the present chapter. France actually is a highly administered country; some say even overadministered. Contrarily, the countries we have in mind cannot afford the luxury of such a vast specialist workforce.

This difference entrains basic variations from the situation obtaining in France even with regard to intermediaries—for the quarrier who ventures into such a country as much as for the machinery manufacturer who advises him and hopes to sell him the machinery for producing aggregates as well as the plant design.

Whereas in France there are local or even regional policies within a national framework for overall direction, there is no such thing in regions where much yet awaits to be created, with new plants for production of aggregates being first and foremost. This situation has both inherent advantages and drawbacks.

Total freedom in selecting a deposit is foremost amongst the advantages. Therefore the task of prospecting (Fig. XX.1) becomes the *total* responsibility of the prospective quarrier whereas in France a part of this task is already accomplished under the aegis of the public authorities.

Fig. XX.1. Prospecting. Crossing of a stream using a rudimentary ferry

To pursue this search along the right lines, the future quarry operator will need to have recourse to a plethora of geological committees and there will often be some difficulty in finding persons sufficiently familiar with the geological details of the region in question. On the other hand, he can freely and completely turn to account the quality of this research, and if the surveys he conducts or commissions lead to an excellent deposit, he exclusively will benefit from this result (Fig. XX.2). He can set himself up and quarry without constraint, at least in principle. He will have no District Commission on Quarries which is likely to raise objections, justifiable though they might be, against his project.

Fig. XX.2. Prospecting. Travelling by a small aircraft. Makeshift landing strip

The major drawback to this absence of a policy on aggregates arises from the fact that the uncertainties are greater. Pierre Massé wrote that the French plan was a reducer of uncertainties. The same could be said of an aggregate policy. This, like the plan in other respects, would include a market survey. The quarrier in France knows fairly well the demands of the zone in which suppliers are already active. He can apprise himself of their annual production etc. without much effort.

This is not so in a newly developing country. One might rightly counter this statement, saying that the question is simple in a number of cases: there is no quarry nearby and the plant would essentially, if not exclusively, aim at supplying to a well-defined job site whose requirements are likewise fairly well defined too. We shall sufficiently develop this point in Sec. 1.3 below and hence not go into it here. But this case of a quarry catering to the market of a particular zone of consumption should be carefully examined since, in all probability, it will undergo expansion in future.

Regardless, there remain other uncertainties, not the least of which are problems of access to the deposit (Fig. XX.3). Perhaps a quarry is opened up in the zone under consideration to feed the job site of a new highway but there is no access road. The machinery would have to be transported under conditions that are not only difficult, but also unfamiliar. Needless to say, transport for delivery of the materials produced would be subject to the same uncertainties. Recruitment of personnel to do the quarrying would again pose other problems, many of which a novice in this profession could never imagine. And what about those to be hired to maintain the quarrying machinery brought in at great expense?

Fig. XX.3: Anachronism. A major aggregate-producing plant in the background. Habits and customs persist, as well they should (Cameroon)

So, a plant for production of aggregates in a newly developing country is truly an adventure. This, perhaps, is its first characteristic, which adds yet another complicating factor to the operation (Fig. XX.4).

First amongst these adventures is identification of the deposit.

1.2 Exploited Deposits

1.2.1 SIMILARITY, YET DIFFERENCES

A French geologist passing through Tropical or Equatorial Africa for the first time would be completely taken in by the similarity between outcrops from an old African basement and rocks in his homeland. No need whatsoever for him to imagine or invent new terminologies to denote differences in genesis or mineralogical structure of the rocks encountered. In visiting the northern part of Gabon, for example, he would note, perhaps with stupefaction, the presence of crystalline rocks similar to those of the Canadian Shield: same age, same structure, same texture.

Further, a prior perusal of the several reports presented in the atlases of newly developing countries would have quickly reassured him that here, as elsewhere in Europe or America, geologists speak the same language: neither terminology nor etymology have changed. Chippings flying from hammers point to the presence of the same families of crystalline or metamorphic rocks, be they granites, basalts, schists, limestones or sandstones. In the depressed zones or plains, he will

Fig. XX.4: Bringing equipment by a towed barge. Unloading at a hastily improvised site

notice the presence of the same deposits of sands polluted to a lesser or greater degree, the same argillaceous silts. In the beds of existing or ancient watercourses and rivers, he might sometimes have to prove the same deposits or terraces of rounded, quartzose or siliceous gravels with here and there identical intercalations of sands that are sometimes reddish, sometimes beige, sometimes whitish. In broken zones, surrounding volcanoes, he will notice the same extrusions of lavas, or pozzolans, some of which have chemical properties greatly similar to those of Puy-de-Dôme or those of Vesuvius.

Truth to say, the surprise of the foreign geologist would lie neither in the difficulty of expressing in appropriate and scientific terms the results of his visual observations, nor the discovery of a multitude of new minerals and unprecedented aggregates. The laterite itself would not astonish him much if his peregrinations had taken him to certain spots in Provence, which are shaded bauxite reddish, the ultimate stage in laterisation according to some authors. No indeed. His astonishment would commence on observing or sensing the landscape and the enveloping environment: high temperatures, density of vegetation cover, alternation of flooding rains and inexorable droughts following well-defined seasons. All this will impress upon him in no time the certitude of considerable action on these rocks, by no means comparable to that of the more temperate climates where he was born, and terms such as weathering, erosion, evolution, decomposition will at once whirl through his mind. In other words, he will realise that any difference from what he is accustomed to seeing stems and results from the profound influence of a particularly aggressive climate, given the heat, humidity and intense biological activity which takes place under the forest cover.

The most striking effect of this climate on the rocks will not escape him: it manifests in the growth and depth of the weathering profiles. He will recall that in France the depth of the wastes overlying the quarries generally remains of the order of several metres, whereas the overburden heights here can reach several tens of metres. And across this depth of weathered materials he will not miss a particular ubiquitous feature: the three-dimensional heterogeneity in the mass, i.e., extreme variety in all directions in respect of colours, textures, geometries, consistencies and hardness.

Nevertheless, whatever be the magnitude and the complexity of the mineralogical and chemical changes these rocks have undergone, compounded by the modifications resulting from morphological evolutions, the mantle of weathering will generally present itself in the following type profile, repeatedly described by geologists interested in the study of soils in intertropical milieus.

— *Horizon A:* Humic horizon, which is greyish and extremely acidic in the upper part, becoming light beige deeper down; alluvial layer subjected to intense leaching in which all the cations including iron and aluminium have been carried away.

— *Horizon B:* Characterised by a high concentration of iron and aluminium, forming nodules, more precisely termed 'ferrallites', which are submerged in a clay, viz., kaolinite.

This gravelly horizon, from which most of the weatherable minerals including silica have very nearly disappeared, perhaps constitutes the most 'authentic' or 'unique' source of aggregates for engineers familiar with these types of climate: these are the lateritic gravels.

— *Horizon C:* This stratum, which establishes the transition with the decayed mother rock of the substratum, is called the zone of departure; clay is more abundant here than iron and aluminium is not individualised; lateritic nodules are scarce.

— *Horizon D:* This is that part of the mother rock which is decomposed to a greater or lesser degree in the upper part; several metres deep lies unaltered rock, which the aggressive action of the overlying ambient atmosphere has not yet reached.

This deep weathering undergone by the rocks will yield aggregates of unique appearance and mechanical properties, besides necessitating special methods of exploration and prospecting, special quarrying techniques and quality standards or applications, some of which differ appreciably from those known, insisted upon and accepted in countries of temperate climate.

Without claiming to be exhaustive, we shall now endeavour to draw up an inventory of the most utilised aggregates side by side with their essential properties and the most common fields of use in the construction industry and public works.

1.2.2 AGGREGATES FROM MASSIVE ROCKS

Aggregates obtained by blasting and crushing of benches in massive rocks are universally used. But in the newly developing countries in tropical or equatorial climate, their exploitation as well as their utilisation follows special methodologies.

Given the generally difficult access, resulting from considerable vegetation cover, absence of roads or tracks, considerable thickness of weathered materials which overlie the deposit, in addition to the very often limited needs which make the unit cost of exploitation particularly high, exploitation is rarely done at depths where truly clean, i.e., unweathered rock is encountered.

Thus blasting is done at levels where the effects of weathering are quite appreciable and mechanical strength tests on rock samples yield relatively low values of hardness. Barring some dolerites or basalts, which are particularly resistant to the action of climate, rare are the rocks in these climates with a Los Angeles coefficient below 30, which is one of their requisite characteristics.

Moreover, the possible presence of highly weathered zones inside the rocky massifs, which can be circumvented only with difficulty at the time of exploitation, affect the cleanliness and homogeneity of the aggregates which, on crushing, give rise to elements that are friable or coated with dust. Of course, sorting is always possible and essayed at the time of actual use but considering the foregoing observations and the moderate levels of service stress of the works in which these aggregates are to be used (buildings of moderate height, roads with light traffic...), the norms and specifications which govern their use are voluntarily made more flexible than those in force in countries where the variety, accessibility and quality of the rocks are better. That is why one will sometimes accept the use of gravels of doubtful shape or having values of Los Angeles coefficient reaching 40, even 45, which would be formally forbidden in France. But the tolerance is not without limits and amongst rocks whose extreme friability often rules out their use in construction the following are generally included: highly weathered sandstones, ampelites, marly limestones, schists and micaschists. Likewise, some crystalophyllic rocks—gneisses, amphibolites—would necessitate better equipped processing plants (use of a tertiary crusher), considering the large number of flakes formed while crushing. There are also some problematical rocks, such as high-sulfur pyrites which cause ruptures in bonds between binders and aggregates; such phenomena are rather exceptional, however.

Substances obtained from the crushing of massive rocks are used in a variety of areas:

— aggregates for concrete,
— crushed sand for sand-asphalt,
— stone for masonry,
— riprap for dock works,
— ballast for railway tracks,
— all-in crusher-run for roadbeds (bases, subbases),
— chippings for wearing courses (surface dressings, bituminous cement).

Lastly, it is rare that one observes a true policy on quarries and quarrying for crushed stone aggregates in newly developing countries, similar to that practised in France. The plants erected at job sites more often close down upon completion of such jobs, given the absence of demand. Only a small number of aggregate-producing plants remain permanently at the periphery of big cities where demand-supply markets justify them. Also, it is advisable to remain highly circumspect with regard to the possibilities of sensible amelioration in the quality of production of these aggregates.

1.2.3 LOOSE MATERIALS FOR ROADWORKS

Under this heading it is in order to differentiate between:

— materials of sedimentary or alluvial origin, most of which resemble in appearance those one comes across in Europe, even though they may have some peculiar geotechnical characteristics in their tropical environment;

— materials of residual origin, which result from a profound transformation, at once chemical and mineralogical, of the mother rock under the action of climate, and whose appearance as well as intrinsic properties constitute phenomena of a remarkable peculiarity for the inexperienced engineer;

— and lastly, materials of varied special origins which differ from the above.

The following may be mentioned under the first group:

— alluvial gravels,

— sedimentary clayey sands of coastal basins or those to be found at the low points or in the eroded beds of streams and rivers,

— sands from lagoons, beaches or dunes.

The second group essentially comprises:

— lateritic gravels,

— lateritic duricrusts,

— sandstone arenites, quartzitic or granitic.

Mention may also be made of those materials which are peculiar to desert zones, namely tuffs, calcareous crusts and gypsum sands, about which the Central Laboratory of Bridges and Roads has contributed some interesting articles.

Under the last group let us note:

— volcanic lavas and scoriae,

— mineral wastes,

— coralline materials,

— shells.

• *Aggregates of sedimentary or alluvial origin*

Alluvial gravels

Coarse alluvia encountered directly in the beds of some rivers or watercourses are generally all-ins of smooth appearance and quartzose or siliceous in nature. Similar in appearance to those found in French gravel pits, they are most often distinguished by a rather low crushing strength (Los Angeles coefficient of the order of 40 to 45) and by a substantial clay content, coming as they do from the muddy or silty waters which carry them along. Varying in depth from several tens of centimeters to several metres, these alluvial gravels are crushed or screened as necessary and used either for making roadbeds or concretes meant for civic constructions. In the former case one might not be happy with their mediocre stability, resulting from the shape of the aggregates and the absence of a plastic mortar, whereas in the latter their adherence to the cement may be only so so. In both cases one should be circumspect about the heterogeneity of the exploitable layers with regard to the nature of the materials (pockets of sand or clay, presence of lumps) as well as their thicknesses (variable thickness of the units or benches constituting the deposit).

In desert zones one will also notice the presence of sizeable deposits of natural gravels, rounded and sandy, termed 'all-ins of regs', used in road-making practice.

Sedimentary sands

These are of two distinct types:

— clayey sands of coastal sedimentary basins, such as those of the Douala or Abidjan region;

— more or less clean sands, found directly in some streams, when it is not at the bottom of lagoons, or quite simply on beaches.

Given the absence of massive rocks in the neighbourhood of these basins, the former often constitute an attractive source for making base and foundation courses for pavements with little traffic, after stabilising with cement. They are occasionally improved by incorporating cleaner sands from nearby lagoons or watercourses to reduce their plasticity and to increase their bearing capacity.

The latter, siliceous or quartzose sands from streams or lagoons, are sometimes used when their granulometry permits, in making sand-asphalt mix meant for the wearing course of paved roads. Blended with beach sands or sands resulting from the crushing of massive rocks, they also serve in making cement mortars and cement concretes.

• Aggregates of residual origin

Lateritic gravelly types

Products of the profound weathering undergone by the rocks under the action of temperature and fluctuations in groundwater, lateritic gravels—mixture of nodules made up of an absolute accumulation and concentration of aluminium and iron hydroxides, embedded in an essentially kaolinitic argillaceous matrix—constitute the most abundant source of loose aggregates in intertropical regions. Under a cover of fine soils 10 cm to 2 or 3 m deep, the gravelly benches per se are generally between some 20 cm to 1 m in height; in some regions of course, they may reach some tens of metres. At the borrow pit the heterogeneity which characterises the deposits is quite often the most frustrating aspect for the quarrier or the user; this heterogeneity encompasses the following:

— height of the overlying wastes (non-uniform overburden);

— height of the benches (again variable);

— geotechnical characteristics of the body: presence of pockets of clay or gross hardened blocks whose dimensions can be of the order of one cubic metre, variation in properties depth-wise (decrease in bearing capacity as one approaches horizon C).

This troublesome characteristic periodically occasions a difficult exploitation, necessitating either the sorting and removal of blocks, or breakage of the deposit into thin layers that are then separated and reassorted, if the pebbly profile is not a chaotic mixture, or outright rejection of the deposit.

Lateritic gravels of uneven granularity, due to low sand content (characteristic property) and variable plasticity, according to whether they are situated in savannah or a forest zone, are used in common practice for laying wearing courses of roads without surface dressings, or foundation and base courses of bituminous roads. Given a bearing capacity, often adequate (instant CBR generally above 50 and CBR after imbibition above 20), incorporation of some 3 to 4% cement imparts to them new properties that enable their withstanding in a base course the stresses of a sometimes aggressive heavy vehicular traffic. Washed to rid them of the clayey gangue coating them, lateritic fine gravels, which are spherical nodules generally below 20 mm in size and blackish in colour (evidence of the presence of iron compounds), are also used as aggregates for concrete or for surface dressings. In addition, some quarries contain an appreciable amount of quartzose elements (un-weathered residues of the mother rock), the largest ones of which, with a diameter between 100 and 150 mm, can be crushed if necessary for use in making cement concretes that yield altogether acceptable performances.

Lastly, mention may be made of pit-run laterite, which on being stabilised mechanically or with cement, can be used for making geoconcrete blocks for low-cost housing.

Lateritic duricrusts and cuirasses

Exposed by erosion phenomena, farruginous cuirasses occur in the form of compact benches that can reach a few metres in thickness; they necessitate recourse to powerful rock-loosening machines, even explosives, because they are very hard. The strongest among them have been used:

— either as all-in from crushing for the metalling of some roads;

— or in the form of blocks or as rubble in masonry: small dwellings, water wings, cribs, breast-walls, riprap, pipeline heads, protective aprons etc.

Arenites

These comprise overlying sandstone rocks, granitic or quartzitic, and in plateau regions sandy deposits resulting from weathering of the substratum, which generally consist of sands that are silty and muddy to a greater or lesser degree. These residual sands when treated with cement and not containing too much mica, are quite often used for making roadbeds of highways with surface dressings and carrying low or moderate traffic.

• Aggregates of diverse origin

Economic considerations together with the moderate degree of stresses to which some structures are subject (low magnitude of loads, low traffic density) have led many newly developing countries to seek and promote a policy of maximal utilisation of local materials, some of which possess unique properties due at once to their special genesis and unusual geotechnical properties. This is notably the case with the following materials whose general use in the road-making industry presents a rather exceptional character.

Volcanic lavas and scoriae

The exclusive presence of these materials in a few regions, such as Cameroon, Zaire, Madagascar, Reunion, Djibouti, La Martinique and Guadelupe, revived an active interest among highway engineers who are using them once again, but with some precautions (*in-situ* crushing or removal of big blocks), in foundations and base courses of roads with low-volume traffic. In certain cases, as in the Douala region, ground pozzolan constitutes an additive in the manufacture of cement. Here again, heterogeneity in the deposits is the rule and exploitation should strive to avoid assimilation or mixing together of elements of variable hardness and disparate particle size.

Mine wastes

The availability in some newly developing countries of tailings or mine wastes which are not saleable, has quite naturally led to envisaging the possible utilisation of these low-cost materials in the construction of highways or other civil engineering structures. Such wastes include manganiferous gravels in the Moanda region in Gabon (fractions rejected from crushed manganese ore), nickel tailings in New Caledonia, chromite ore tailings not far from Tamatave in Madagascar, not to speak of the slags of Algeria. These materials, often possessing very unique properties—high density, porosity, dubious shape—are utilised with varying degrees of success depending on the precautions taken at the stage of their processing and their ultimate use. Improvement with cement and selection of the fine elements from tailings derived from the exploitation of nickel in New Caledonia, has been a necessary and prior condition for their utilisation. These aggregates, most often confined to a well-demarcated zone, are cited solely as a reminder here because, despite their special behaviour, which in some cases is as yet poorly understood, they offer interesting potential utilisation when the more conventional sources of materials are scarce.

Coralline and shell materials

This account would be incomplete if mention were not made of the use in some regions (Mauritania, French Polynesia) of corals and shells for making the base and foundation courses of roads, even wearing courses (Mauritanian sand-asphalt made with shell-sand mixture). Well-reasoned reports have been published on the characteristics of coralline materials and their use in road-making practice in those regions where the more conventional sources of aggregates either do not exist or are too far away from construction sites.

Conclusions

Profoundly weathered by the aggressive action of atmospheric agents obtaining in a tropical milieu, the rocks in newly developing countries situated in such an environment are distinguished by properties of hardness, cleanliness and shape that are far less acceptable than those of the same age but situated in less aggressive temperate climates.

Since economic constraints and limited needs generally do not justify extending the working faces deeper, where the clean portions of crystalline structures lie hidden, a certain margin of tolerance is permitted in the processing norms and specifications for application of aggregates, so long as this does not compromise the quality of the structures most often subjected to moderate stresses.

Barring laterites, which constitute the purest and most unusual amongst the new aggregates generated under the action of these climates, the other types of loose materials, be they sedimentary, alluvial or residual in origin, will hardly astonish the engineer who observes them for the first time. The only surprise for him would be the methods of exploitation and utilisation, sometimes rudimentary but always ingenious, which perforce arise from the limited financial resources of these young countries.

In the years to come, due to advances resulting from development, manufacturing procedures and quality criteria will become more rigorous and minor differences will gradually disappear. But these magnificent weathered profiles, these marvellous mantles of soils dark coloured or sparkling red, will forever remain telltale signs of a past and an evolution which nothing, no one—not even Providence—can change, and herein lies and will continue to lie for a long time to come, a difference that is undeniable.

1.3 Conditions for Initiating an Exploitation

1.3.1 END-USE OF THE AGGREGATES TO BE PRODUCED

That a quarrier would venture into a region of a newly developing country with the intention of catering to a zone of diversified consumption comprising highways, buildings, highway structures, concretes etc., belies credibility. A projected plant is generally destined to supply a well-defined job site for which a tender inviting offers has been issued.

Fortunately, most of the contenders would be familiar with this kind of job site and have already successfully acquitted themselves under similar conditions, some even within the very country issuing the tender. However, it has happened that entrepreneurs with no idea of the particular conditions of work have secured prior authorisation to tender and even declared the successful bidder. Obviously, it would be better that the bidder not be someone with little understanding of the problems related to distance, the conditions of work and the various factors peculiar to the country in question. When this happens, the subcontractor is caught in an awkward situation since his client, the successful bidder for the works at an unrealistic cost, will tend to impose very difficult conditions upon him. We feel obliged to call attention to this highly problematic situation but shall not expatiate upon it further.

Let us consider instead only the normal case of a contract concluded under proper conditions.

We shall not repeat here what has already been covered in Chapter XV on project engineering, which deals in particular with an example of setting up a plant dedicated to a large job site.

1.3.2 OPENING THE QUARRY

Let us suppose that a new deposit is to be exploited. Notwithstanding all the geological investigations which could have been conducted by highly competent professional services, nor the existence of relatively precise geological maps, unexpected contingencies may well arise.

Hence a detailed exploration is mandatory. Test boreholes have to be drilled. One can never take too many precautions.

In any case, from the moment of attacking a rocky mass, even if the rock should subsequently prove to be homogeneous, one is on the alert, as mentioned earlier, for a zone weathered by air and the streaming action of water.

In some countries precipitation is often copious and hard-hitting; the working of a deposit should take this into account. It is equally necessary to make certain that a plant for the treatment of alluvial materials is not located in the flood zones of a watercourse whose flow rises immeasurably in just a few hours.

1.3.3 COMPRESSED TIME SCHEDULES

Time elapses between the submission of tenders and selection of the bidder so the signal for starting the works is slow to come. Yet the government of the country interested in construction of such or such highways, ports, aerodromes, or administrative buildings, insists upon completion of these works by a definite date, usually one that corresponds to a commemoration, an anniversary in the country's history, leaving the executors of these tasks little time in hand. As a result, plants are paradoxically oversized for jobs which sometimes do not materialise and, in any case, for operations that cannot recur.

Occasionally, complications crop up which postpone commissioning of the site. Complications arise either from overlapping of the authorities nominated or redundancy of inspection functions. The department for inspection of the materials rejects them and the contractor-in-charge takes refuge behind this technical aspect. Result, a deadlock. Non-acceptance of the first samples of materials can stem from the fact that a newly opened deposit lacks the characteristics envisaged in the specifications etc.

Customs procedures are often sticky even though an exemption from customs duty may have been sanctioned in the terms of the contract drawn up under international financing for machinery under temporary import.

1.3.4 SHORTAGE OF CREDITS

Often, time elapses after submission of the offers before the governments and the financiers actually get together. Occasionally, the time limit for the validity of the offer is reached or exceeded. Extension of these periods is then sought. Meanwhile prices have escalated. But funding is not augmented, nor is it likely to be.

Yet the project has to be completed since stopping midway is counter to fulfilling the policy objective. Technical concessions, prejudicial to the service life of the structure, are made.

True, this is not a common situation but nevertheless is one which could very well be encountered.

1.3.5 RIGOROUS TECHNICAL SPECIFICATIONS

In the preparation of materials, we have sometimes had to conform to norms that are not customary in European countries (Figs. XX.5 and XX.6).

There is indeed a tendency towards overspecification on the part of design bureaus which draw up the tender calls. This could very well be a safeguard against uncertainties, the unforeseeable; for example, sometimes the particle size ranges are narrower than those customarily called for.

We also saw in Sec. 1.2 that the materials in these countries often have Los Angeles coefficients higher than those normally recorded in France. But the sup-

Fig. XX.5. The problem of overloading. This type of loading is not uncommon

Fig. XX.6. Canadian construction execution in Cameroon. French crushing, grinding, screening...etc. equipment. NORDBERG. An excellent road structure

plier of materials cannot do much about this. To modify the geology is obviously beyond him.

1.4 Operation of Job Site

1.4.1 PRIORITY TO MATERIALS PRODUCTION

In practice, in the operation of a job site, especially one of roadworks, materials are needed very early (Figs. XX.7 and XX.8).

Certainly earthwork starts first but the bed has to be protected by placement of the first materials which serve to resist severe and violent spells of bad weather.

A plant for the production of crushed, ground, screened materials calls for an investigation, fabrication, transport and assembly.

This is one aspect in the succession of operations which should not be ignored. The orders for machinery should be placed expeditiously by the successful entrepreneurs, the bidders; however, they can only do this upon receipt of guarantees, as stipulated in the contract, that the machinery will be delivered in good and working condition without reservations.

1.4.2 CONDITIONS OF COMMISSIONING, MAINTENANCE AND EMERGENCY REPAIRS

The problem of expropriations in countries where the unwritten law sanctioned by convention is sometimes overriding, is not without its element of surprise and a certain folklorish twist, which the business of public works could well do without, having already been subjected to delays (see above). We know of a case wherein a whole machinery unit, delivered within schedule, was transported to the vicinity of assembly and remained there unpacked, due to negotiations for the last expropriation being unresolved.

In practice, the time period (originally envisaged) for producing all the materials must necessarily quite often be shortened, and the total quantity of materials needed often tends to exceed that projected.

Furthermore, for a given target the entrepreneur quite logically provides for a minimum of machinery and sometimes a three-shift utilisation.

Regardless, it is evident that maintenance must be sound. Wearing components should be changed before the normal stage of use is over.

At the time of installing a plant, one spare lot of wearing components is supplied for replacement at the appropriate time and it is here that the experience of the person in charge of quarry and sand-pit machinery assumes great significance. Based on his experience, he specifies with adequate precision which are the wearing components that will need to be replaced during the duration of the job.

Then there is also the problem of emergency repairs.

In practice, any mechanism, howsoever well designed and well manufactured it may be, can break down.

Storing at the job site a certain number of spares to forestall possible stoppage of a machine is rather illusory since unfortunately it is not always the piece one has stocked that will need repair.

Fig. XX.7. After surfacing: Ivory Coast. Jean Lefebvre firm

Fig. XX.8. Overlaying of 0/31.5 protects the roadbase after earthworks in the case of heavy rains: common situation

Furthermore, only an equipment manufacturer who has a large number of installations in service within a given geographical zone can reasonably maintain a huge inventory of spare parts for the benefit of his clients.

Now let us speak of emergency mechanical spares rather than wearing components.

The most rational and most reasonable solution is that of swift emergency repairs from the fabricator's factory, and this with due regard to the frequency of airline services.

Thus it is necessary for the entrepreneur to approach a sufficiently major equipment manufacturer, who should also be financially quite sound we might add, i.e., able to readily stock most of the components that might be required.

If the component is in stock, it can be brought to the airport within 48 hours adequately packaged. The component will be conveyed to its destination within a few hours.

In some cases, and for ticklish problems, technicians from the equipment manufacturer's plant will have to accompany the components to ensure proper emergency repairs.

The best solution for job sites in newly developing countries certainly consists of using none but the best, most reliable machinery; even then it is preferable to have a supplier capable of immediately expediting by air any component required and, if need be, competent personnel.

1.4.3 INSPECTIONS OFTEN REQUIRING STRICT CONFORMITY TO SPECIFICATIONS NOT ALWAYS ADAPTED TO THE SITUATION

We have already stated that the norms imposed are sometimes stricter than those one is accustomed to taking into consideration.

Specifications are often laid down far away from the place of exploitation of the materials nature has placed at the disposal of man. Hence it sometimes happens that the specifications are wholly incompatible with the actual characteristics of the available rocks.

But, are these technical imperatives always adhered to?

In most cases, yes. Therefore they have to be taken very seriously. This leads to situations hopelessly deadlocked, which in turn lead to delays in operational progress with the attendant threats of penalties that are sometimes quite hefty.

This happened to us in an overlay project of runways in a major aerodrome and it became necessary to modify an aggregate-producing plant in record time. The plant in question had been very meticulously designed on the basis of grading curves of pit-runs, which proved altogether different from the curves of the materials that were actually extracted.

Given such unforeseen factors, there is every need for flexible plants whose material-handling circuits can be readily modified.

It is often also preferable to have multiple plants with medium-size equipment, rather than one plant with the largest capacity equipment since the latter will be the heaviest too.

At this stage, one should think in terms of the adaptability of the plant and we shall discuss this point right now.

2. PLANTS FOR OPERATING UNDER THE CONDITIONS DESCRIBED ABOVE

Reference may be made with advantage to Chapter XV (project engineering) which is based on an example of plants meant to cater to a particular job site.

2.1 Mobile Installations

It is obvious that at the first setting up, it is paramount for the machinery to reach the exploitation site as fast as possible and be assembled in the shortest time (Fig. XX.9).

From this point of view, mobile plants engender an advantage since trailers, tank transporters suited for heavy loads, are not always available on location and sometimes the distances to be travelled are long. We have demonstrated this in the construction of the highway for uranium exploitation in Niger.

Mobile units are profitable to design and construct in direct proportion to the long distances to be covered on tracks. The clearances under the chassis should be liberal and bolts well fastened since otherwise part of the equipment will rest on the tracks. As roads are not always laid in advance and sometimes to success-fully lay a few hundred kilometres, thousands of kilometres have first to be traversed on tracks made of corrugated sheets.

Fig. XX.9. Douala-Yaoundé road (Cameroon). Mobile plant with NORDBERG machinery

2.2 Special Features

2.2.1 SCALPING MATERIALS

Since quite often a new deposit is taken up, some percentage of weathered materials, friable materials and even earth materials is present initially in the feed for the crushing, grinding, screening etc. machinery. Effective scalping is therefore imperative prior to the jaw-crusher, to remove elements of particle size between 0 and 40 mm for instance.

In some plant sites, especially in the Ivory Coast, a second screening between the primary and the secondary has been made obligatory; here again the 0/40 fraction is eliminated.

The illustrations appearing in this chapter depict plants or block diagrams of plants which implement these successive eliminations of weathered products.

2.2.2 THE PREFERENTIAL CIRCUIT

This term, preferential circuit, originated from the fact that with the granites of the Ivory Coast or Cameroon it became necessary to improve the Los Angeles coefficient. This was achieved in particular for the construction of the Douala-Yaoundé highway by proceeding in the following manner. These special schemes were then gradually widely adopted (Fig. XX.10).

The plant scalped the materials prior to the primary jaw-crusher (Fig. XX.11), followed by, whenever necessary an elimination of the 0/10, 0/30 or 0/40 elements between the primary and the secondary crusher.

After the secondary crushing, a screening termed the dispatch screening, was done as follows: the screen incorporated a deck for a 6-mm cut-off, one for 35-mm and yet another for protection with a grid of 60-mm mesh, for example.

Accordingly, this screen received the materials coming from the primary crusher, generally of the jaw type, the secondary crusher, generally of the cone type, and hence received materials of very average, if not downright poor cubicity.

In fact, the primary jaw-crusher is not designed for producing cubic materials and the primary cone crusher in a size reduction circuit is not exactly the type which gives the best results in respect of shape.

In reality, the 0/6 materials, constituting a small proportion of the whole, that is, nearly 10% of the total products arriving on the dispatch screen, are conveyed to the finishing screen.

The 6/35 material, which therefore is not poor in flat elements, needle-shaped elements, flakes etc., is conveyed to an autogenous crusher. This is an equipment which gives excellent results from the viewpoint of fines proportion in the output material, as well as cubicity of the elements obtained.

At the end, the + 35 is sent to the cone crusher downstream; generally this + 35 represents the rock's core, that is, material which is most resistant to the crushing tests already passed.

With installations of this type (Fig. XX.12), we have obtained an improvement of up to 4 points in the Los Angeles coefficient for medium quality granites.

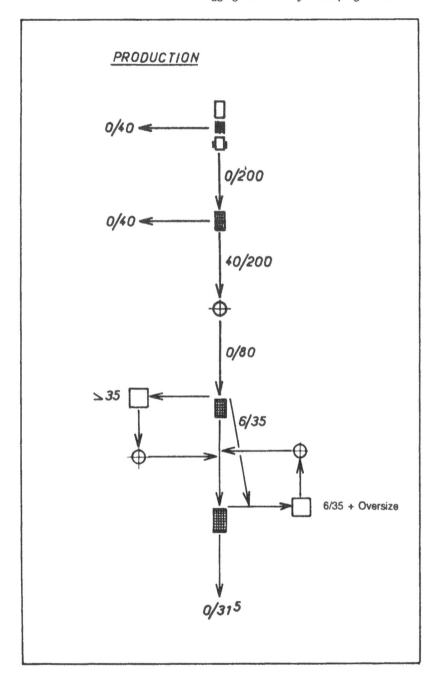

Fig. XX.10. Production of '0/31.5 grading envelope' starting with polluted granite with a high Los Angeles coefficient

Fig. XX.11. Primary station with scalping. Precrushed buffer stock

2.2.3 SELECTIVE CRUSHING: EQUIPMENT FOR FRAGMENTATION BY AUTOGENOUS GRINDING

Used in the position of tertiaries, it is these equipment which have to produce cubic materials, perfect the Los Angeles factor, recycle the excess elements etc.

There was a time when a GYRADISC equipment was best suited to carry out this task for the construction of the Douala-Yaoundé road. All sites where materials were prepared were equipped with this apparatus. There are now a certain number of crushers which vie with one another for the best cubicity, the best resistance to wear etc. These are now the new generation equipment and they give good results.

2.2.4 OPEN STORAGE OF MATERIALS

Plants with batteries of hoppers or silos are not common in newly developing countries.

Generally, the finished products and the processed products are stacked in the open, in some cases in recovery tunnels, but commonly in stacking areas from where they are recovered by loaders and transported by trucks.

2.2.5 SEGREGATION

The output is generally divided into a size fraction of 0/31.5 for the first layer, if necessary a size fraction of 0/20 for an intermediate layer, and then either a size fraction of 0/12.5 for a bituminous cement, or a production of 2 fine gravels of

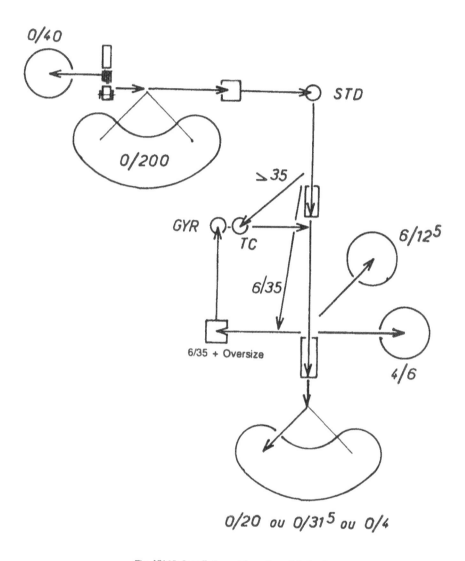

Fig. XX.12. Installation with preferential circuit

continuous or discontinuous granulometry for making a two-layer dressing, for example: 4/6 and 10/14.

Regarding storage of the 0/31.5 or 0/20 fraction, and more so with respect to the 0/31.5 fraction, which constitutes the largest mass, a problem of segregation arises.

The best solution is to use a conveyor of adjustable direction and inclination. Throw-down and fall of materials from a large height will thereby be avoided.

Fig. XX.13. Plant being assembled: preferential circuit and crushers function under autogenous grinding

Dumping at a moderate height as the conveyor moves along will result in separate bean-shaped piles of each fraction.

2.2.6 MOISTURISATION

In many cases we have incorporated a moisturising device in the stacking equipment. Given a regular supply of water, the transported materials reach the place of laying and compacting with a moisture content that is wholly sufficient, say of the order of 6 to 8%. Spraying water simultaneous with the stacking operation ensures a much more uniform distribution within the bulk of the materials than can be achieved by watering a layer spread on the road. Losses due to solar radiation, thus evaporation, are thereby minimised.

Furthermore, this moisturising device prevents loss of fillers when the wind is strong.

3. WINDING UP THE SITE

Three different situations are possible:

3.1 Plant Remains Idle Where It Is

This happens:

— Either because the construction site was a very large affair and everything so planned that the plant was operated up to its extreme wear. Accounts are rendered and the facility amortised;

— Or it was envisaged in the contract that upon winding up the site, the plant would remain the property of the country in which the works were executed and this country ceases operation.

3.2 Plant Remains Where It Is for an Identical or Different Utilisation: Adaptability

Obviously, reuse of a plant set up for a particular job site is the more common practice.

If the plant need not be moved out and if it can continue the same production, there is no problem. Instead of working three shifts, the plant works one per day. This is indeed an ideal solution. But quite often, remaining *in situ*, the plant is expected to ensure different outputs. For example, utilised initially for the production of ballast, it now switches to production of road-making materials.

Inevitably, the circuits of material will not be the same, nor the settings of crushers; furthermore, the crushers per se should be differently equipped.

In practice, every problem should be taken up individually and one should evaluate equipment which is versatile and that which can be fitted with different accessories.

In switching over from the production of ballast to road-making materials, one or two stages of crushing should necessarily be added.

If the same plant is to concomitantly produce materials for making roads and those for buildings or prefabrication, it is on the screening side that new equipment will need to be incorporated.

3.3 Plant Installed Elsewhere: Mobility Plus Adaptability

What now remains is the case of a plant dismantled, then reassembled a few kilometres or a few thousand kilometres away for resumption of operation.

In this case preferably the plant should be mobile and compact since generally it is the volume more than the weight which is penalised in transport costs.

In practice, the problem should be examined at the time of investment and the equipment manufacturer apprised of future disposition of the plant once the initial site is wound up.

4. CONCLUSION

We have perhaps painted a gloomy picture by highlighting the problems incidental to crusher sites in newly developing countries.

European and French contracting firms have surmounted these problems and France's position in the matter of public works has always been highly enviable, ranking third or fourth in the volume of works executed outside the country. The list of prestigious projects completed by leading French firms is quite long.

We have selected a few very large crusher sites in various domains and included photographs of these executions, to constitute, as it were, the conclusion of this chapter and this book.

These photographs pay homage to the public works and construction profession and its French representatives (Figs. XX.14 to XX.18).

Fig. XX.14. Mayotte Island: Congori Port

Fig. XX.15. Katze dam, Lesotho—Doc. Bouygues

Fig. XX.16. Buyo dam, Ivory Coast

Fig. XX.17. Jakarta Aerodrome—Colas

Fig. XX.18. James Bay dam—Doc Bauygues

Printed and bound by CPI Group (UK) Ltd, Croydon, CR0 4YY

23/10/2024

01777686-0002